Series in Real Analysis - Vol. 12

HENSTOCK–KURZWEIL INTEGRATION ON EUCLIDEAN SPACES

SERIES IN REAL ANALYSIS

Published

Vol. 1: Lectures on the Theory of Integration
R Henstock

Vol. 2: Lanzhou Lectures on Henstock Integration
Lee Peng Yee

Vol. 3: The Theory of the Denjoy Integral & Some Applications
V G Celidze & A G Dzvarseisvili
translated by P S Bullen

Vol. 4: Linear Functional Analysis
W Orlicz

Vol. 5: Generalized ODE
S Schwabik

Vol. 6: Uniqueness & Nonuniqueness Criteria in ODE
R P Agarwal & V Lakshmikantham

Vol. 7: Henstock–Kurzweil Integration: Its Relation to Topological Vector Spaces
Jaroslav Kurzweil

Vol. 8: Integration between the Lebesgue Integral and the Henstock–Kurzweil Integral: Its Relation to Local Convex Vector Spaces
Jaroslav Kurzweil

Vol. 9: Theories of Integration: The Integrals of Riemann, Lebesgue, Henstock–Kurzweil, and McShane
Douglas S Kurtz & Charles W Swartz

Vol. 10: Topics in Banach Space Integration
Štefan Schwabik & Ye Guoju

Vol. 12: Henstock–Kurzweil Integration on Euclidean Spaces
Tuo Yeong Lee

Series in Real Analysis - Vol. 12

HENSTOCK–KURZWEIL INTEGRATION ON EUCLIDEAN SPACES

Lee Tuo Yeong

Nanyang Technological University, Singapore

World Scientific

NEW JERSEY • LONDON • SINGAPORE • BEIJING • SHANGHAI • HONG KONG • TAIPEI • CHENNAI

Published by

World Scientific Publishing Co. Pte. Ltd.

5 Toh Tuck Link, Singapore 596224

USA office: 27 Warren Street, Suite 401-402, Hackensack, NJ 07601

UK office: 57 Shelton Street, Covent Garden, London WC2H 9HE

British Library Cataloguing-in-Publication Data
A catalogue record for this book is available from the British Library.

HENSTOCK–KURZWEIL INTEGRATION ON EUCLIDEAN SPACES
Series in Real Analysis — Vol. 12

ISBN-13 978-981-4324-58-8
ISBN-10 981-4324-58-2

Printed in Singapore by World Scientific Printers.

Preface

The Riemann integral is a useful tool for solving many mathematical problems in elementary calculus. However, at the end of nineteenth century, mathematicians found that the Riemann integral has the following shortcomings:

(A) If a function is Riemann integrable on a bounded interval $[a, b]$ of the real line, then the function is bounded there.
(B) There is a bounded non-Riemann integrable derivative.
(C) There is a sequence $(f_n)_{n=1}^{\infty}$ of uniformly bounded, Riemann integrable functions converging *pointwise* to a function f on a bounded interval $[a, b]$ of the real line, yet f *is not* Riemann integrable on $[a, b]$.

In 1902, Lebesgue defined a more general integral to remove the above-mentioned defects (A), (B), and (C). On the other hand, the Lebesgue integral is not powerful enough to integrate all *finite* derivatives. Thus, it is natural for Lebesgue to ask for another general integral that would overcome this drawback.

In 1912, Denjoy extended the one-dimensional Lebesgue integral so that the resulting integral can integrate all finite deriviatves. Two years later, Perron employed a different method to define his integral and used it in the study of differential equations. It is somewhat surprising that the above integrals of Denjoy and Perron coincide, and this integral is now commonly known as the Denjoy-Perron integral. On the other hand, the classical definitions of Riemann, Lebesgue, Denjoy, and Perron integrals have very little in common.

In the 1950's, Henstock and Kurzweil gave a slight but ingenious modification of the classical Riemann integral to obtain the Henstock-Kurzweil integral, which is a Riemann-type definition of the Denjoy-Perron integral. Since then, various higher-dimensional Riemann-type integrals have been defined and studied by several authors; see, for example, [58,63–66,74,78,79,81,82,88,120,137,140]. Since the one-dimensional Henstock-Kurzweil integral is equivalent to the Denjoy-Perron integral, it is not so surprising that the multiple Henstock-Kurzweil integral is equivalent to the classical multiple Perron integral.

This book gives an elementary treatment of the Henstock-Kurzweil integration on Euclidean spaces; most of this book can be understood with a prerequisite of advanced calculus. Furthermore, we present some recent advancement of the theory.

In Chapter 1, we present some basic theory of the one-dimensional Henstock-Kurzweil integral. The definition of this integral is very similar to that of the classical Riemann integral, and it encompasses improper Riemann integrals. Moreover, the relatively new integral is powerful enough to integrate *every* finite derivative.

Chapter 2 deals with the multiple Henstock-Kurzweil integral; in particular, Fubini's theorem is established for this integral. Unlike the corresponding result in the classical Lebesgue theory, neither *measure theory* nor *topology* is needed for the proof.

In the long Chapter 3 we use the Henstock-Kurzweil integral to define the Lebesgue integral. Consequently, we recover all the standard results in the Lebesgue theory.

In Chapters 4 and 5, we obtain further properties of Henstock-Kurzweil integrable functions via the Henstock variational measure.

Chapter 6 is devoted to the proof of Kurzweil's multiple integration by parts formula, and we show that Kurzweil's result cannot be improved.

In Chapter 7, we employ a new generalized Dirichlet test to study certain classes of one-dimensional trigonometric series. Finally, in Chapter 8, we employ the above chapters to deduce some delicate convergence theorems for Henstock-Kurzweil integrals and double trigonometric series. Some open problems are also given in Section 8.11.

I would like to thank Professor Lee Peng Yee, Professor Chew Tuan Seng, my family members, and friends for their support and help throughout the writing, without which this book would not have been completed.

Contents

Chapter 1

The one-dimensional Henstock-Kurzweil integral

1.1 Introduction and Cousin's Lemma

The purpose of this monograph is to study multiple Henstock-Kurzweil integrals. In the present chapter, we shall first present and prove certain results for the one-dimensional Henstock-Kurzweil integral.

Unless mentioned otherwise, the following conventions and notations will be used throughout this monograph. $\mathbb{R}$, $\mathbb{R}^+$, and $\mathbb{N}$ denote the real line, the positive real line, and the set of positive integers respectively. An interval in $\mathbb{R}$ is a set of the form $[\alpha, \beta]$, where $-\infty < \alpha < \beta < \infty$, and $[a, b]$ denotes a fixed interval in $\mathbb{R}$.

Definition 1.1.1.

(i) Two intervals $[u, v]$, $[s, t]$ in $\mathbb{R}$ are said to be *non-overlapping* if $(u, v) \cap (s, t) = \emptyset$.

(ii) If $\{[u_1, v_1], \ldots, [u_p, v_p]\}$ is a finite collection of *pairwise non-overlapping* subintervals of $[a, b]$ such that $[a, b] = \bigcup_{k=1}^{p} [u_k, v_k]$, we say that $\{[u_1, v_1], \ldots, [u_p, v_p]\}$ is a *division* of $[a, b]$.

(iii) A *point-interval pair* $(t, [u, v])$ consists of a point $t \in \mathbb{R}$ and an interval $[u, v]$ in $\mathbb{R}$. Here t is known as the *tag* of $[u, v]$.

(iv) A *Perron partition* of $[a, b]$ is a finite collection $\{(t_1, [u_1, v_1]), \ldots, (t_p, [u_p, v_p])\}$ of point-interval pairs, where $\{[u_1, v_1], \ldots, [u_p, v_p]\}$ is a division of $[a, b]$, and $t_k \in [u_k, v_k]$ for $k = 1, \ldots, p$.

(v) A function $\delta : [a, b] \longrightarrow \mathbb{R}^+$ is known as a *gauge* on $[a, b]$.

(vi) Let δ be a gauge on $[a, b]$. A Perron partition $\{(t_1, [u_1, v_1]), \ldots, (t_p, [u_p, v_p])\}$ of $[a, b]$ is said to be *δ-fine* if $[u_k, v_k] \subset (t_k - \delta(t_k), t_k + \delta(t_k))$ for $k = 1, \ldots, p$.

The following example shows that tags play an important role in our study.

Example 1.1.2. We define a gauge δ on $[0,1]$ by setting

$$\delta(t) = \begin{cases} t & \text{if } 0 < t \leq 1, \\ \frac{1}{2} & \text{if } t = 0. \end{cases}$$

Then the following statements are true.

(i) $\{[0,\frac{1}{3}],[\frac{1}{3},\frac{1}{2}],[\frac{1}{2},1]\}$ is a division of $[0,1]$.
(ii) $\{(0,[0,\frac{1}{3}]),(\frac{1}{2},[\frac{1}{3},\frac{1}{2}]),(1,[\frac{1}{2},1])\}$ is a δ-fine Perron partition of $[0,1]$.
(iii) $\{(0,[0,\frac{1}{3}]),(\frac{1}{2},[\frac{1}{3},\frac{1}{2}]),(\frac{1}{2},[\frac{1}{2},1])\}$ is *not* a δ-fine Perron partition of $[0,1]$.

The following natural question arises from Example 1.1.2.

Question 1.1.3. If δ is an *arbitrary* gauge on $[a,b]$, is it possible to find a δ-fine Perron partition of $[a,b]$?

In order to proceed further, we need the following result.

Lemma 1.1.4. *Let δ be a gauge on $[a,b]$ and let $a < c < b$. If P_1 and P_2 are δ-fine Perron partitions of $[a,c]$ and $[c,b]$ respectively, then $P_1 \cup P_2$ is a δ-fine Perron partition of $[a,b]$.*

Proof. Exercise. □

The following theorem gives an affirmative answer to Question 1.1.3.

Theorem 1.1.5 (Cousin's Lemma). *If δ is a gauge on $[a,b]$, then there exists a δ-fine Perron partition of $[a,b]$.*

Proof. Proceeding towards a contradiction, suppose that $[a,b]$ does not have a δ-fine Perron partition. We divide $[a,b]$ into $[a,\frac{a+b}{2}]$ and $[\frac{a+b}{2},b]$ so that $[a,b]$ is the union of two non-overlapping intervals in $\mathbb{R}$. In view of Lemma 1.1.4, we can choose an interval $[a_1,b_1]$ from the set $\{[a,\frac{a+b}{2}],[\frac{a+b}{2},b]\}$ so that $[a_1,b_1]$ does not have a δ-fine Perron partition. Using induction, we construct intervals $[a_1,b_1],[a_2,b_2],\ldots$ in $\mathbb{R}$ so that the following properties are satisfied for every $n \in \mathbb{N}$:

(i) $[a_n,b_n] \supset [a_{n+1},b_{n+1}]$;
(ii) no δ-fine Perron partition of $[a_n,b_n]$ exists;
(iii) $b_n - a_n = \frac{b-a}{2^n}$.

Since properties (i) and (iii) hold for every $n \in \mathbb{N}$, it follows from the Nested Interval Theorem [6, Theorem 2.5.3] that

$$\bigcap_{k=1}^{\infty} [a_k, b_k] = \{t_0\}$$

for some $t_0 \in \mathbb{R}$. On the other hand, since $\bigcap_{k=1}^{\infty} [a_k, b_k] = \{t_0\}$ and $\delta(t_0) > 0$, it follows from property (iii) that there exists $N \in \mathbb{N}$ such that $\{(t_0, [a_N, b_N])\}$ is a δ-fine Perron partition of $[a_N, b_N]$, a contradiction to (ii). This contradiction completes the proof. □

Let $C[a, b]$ denote the space of real-valued continuous functions on $[a, b]$. A simple application of Theorem 1.1.5 gives the following classical result.

Theorem 1.1.6. *If $f \in C[a, b]$, then f is uniformly continuous on $[a, b]$.*

Proof. Let $\varepsilon > 0$ be given. Using the continuity of f on $[a, b]$, for each $x_0 \in [a, b]$ there exists $\delta_0(x_0) > 0$ such that

$$|f(x) - f(x_0)| < \frac{\varepsilon}{2}$$

whenever $x \in (x_0 - \delta_0(x_0), x_0 + \delta_0(x_0)) \cap [a, b]$. We want to prove that there exists $\eta > 0$ with the following property:

$$s, t \in [a, b] \text{ with } |s - t| < \eta \implies |f(s) - f(t)| < \varepsilon.$$

Define a gauge δ on $[a, b]$ by setting $\delta = \frac{1}{2}\delta_0$. In view of Cousin's Lemma, we may select and fix a δ-fine Perron partition $\{(t_1, [u_1, v_1]), \ldots, (t_p, [u_p, v_p])\}$ of $[a, b]$. If $s, t \in [a, b]$ with $|t - s| < \eta := \min\{\delta(t_i) : i = 1, \ldots, p\}$, then there exists $j \in \{1, \ldots, p\}$ such that $|t - t_j| < \delta(t_j)$ and so $|s - t_j| \leq |s - t| + |t - t_j| < 2\delta(t_j) = \delta_0(t_j)$. Thus

$$|f(t) - f(s)| \leq |f(t) - f(t_j)| + |f(t_j) - f(s)| < \varepsilon.$$

Therefore, f is uniformly continuous on $[a, b]$. □

Following the proof of Theorem 1.1.6, we obtain the following corollary.

Corollary 1.1.7. *If $f \in C[a, b]$, then f is bounded on $[a, b]$.*

1.2 Definition of the Henstock-Kurzweil integral

Let $P = \{(t_1, [u_1, v_1]), \ldots, (t_p, [u_p, v_p])\}$ be a Perron partition of $[a, b]$. If f is a real-valued function defined on $\{t_1, \ldots, t_p\}$, we write

$$S(f, P) = \sum_{i=1}^{p} f(t_i)(v_i - u_i).$$

We first define the Riemann integral.

Definition 1.2.1. A function $f : [a, b] \longrightarrow \mathbb{R}$ is said to be *Riemann integrable* on $[a, b]$ if there exists $A_0 \in \mathbb{R}$ with the following property: given $\varepsilon > 0$ there exists a <u>constant</u> gauge δ on $[a, b]$ such that

$$|S(f, P) - A_0| < \varepsilon \tag{1.2.1}$$

for each δ-fine Perron partition P of $[a, b]$.

The collection of all functions that are Riemann integrable on $[a, b]$ will be denoted by $R[a, b]$.

Once we omit the word "constant" from Definition 1.2.1, we obtain the following modification of the Riemann integral.

Definition 1.2.2. A function $f : [a, b] \longrightarrow \mathbb{R}$ is said to be *Henstock-Kurzweil integrable* on $[a, b]$ if there exists $A \in \mathbb{R}$ with the following property: given $\varepsilon > 0$ there exists a gauge δ on $[a, b]$ such that

$$|S(f, P) - A| < \varepsilon \tag{1.2.2}$$

for each δ-fine Perron partition P of $[a, b]$.

The collection of all functions that are Henstock-Kurzweil integrable on $[a, b]$ will be denoted by $HK[a, b]$.

It is easy to deduce from Definitions 1.2.1 and 1.2.2 that if $f \in R[a, b]$, then $f \in HK[a, b]$. In this case, Cousin's Lemma shows that there is a unique number satisfying Definitions 1.2.1 and 1.2.2.

Theorem 1.2.3. *There is at most one number A satisfying Definition 1.2.2.*

Proof. Suppose that A_1 and A_2 satisfy Definition 1.2.2. We claim that $A_1 = A_2$.

Let $\varepsilon > 0$ be given. Since A_1 satisfies Definition 1.2.2, there exists a gauge δ_1 on $[a, b]$ such that

$$|S(f, P_1) - A_1| < \frac{\varepsilon}{2}$$

for each δ_1-fine Perron partition P_1 of $[a, b]$. Similarly, there exists a gauge δ_2 on $[a, b]$ such that

$$|S(f, P_2) - A_2| < \frac{\varepsilon}{2}$$

for each δ_2-fine Perron partition P_2 of $[a, b]$.

Define a gauge δ on $[a, b]$ by setting

$$\delta(x) = \min\{\delta_1(x), \delta_2(x)\}. \tag{1.2.3}$$

According to Cousin's Lemma (Theorem 1.1.5), we may fix a δ-fine Perron partition P of $[a, b]$. Since (1.2.3) implies that the δ-fine Perron partition P is both δ_1-fine and δ_2-fine, it follows from the triangle inequality that

$$|A_1 - A_2| \leq |S(f, P) - A_1| + |S(f, P) - A_2| < \varepsilon.$$

Since $\varepsilon > 0$ is arbitrary, we conclude that $A_1 = A_2$. □

Theorem 1.2.3 tells us that if $f \in HK[a, b]$, then there is a unique number A satisfying Definition 1.2.2. In this case the number A, denoted by $(HK) \int_a^b f$, $(HK) \int_a^b f(x)\, dx$ or $(HK) \int_a^b f(t)\, dt$, is known as the *Henstock-Kurzweil integral* of f over $[a, b]$.

It is clear that if $f \in R[a, b]$, then $f \in HK[a, b]$ and the number $(HK) \int_a^b f$ satisfies Definition 1.2.1. In this case the unique number $(HK) \int_a^b f$, denoted by $\int_a^b f$, $\int_a^b f(x)\, dx$ or $\int_a^b f(t)\, dt$, is known as the Riemann integral of f over $[a, b]$. The following example shows that the inclusion $R[a, b] \subsetneq HK[a, b]$ is *proper*.

Example 1.2.4. Let $\mathbb{Q}$ be the set of all rational numbers, and define the function $f : [0, 1] \longrightarrow \mathbb{R}$ by setting

$$f(x) = \begin{cases} 1 & \text{if } x \in [0, 1] \cap \mathbb{Q}, \\ 0 & \text{otherwise.} \end{cases}$$

Then $f \in HK[0, 1] \setminus R[0, 1]$.

Proof. Let $(r_n)_{n=1}^{\infty}$ be an enumeration of $[0, 1] \cap \mathbb{Q}$ and let $\varepsilon > 0$. We define a gauge δ on $[0, 1]$ by setting

$$\delta(x) = \begin{cases} \frac{\varepsilon}{2^{n+1}} & \text{if } x = r_n \text{ for some } n \in \mathbb{N}, \\ 1 & \text{if } x \in [0, 1] \setminus \mathbb{Q}. \end{cases}$$

If P is a δ-fine Perron partition of $[0,1]$, then

$$\begin{aligned}|S(f,P)-0| &= \left|\sum_{\substack{(t,[u,v])\in P\\ t\in\mathbb{Q}\cap[0,1]}} f(t)(v-u) + \sum_{\substack{(t,[u,v])\in P\\ t\in[0,1]\setminus\mathbb{Q}}} f(t)(v-u)\right| \\ &= \sum_{\substack{(t,[u,v])\in P\\ t\in\mathbb{Q}\cap[0,1]}} f(t)(v-u) \\ &< \sum_{k=1}^{\infty}\frac{\varepsilon}{2^k} \\ &= \varepsilon.\end{aligned}$$

Since $\varepsilon > 0$ is arbitrary, we conclude that $f \in HK[0,1]$ and $(HK)\int_0^1 f = 0$.

It remains to prove that $f \notin R[0,1]$. Proceeding towards a contradiction, suppose that $f \in R[0,1]$. Since $f \in R[0,1] \subseteq HK[0,1]$ and $(HK)\int_0^1 f = 0$, we have $\int_0^1 f = 0$. Hence for $\varepsilon = 1$ there exists a constant gauge δ_1 on $[0,1]$ such that

$$|S(f,P_1)| < 1$$

for each δ_1-fine Perron partition P_1 of $[0,1]$. If q is a positive integer satisfying $q^{-1} < \delta_1$, then

$$P_2 := \left\{\left((k_1-1)q^{-1}, [(k_1-1)q^{-1}, k_1q^{-1}]\right) : k_1 = 1,\ldots,q\right\}$$

is a δ_1-fine Perron partition of $[0,1]$. A contradiction follows:

$$1 > S(f,P_2) = \sum_{k=1}^{q}(kq^{-1} - (k-1)q^{-1}) = 1.$$

□

The following theorem shows that the one-dimensional Henstock-Kurzweil integral is useful for formulating a Fundamental Theorem of Calculus.

Theorem 1.2.5. *Let $f : [a,b] \longrightarrow \mathbb{R}$ and let $F \in C[a,b]$. If F is differentiable on (a,b) and $F'(x) = f(x)$ for all $x \in (a,b)$, then $f \in HK[a,b]$ and*

$$(HK)\int_a^b f = F(b) - F(a).$$

Proof. Let $\varepsilon > 0$ be given. Since F is continuous on $[a, b]$, for each $x \in [a, b]$ there exists $\delta_1(x) > 0$ such that

$$|F(x) - F(y)| < \frac{\varepsilon}{6}$$

whenever $y \in [a, b] \cap (x - \delta_1(x), x + \delta_1(x))$.

Since F is differentiable on (a, b), for each $x \in (a, b)$ there exists $\delta_2(x) > 0$ such that

$$|F'(x)(v - u) - (F(v) - F(u))| \le \frac{\varepsilon(v - u)}{3(b - a)}$$

whenever $x \in [u, v] \subseteq (a, b) \cap (x - \delta_2(x), x + \delta_2(x))$.

Define a gauge δ on $[a, b]$ by the formula

$$\delta(x) = \begin{cases} \min\left\{\delta_1(x), \delta_2(x), \frac{1}{2}(x - a), \frac{1}{2}(b - x)\right\} & \text{if } a < x < b, \\ \dfrac{\varepsilon}{6(|f(a)| + |f(b)| + 1)} & \text{if } x \in \{a, b\}. \end{cases}$$

If P is a δ-fine Perron partition of $[a, b]$, then

$$\begin{aligned}
&|S(f, P) - (F(b) - F(a))| \\
&= \left| \sum_{(t,[u,v]) \in P} \left\{ f(t)(v - u) - \big(F(v) - F(u)\big) \right\} \right| \\
&\le \sum_{\substack{(t,[u,v]) \in P \\ t \in \{a,b\}}} |f(t)(v - u) - (F(v) - F(u))| \\
&+ \sum_{\substack{(t,[u,v]) \in P \\ a<t<b}} |f(t)(v - u) - (F(v) - F(u))| \\
&\le \sum_{\substack{(t,[u,v]) \in P \\ t \in \{a,b\}}} \big(|f(t)|\,(v - u) + |F(v) - F(u)| \big) + \sum_{\substack{(t,[u,v]) \in P \\ a<t<b}} \frac{\varepsilon(v - u)}{3(b - a)} \\
&\le \sum_{\substack{(t,[u,v]) \in P \\ t \in \{a,b\}}} |f(t)|\,(v - u) + \sum_{\substack{(t,[u,v]) \in P \\ t \in \{a,b\}}} |F(v) - F(u)| + \sum_{\substack{(t,[u,v]) \in P \\ a<t<b}} \frac{\varepsilon(v - u)}{3(b - a)} \\
&< \frac{\varepsilon(|f(a)| + |f(b)|)}{6(|f(a)| + |f(b)| + 1)} + \frac{2\varepsilon}{6} + \sum_{\substack{(t,[u,v]) \in P \\ a<t<b}} \frac{\varepsilon(v - u)}{3(b - a)} \\
&< \varepsilon.
\end{aligned}$$

Since $\varepsilon > 0$ is arbitrary, we conclude that $f \in HK[a, b]$ and

$$(HK) \int_a^b f = F(b) - F(a).$$

□

The following examples are special cases of Theorem 1.2.5.

Example 1.2.6. Let

$$F(x) = \begin{cases} x^2 \sin \dfrac{1}{x^2} & \text{if } 0 < x \leq 1, \\ 0 & \text{if } x = 0. \end{cases}$$

Then F is differentiable on $[0, 1]$. In particular, $F' \in HK[0, 1]$ and

$$(HK) \int_0^1 F' = \sin 1.$$

Example 1.2.7. Let

$$f(x) = \begin{cases} \dfrac{2}{x^3} - \dfrac{2\cos x}{\sin^3 x} & \text{if } 0 < x \leq \frac{\pi}{2}, \\ 0 & \text{if } x = 0. \end{cases}$$

Then $f \in HK[0, \frac{\pi}{2}]$ and

$$(HK) \int_0^{\frac{\pi}{2}} f = \frac{2}{3} - \frac{4}{\pi^2}.$$

The following theorem is a consequence of Theorem 1.2.5.

Theorem 1.2.8. *If $f : [a, b] \longrightarrow \mathbb{R}$ is a derivative on $[a, b]$, then $f \in HK[a, b]$.*

We remark that the converse of Theorem 1.2.8 is not true. More details will be given in Section 4.5.

1.3 Simple properties

The aim of this section is to prove some basic properties of the Henstock-Kurzweil integral via Definition 1.2.2.

Theorem 1.3.1. *If $f, g \in HK[a, b]$, then $f + g \in HK[a, b]$ and*

$$(HK) \int_a^b (f + g) = (HK) \int_a^b f + (HK) \int_a^b g.$$

Proof. Let $\varepsilon > 0$ be given. Since $f \in HK[a,b]$, there exists a gauge δ_1 on $[a,b]$ such that

$$\left| S(f, P_1) - (HK)\int_a^b f \right| < \frac{\varepsilon}{2}$$

for each δ_1-fine Perron partition P_1 of $[a,b]$. Similarly, there exists a gauge δ_2 on $[a,b]$ such that

$$\left| S(g, P_2) - (HK)\int_a^b g \right| < \frac{\varepsilon}{2}$$

for each δ_2-fine Perron partition P_2 of $[a,b]$.

Define a gauge δ on $[a,b]$ by setting

$$\delta(x) = \min\{\delta_1(x), \delta_2(x)\}, \tag{1.3.1}$$

and let P be a δ-fine Perron partition of $[a,b]$. Since (1.3.1) implies that the δ-fine Perron partition P is both δ_1-fine and δ_2-fine, the identity

$$S(f+g, P) = S(f, P) + S(g, P)$$

and the triangle inequality yield

$$\begin{aligned} &\left| S(f+g, P) - \left\{ (HK)\int_a^b f + (HK)\int_a^b g \right\} \right| \\ &\leq \left| S(f, P) - (HK)\int_a^b f \right| + \left| S(g, P) - (HK)\int_a^b g \right| \\ &< \varepsilon. \end{aligned}$$

Since $\varepsilon > 0$ is arbitrary, we conclude that $f + g \in HK[a,b]$ and

$$(HK)\int_a^b (f+g) = (HK)\int_a^b f + (HK)\int_a^b g.$$

□

Theorem 1.3.2. *If $f \in HK[a,b]$ and $c \in \mathbb{R}$, then $cf \in HK[a,b]$ and*

$$(HK)\int_a^b cf = c\left\{ (HK)\int_a^b f \right\}.$$

Proof. Let $\varepsilon > 0$ be given. Since $f \in HK[a,b]$, there exists a gauge δ on $[a,b]$ such that

$$\left| S(f, P_1) - (HK)\int_a^b f \right| < \frac{\varepsilon}{|c|+1}$$

for each δ-fine Perron partition P_1 of $[a, b]$. If P is a δ-fine Perron partition of $[a, b]$, then

$$\left|S(cf, P) - c\left\{(HK)\int_a^b f\right\}\right| = |c|\left|S(f, P) - (HK)\int_a^b f\right| < \frac{|c|\,\varepsilon}{|c|+1} < \varepsilon.$$

Since $\varepsilon > 0$ is arbitrary, we conclude that $cf \in HK[a, b]$ and

$$(HK)\int_a^b cf = c\left\{(HK)\int_a^b f\right\}.$$

□

Theorem 1.3.3. *If $f, g \in HK[a, b]$ and $f(x) \leq g(x)$ for all $x \in [a, b]$, then*

$$(HK)\int_a^b f \leq (HK)\int_a^b g.$$

Proof. Let $\varepsilon > 0$ be given. Since $f \in HK[a, b]$, there exists a gauge δ_1 on $[a, b]$ such that

$$\left|S(f, P_1) - (HK)\int_a^b f\right| < \frac{\varepsilon}{2}$$

for each δ_1-fine Perron partition P_1 of $[a, b]$. Similarly, there exists a gauge δ_2 on $[a, b]$ such that

$$\left|S(g, P_2) - (HK)\int_a^b g\right| < \frac{\varepsilon}{2}$$

for each δ_2-fine Perron partition P_2 of $[a, b]$.

Define a gauge δ on $[a, b]$ by setting

$$\delta(x) = \min\{\delta_1(x), \delta_2(x)\}, \tag{1.3.2}$$

and we apply Cousin's Lemma to fix a δ-fine Perron partition P_0 of $[a, b]$. Since (1.3.2) implies that the δ-fine Perron partition P_0 is both δ_1-fine and δ_2-fine, it follows from the inequality $S(f, P_0) \leq S(g, P_0)$ that

$$(HK)\int_a^b f < S(f, P_0) + \frac{\varepsilon}{2} \leq S(g, P_0) + \frac{\varepsilon}{2} < (HK)\int_a^b g + \varepsilon,$$

and the desired inequality follows from the arbitrariness of ε. □

The following theorem gives a useful necessary and sufficient condition for a function f to be Henstock-Kurzweil integrable on $[a, b]$.

Theorem 1.3.4 (Cauchy Criterion). *A function $f : [a,b] \longrightarrow \mathbb{R}$ is Henstock-Kurzweil integrable on $[a,b]$ if and only if given $\varepsilon > 0$ there exists a gauge δ on $[a,b]$ such that*

$$|S(f,P) - S(f,Q)| < \varepsilon \tag{1.3.3}$$

for each pair of δ-fine Perron partitions P and Q of $[a,b]$.

Proof. ($\Longrightarrow$) Let $\varepsilon > 0$ be given. Since $f \in HK[a,b]$, there exists a gauge δ on $[a,b]$ such that

$$\left| S(f,P_0) - (HK)\int_a^b f \right| < \frac{\varepsilon}{2} \tag{1.3.4}$$

for each δ-fine Perron partition P_0 of $[a,b]$. If P and Q are two δ-fine Perron partitions of $[a,b]$, the triangle inequality and (1.3.4) yield

$$|S(f,P) - S(f,Q)| \leq \left| S(f,P) - (HK)\int_a^b f \right| + \left| S(f,Q) - (HK)\int_a^b f \right| < \varepsilon.$$

($\Longleftarrow$) For each $n \in \mathbb{N}$ we let δ_n be a gauge on $[a,b]$ so that

$$|S(f,Q_n) - S(f,R_n)| < \frac{1}{n}$$

for each pair of δ_n-fine Perron partitions Q_n and R_n of $[a,b]$. Next we define a gauge Δ_n on $[a,b]$ by setting

$$\Delta_n(x) = \min\{\delta_1(x), \dots, \delta_n(x)\},$$

and apply Cousin's Lemma to fix a Δ_n-fine Perron partition P_n of $[a,b]$.

We claim that $(S(f,P_n))_{n=1}^{\infty}$ is a Cauchy sequence of real numbers. Let $\varepsilon > 0$ be given and choose a positive integer N so that $\frac{1}{N} < \varepsilon$. If n_1 and n_2 are positive integers such that $\min\{n_1,n_2\} \geq N$, then P_{n_1} and P_{n_2} are both $\Delta_{\min\{n_1,n_2\}}$-fine Perron partitions of $[a,b]$ and so

$$|S(f,P_{n_1}) - S(f,P_{n_2})| < \frac{1}{\min\{n_1,n_2\}} \leq \frac{1}{N} < \varepsilon.$$

Consequently, $(S(f,P_n))_{n=1}^{\infty}$ is a Cauchy sequence of real numbers. Since $\mathbb{R}$ is complete, the sequence $(S(f,P_n))_{n=1}^{\infty}$ converges to some real number A.

It remains to prove that $f \in HK[a,b]$ and $A = (HK)\int_a^b f$. Let P be a Δ_N-fine Perron partition of $[a,b]$. Since our construction implies that the sequence $(\Delta_n)_{n=1}^{\infty}$ of gauges is non-increasing, we see that the Δ_n-fine Perron partition P_n is Δ_N-fine for every integers $n \geq N$. Thus

$$|S(f,P) - A| = \lim_{n\to\infty} |S(f,P) - S(f,P_n)| \leq \frac{1}{N} < \varepsilon.$$

Since $\varepsilon > 0$ is arbitrary, we conclude that $f \in HK[a,b]$ and $A = (HK)\int_a^b f$.

□

We are now ready to give an important class of Henstock-Kurzweil integrable functions.

Theorem 1.3.5. *If $f \in C[a,b]$, then $f \in HK[a,b]$.*

Proof. Let $\varepsilon > 0$ be given. Since f is continuous on $[a,b]$, for each $x \in [a,b]$ there exists $\delta(x) > 0$ such that

$$|f(y) - f(x)| < \frac{\varepsilon}{2(b-a)}$$

whenever $y \in [a,b] \cap (x - \delta(x), x + \delta(x))$.

Clearly, the function $x \mapsto \delta(x)$ is a gauge on $[a,b]$. Let $P = \{(t_1,[u_1,v_1]),\ldots,(t_p,[u_p,v_p])\}$ and $Q = \{(w_1,[x_1,y_1]),\ldots,(w_q,[x_q,y_q])\}$ be two δ-fine Perron partitions of $[a,b]$. If $[u_j,v_j] \cap [x_k,y_k]$ is non-empty for some $j \in \{1,\ldots,p\}$ and $k \in \{1,\ldots,q\}$, we select and fix a point $z_{j,k} \in [u_j,v_j] \cap [x_k,y_k]$. On the other hand, if $[u_r,v_r] \cap [x_s,y_s]$ is empty for some $r \in \{1,\ldots,p\}$ and $s \in \{1,\ldots,q\}$, we set $z_{r,s} = a$. Let $\mu_1(\emptyset) = 0$ and let $\mu_1([\alpha,\beta]) = \beta - \alpha$ for each pair of real numbers α and β satisfying $\alpha \leq \beta$. By the triangle inequality,

$$\begin{aligned}
&|S(f,P) - S(f,Q)| \\
&= \left|\sum_{j=1}^{p} f(t_j)(v_j - u_j) - \sum_{k=1}^{q} f(w_k)(y_k - x_k)\right| \\
&= \left|\sum_{j=1}^{p}\sum_{k=1}^{q} f(t_j)\mu_1([u_j,v_j] \cap [x_k,y_k]) - \sum_{k=1}^{q}\sum_{j=1}^{p} f(w_k)\mu_1([u_j,v_j] \cap [x_k,y_k])\right| \\
&\leq \left|\sum_{j=1}^{p}\sum_{k=1}^{q} (f(t_j) - f(z_{j,k}))\mu_1([u_j,v_j] \cap [x_k,y_k])\right| \\
&+ \left|\sum_{j=1}^{p}\sum_{k=1}^{q} (f(w_k) - f(z_{j,k}))\mu_1([u_j,v_j] \cap [x_k,y_k])\right| \\
&< \varepsilon.
\end{aligned}$$

An application of Theorem 1.3.4 completes the proof. □

The following theorem is also a consequence of Theorem 1.3.4.

Theorem 1.3.6. *If $f \in HK[a,b]$, then $f \in HK[c,d]$ for each interval $[c,d] \subset [a,b]$.*

Proof. Let $[c,d]$ be a proper subinterval of $[a,b]$. For each $\varepsilon > 0$ we use Theorem 1.3.4 to select a gauge δ on $[a,b]$ such that

$$|S(f,P) - S(f,Q)| < \varepsilon$$

for each pair of δ-fine Perron partitions P and Q of $[a,b]$. Since $[c,d]$ is a proper subinterval of $[a,b]$, there exists a finite collection $\{[u_1,v_1],\ldots,[u_N,v_N]\}$ of pairwise non-overlapping subintervals of $[a,b]$ such that $[c,d] \notin \{[u_1,v_1],\ldots,[u_N,v_N]\}$ and

$$[a,b] = [c,d] \cup \bigcup_{k=1}^{N} [u_k,v_k].$$

For each $k \in \{1,\ldots,N\}$ we fix a δ-fine Perron partition P_k of $[u_k,v_k]$. If $P_{[c,d]}$ and $Q_{[c,d]}$ are δ-fine Perron partitions of $[c,d]$, then it is clear that $P_{[c,d]} \cup \bigcup_{k=1}^{N} P_k$ and $Q_{[c,d]} \cup \bigcup_{k=1}^{N} P_k$ are δ-fine Perron partitions of $[a,b]$. Thus

$$\begin{aligned}
&\left|S(f,P_{[c,d]}) - S(f,Q_{[c,d]})\right| \\
&= \left|S(f,P_{[c,d]}) + \sum_{k=1}^{N} S(f,P_k) - S(f,Q_{[c,d]}) - \sum_{k=1}^{N} S(f,P_k)\right| \\
&= \left|S(f,P_{[c,d]} \cup \bigcup_{k=1}^{N} P_k) - S(f,Q_{[c,d]} \cup \bigcup_{k=1}^{N} P_k)\right| \\
&< \varepsilon.
\end{aligned}$$

By Theorem 1.3.4, $f \in HK[c,d]$. □

Remark 1.3.7. If $f \in HK[a,b]$ and $c \in [a,b]$, we define the Henstock-Kurzweil integral of f over $\{c\}$ to be zero.

Theorem 1.3.8. *Let $f : [a,b] \longrightarrow \mathbb{R}$ and let $a < c < b$. If $f \in HK[a,c] \cap HK[c,b]$, then $f \in HK[a,b]$ and*

$$(HK)\int_a^b f = (HK)\int_a^c f + (HK)\int_c^b f.$$

Proof. Given $\varepsilon > 0$ there exists a gauge δ_a on $[a,c]$ such that

$$\left|S(f,P_a) - (HK)\int_a^c f\right| < \frac{\varepsilon}{2}$$

whenever P_a is a δ_a-fine Perron partition of $[a,c]$. A similar argument shows that there exists a gauge δ_b on $[c,b]$ such that

$$\left|S(f,P_b) - (HK)\int_c^b f\right| < \frac{\varepsilon}{2}$$

for each δ_b-fine Perron partition P_b of $[c, b]$.

Define a gauge δ on $[a, b]$ by setting

$$\delta(x) = \begin{cases} \min\{\delta_a(x), c - x\} \text{ if } a \le x < c, \\ \min\{\delta_a(x), \delta_b(x)\} \text{ if } x = c, \\ \min\{\delta_b(x), x - c\} \text{ if } c < x \le b, \end{cases}$$

and let $P = \{(t_1, [u_1, v_1]), \ldots, (t_p, [u_p, v_p])\}$ be a δ-fine Perron partition of $[a, b]$. Since our choice of δ implies that $c = u_j = v_k$ for some $j, k \in \{1, \ldots, p\}$, we conclude that $P = P_1 \cup P_2$ for some δ-fine Perron partitions P_1, P_2 of $[a, c]$ and $[c, b]$ respectively. Consequently,

$$\begin{aligned} &\left| S(f, P) - \left\{ (HK) \int_a^c f + (HK) \int_c^b f \right\} \right| \\ &\le \left| S(f, P_1) - (HK) \int_a^c f \right| + \left| S(f, P_2) - (HK) \int_c^b f \right| \\ &< \varepsilon. \end{aligned}$$

Since $\varepsilon > 0$ is arbitrary, the theorem follows. □

Exercise 1.3.9.

(i) Prove that if $f, g \in R[a, b]$, then $f + g \in R[a, b]$ and

$$\int_a^b (f + g) = \int_a^b f + \int_a^b g.$$

(ii) Prove that if $f \in R[a, b]$ and $c \in \mathbb{R}$, then $cf \in R[a, b]$ and

$$\int_a^b cf = c \int_a^b f.$$

(iii) Let $f : [a, b] \longrightarrow \mathbb{R}$ and let $a < c < b$. Prove that if $f \in R[a, c] \cap R[c, b]$, then $f \in R[a, b]$ and

$$\int_a^b f = \int_a^c f + \int_c^b f.$$

1.4 Saks-Henstock Lemma

The aim of this section is to establish the important Saks-Henstock Lemma (Theorem 1.4.4) for the Henstock-Kurzweil integral. As a result, we deduce that there are no improper Henstock-Kurzweil integrals (Theorems 1.4.6 and 1.4.8). We begin with the following definitions.

Definition 1.4.1. A finite collection $\{(t_1, [u_1, v_1]), \ldots, (t_p, [u_p, v_p])\}$ of point-interval pairs is said to be a *Perron subpartition* of $[a, b]$ if $t_i \in [u_i, v_i]$ for $i = 1, \ldots, p$, and $\{[u_1, v_1], \ldots, [u_p, v_p]\}$ is a finite collection of non-overlapping subintervals of $[a, b]$.

Definition 1.4.2. Let $\{(t_1, [u_1, v_1]), \ldots, (t_p, [u_p, v_p])\}$ be a Perron subpartition of $[a, b]$ and let δ be a gauge on $\{t_1, \ldots, t_p\}$. The Perron subpartition $\{(t_1, [u_1, v_1]), \ldots, (t_p, [u_p, v_p])\}$ is said to be *δ-fine* if $[u_i, v_i] \subset (t_i - \delta(t_i), t_i + \delta(t_i))$ for $i - 1, \ldots, p$.

By replacing δ-fine Perron partitions by δ-fine Perron subpartitions in Definition 1.2.2, we obtain the following

Lemma 1.4.3. *Let $f \in HK[a, b]$ and let $\varepsilon > 0$. If δ is a gauge on $[a, b]$ such that*

$$\left| \sum_{(x,[y,z]) \in Q} f(x)(z - y) - (HK) \int_a^b f \right| < \varepsilon$$

for each δ-fine Perron partition Q of $[a, b]$, then

$$\left| \sum_{(t,[u,v]) \in P} \left\{ f(t)(v - u) - (HK) \int_u^v f \right\} \right| \leq \varepsilon \tag{1.4.1}$$

for each δ-fine Perron subpartition P of $[a, b]$.

Proof. Let P be a δ-fine Perron subpartition of $[a, b]$. If $\bigcup_{(t,[u,v]) \in P} [u, v] = [a, b]$, then (1.4.1) follows from Theorem 1.3.8. Henceforth we assume that $\bigcup_{(t,[u,v]) \in P} [u, v] \subset [a, b]$; in this case, we choose non-overlapping intervals $[x_1, y_1], \ldots, [x_q, y_q]$ such that

$$[a, b] \setminus \bigcup_{(t,[u,v]) \in P} (u, v) = \bigcup_{k=1}^{q} [x_k, y_k].$$

For each $k \in \{1, \ldots, q\}$ we infer from Theorem 1.3.6 that $f \in HK[x_k, y_k]$. Hence for each $\eta > 0$ there exists a gauge δ_k on $[x_k, y_k]$ such that

$$\left| S(f, P_k) - (HK) \int_{x_k}^{y_k} f \right| < \frac{\eta}{q}$$

for each δ_k-fine Perron partition P_k of $[x_k, y_k]$. Since δ and δ_k are gauges on $[u_k, v_k]$, we can apply Cousin's Lemma to select and fix a $\min\{\delta, \delta_k\}$-fine Perron partition Q_k of $[x_k, y_k]$.

Following the proof of Lemma 1.1.4, we conclude that $Q := P \cup \bigcup_{k=1}^{q} Q_k$ is a δ-fine Perron partition of $[a,b]$ such that

$$S(f,Q) = S(f,P) + \sum_{k=1}^{q} S(f,Q_k)$$

and

$$(HK)\int_a^b f = \sum_{(t,[u,v])\in P} (HK)\int_u^v f + \sum_{k=1}^{q} (HK)\int_{x_k}^{y_k} f.$$

Consequently,

$$\begin{aligned}
&\left|\sum_{(t,[u,v])\in P} \left\{ f(t)(v-u) - (HK)\int_u^v f \right\}\right| \\
&= \left| \left\{ S(f,Q) - \sum_{k=1}^{q} S(f,Q_k) \right\} - \left\{ (HK)\int_a^b f - \sum_{k=1}^{q} (HK)\int_{x_k}^{v_k} f \right\} \right| \\
&\leq \left| S(f,Q) - (HK)\int_a^b f \right| + \sum_{k=1}^{q} \left| S(f,Q_k) - (HK)\int_{x_k}^{v_k} f \right| \\
&< \varepsilon + \eta.
\end{aligned}$$

Since $\eta > 0$ is arbitrary, the lemma is proved. □

We are now ready to state and prove the following crucial Saks-Henstock Lemma, which plays an important role for the rest of this chapter.

Theorem 1.4.4 (Saks-Henstock). *If $f \in HK[a,b]$, then for each $\varepsilon > 0$ there exists a gauge on δ on $[a,b]$ such that*

$$\sum_{(t,[u,v])\in P} \left| f(t)(v-u) - (HK)\int_u^v f \right| < \varepsilon \tag{1.4.2}$$

for each δ-fine Perron subpartition P of $[a,b]$.

Proof. Since f is Henstock-Kurzweil integrable on $[a,b]$, it follows from Lemma 1.4.3 that there exists a gauge δ on $[a,b]$ such that

$$\left|\sum_{(z,[x,y])\in Q} \left\{ f(z)(y-x) - (HK)\int_x^y f \right\}\right| < \frac{\varepsilon}{2} \tag{1.4.3}$$

for each δ-fine Perron subpartition Q of $[a,b]$. Let P be a δ-fine Perron subpartition of $[a,b]$, let

$$P^+ = \left\{ (t,[u,v]) \in P : f(t)(v-u) - (HK)\int_u^v f \geq 0 \right\}$$

and let $P^- = P \backslash P^+$. Then $P^+ \cup P^-$ is a δ-fine Perron subpartition of $[a, b]$ and hence the desired result follows from (1.4.3):

$$\begin{aligned}
&\sum_{(t,[u,v]) \in P} \left| f(t)(v-u) - (HK)\int_u^v f \right| \\
&= \sum_{(t,[u,v]) \in P^+} \left\{ f(t)(v-u) - (HK)\int_u^v f \right\} \\
&- \sum_{(t,[u,v]) \in P^-} \left\{ f(t)(v-u) - (HK)\int_u^v f \right\} \\
&< \varepsilon.
\end{aligned}$$

□

Theorem 1.4.5. *Let $f \in HK[a, b]$ and let $F(x) = (HK)\int_a^x f$ for each $x \in [a, b]$. Then F is continuous on $[a, b]$.*

Proof. For each $\varepsilon > 0$ we apply the Saks-Henstock Lemma to select a gauge δ on $[a, b]$ such that

$$\sum_{(t,[u,v]) \in P} \left| f(t)(v-u) - (HK)\int_u^v f \right| < \frac{\varepsilon}{2}$$

for each δ-fine Perron subpartition P of $[a, b]$. By making δ smaller, we may assume that $\delta(t) \leq \frac{\varepsilon}{2(1+|f(t)|)}$ for all $t \in [a, b]$. If $x \in [a, b]$, then

$$|F(y) - F(x)| \leq |f(x)(y-x) - (F(y) - F(x))| + |f(x)(y-x)| < \varepsilon$$

whenever $y \in (x - \delta(x), x + \delta(x)) \cap [a, b]$. Since $x \in [a, b]$ is arbitrary, we conclude that F is continuous on $[a, b]$. □

Theorem 1.4.6 (Cauchy extension). *A function $f : [a, b] \longrightarrow \mathbb{R}$ is Henstock-Kurzweil integrable on $[a, b]$ if and only if for each $c \in (a, b)$ the function $f|_{[a,c]}$ is Henstock-Kurzweil integrable on $[a, c]$ and*

$$\lim_{c \to b^-} (HK)\int_a^c f \quad \textit{exists.} \tag{1.4.4}$$

In this case, $(HK)\int_a^b f = \lim_{c \to b^-} (HK)\int_a^c f$.

Proof. ($\Longrightarrow$) This follows from Theorem 1.4.5.

($\Longleftarrow$) Let $\varepsilon > 0$ and let $(c_n)_{n=0}^\infty$ be a strictly increasing sequence of real numbers such that $c_0 = a$ and $\sup_{n \in \mathbb{N}} c_n = b$. Since (1.4.4) holds, there exists a positive integer N such that

$$\left| (HK)\int_a^x f - \lim_{c \to b^-} (HK)\int_a^c f \right| < \frac{\varepsilon}{4} \tag{1.4.5}$$

whenever $x \in [c_N, b)$.

For each $k \in \mathbb{N}$ we let δ_k be a gauge on $[c_{k-1}, c_k]$ so that the inequality

$$\sum_{(z,[x,y])\in Q_k} \left| f(z)(y-x) - (HK)\int_x^y f \right| < \frac{\varepsilon}{4(2^k)} \tag{1.4.6}$$

holds for each δ_k-fine Perron subpartition Q_k of $[c_{k-1}, c_k]$.

Define a gauge δ on $[a, b]$ by setting

$$\delta(x) = \begin{cases} \frac{1}{2}(c_1 - c_0) & \text{if } x = c_0, \\ \min\{\delta_k(c_k), \delta_{k+1}(c_k), \frac{1}{2}(c_k - c_{k-1}), \frac{1}{2}(c_{k+1} - c_k)\} & \\ \qquad\qquad \text{if } x = c_k \text{ for some } k \in \mathbb{N}, & \\ \min\{\delta_k(x), \frac{1}{2}(x - c_{k-1}), \frac{1}{2}(c_k - x)\} & \\ \qquad\qquad \text{if } x \in (c_{k-1}, c_k) \text{ for some } k \in \mathbb{N}, & \\ \min\{b - c_N, \frac{\varepsilon}{4(|f(b)|+1)}\} & \text{if } x = b, \end{cases}$$

and let $P = \{(t_1, [x_0, x_1]), \ldots, (t_p, [x_{p-1}, x_p])\}$ be a δ-fine Perron partition of $[a, b]$. After a suitable reordering, we may assume that $a = x_0 < x_1 < \cdots < x_p = b$. Since $b \notin \bigcup_{k=1}^{\infty} [c_{k-1}, c_k]$, our choice of δ implies that $t_p = b$ and $x_{p-1} \in (c_r, c_{r+1}]$ for some unique positive integer r. We also observe that if $k \in \{1, \ldots, r\}$, our choice of δ implies that

$$\{(t, [u, v]) \in P : [u, v] \subseteq [c_{k-1}, c_k]\}$$

is a δ_k-fine Perron partition of $[c_{k-1}, c_k]$. Thus

$$\begin{aligned} &\left| S(f, P) - \lim_{c \to b^-} (HK)\int_a^c f \right| \\ &\leq \left| \sum_{k=1}^{r} \left\{ \sum_{\substack{(t,[u,v])\in P \\ [u,v]\subseteq[c_{k-1},c_k]}} f(t)(v-u) - (HK)\int_{c_{k-1}}^{c_k} f \right\} \right| \\ &\quad + \left| \sum_{\substack{(t,[u,v])\in P \\ [u,v]\subseteq[c_r,x_{p-1}]}} f(t)(v-u) - (HK)\int_{c_r}^{x_{p-1}} f \right| + |f(b)|\,(b - x_{p-1}) \\ &\quad + \left| (HK)\int_a^{x_{p-1}} f - \lim_{c \to b^-} (HK)\int_a^c f \right| \\ &< \varepsilon. \end{aligned}$$

□

Corollary 1.4.7. *Let $f : [a,b] \longrightarrow \mathbb{R}$ and suppose that $f|_{[a,c]}$ is Riemann integrable on $[a,c]$ for every $c \in (a,b)$. If $\lim_{c\to b^-} \int_a^c f(t)\, dt$ exists, then $f \in HK[a,b]$ and*

$$\lim_{c\to b^-} \int_a^c f = (HK) \int_a^b f.$$

Likewise, we have the following modification of Theorem 1.4.6.

Theorem 1.4.8 (Cauchy extension). *A function $f : [a,b] \longrightarrow \mathbb{R}$ is Henstock-Kurzweil integrable on $[a,b]$ if and only if for each $c \in (a,b)$ the function $f|_{[c,b]}$ is Henstock-Kurzweil integrable on $[c,b]$ and*

$$\lim_{c\to a^+} (HK) \int_c^b f \text{ exists.}$$

In this case, $(HK) \int_a^b f = \lim_{c\to a^+} (HK) \int_c^b f$.

The following corollary is an immediate consequence of Theorem 1.4.8.

Corollary 1.4.9. *Let $f : [a,b] \longrightarrow \mathbb{R}$ and suppose that $f|_{[c,b]}$ is Riemann integrable on $[c,b]$ for every $c \in (a,b)$. If $\lim_{c\to a^+} \int_c^b f(t)\, dt$ exists, then $f \in HK[a,b]$ and*

$$\lim_{c\to b^-} \int_a^c f = (HK) \int_a^b f.$$

Example 1.4.10. Let

$$f(x) = \begin{cases} \frac{1}{\sqrt{1-x}} & \text{if } 0 \le x < 1, \\ 0 & \text{if } x = 0. \end{cases}$$

Since f is Henstock-Kurzweil integrable on $[0,c]$ for all $c \in [0,1)$ and

$$\lim_{c\to 1^-} (HK) \int_0^c \frac{1}{\sqrt{1-t}}\, dt = \lim_{c\to 1^-} \left(2 - 2\sqrt{1-c}\right) = 2,$$

it follows from Theorem 1.4.6 that $f \in HK[0,1]$ and

$$(HK) \int_0^1 f = 2.$$

Example 1.4.11. Let

$$h(x) = \begin{cases} \frac{1}{x^2} \sin \frac{1}{x} & \text{if } 0 < x \leq 1, \\ 0 & \text{if } x = 0. \end{cases}$$

It is clear that h is Henstock-Kurzweil integrable on $[c, 1]$ for all $c \in (0, 1)$. On the other hand, since the limit

$$\lim_{c \to 0^+} (HK) \int_c^1 \frac{1}{x^2} \sin \frac{1}{x} \, dx = \lim_{c \to 0^+} \left(\cos 1 - \cos \frac{1}{c} \right)$$

does not exist, an application of Theorem 1.4.8 shows that h is not Henstock-Kurzweil integrable on $[0, 1]$.

Exercise 1.4.12. Show that the Saks-Henstock Lemma remains true for the Riemann integral.

Further applications of the Saks-Henstock Lemma will be given in the subsequent chapters.

1.5 Notes and Remarks

Cousin's Lemma has been used by Gordon [46] to prove some classical results in analysis. Theorem 1.1.6 is also due to Gordon [46].

It is known that $R[a, b]$ is a linear space. Further properties of the Riemann integral can be found in [88, Sections 1.3-1.5] or [6, Chapter 7].

1n 1957, Kurzweil [71] gave a slight but ingenious modification of the classical Riemann integral and used it in his work on differential equations. Later, Henstock [55] discovered the integral independently and developed the theory further. This integral, which is now commonly known as the Henstock-Kurzweil integral, is also known as the Henstock integral, the Kurzweil-Henstock integral, or the generalized Riemann integral; see, for example, [4–6, 85, 88]. In dimension one, this integral is equivalent to the Perron integral in the following sense: a function which is integrable in one sense is integrable in the other sense and both integrals coincide; a proof of this result is given in [44, Chapter 11]. A good overall view of the theory can be found in Bongiorno [10] and Lee [86]. See also [133].

Chapter 2

The multiple Henstock-Kurzweil integral

The goal of this chapter is to prove some basic properties of the m-dimensional Henstock-Kurzweil integral, where m is a positive integer. While most results are similar to that of Chapter 1, certain proofs require different techniques.

2.1 Preliminaries

Let $\mathbb{R}^m$ be the m-dimensional Euclidean space. A typical element $(x_1, \ldots, x_m)$ in $\mathbb{R}^m$ will be denoted by $\boldsymbol{x}$.

An *interval* in $\mathbb{R}^m$ is a set of the form $[\boldsymbol{u}, \boldsymbol{v}] := \prod_{i=1}^m [u_i, v_i]$, where $-\infty < u_i < v_i < \infty$ for $i = 1, \ldots, m$. On the other hand, the set $\prod_{i=1}^m [s_i, t_i] \subset \mathbb{R}^m$ is known as a *degenerate* interval if $s_i = t_i$ for some $i \in \{1, \ldots, m\}$. The aim of this section is to establish some basic properties of intervals in $\mathbb{R}^m$.

Definition 2.1.1. Two intervals $[\boldsymbol{s}, \boldsymbol{t}]$, $[\boldsymbol{u}, \boldsymbol{v}]$ in $\mathbb{R}^m$ are said to be *non-overlapping* if $\prod_{i=1}^m (s_i, t_i) \cap \prod_{i=1}^m (u_i, v_i)$ is empty.

We remark that it is not difficult to check that if $[s, t]$ and $[u, v]$ are non-overlapping intervals in $\mathbb{R}$, then $[s, t] \cup [u, v]$ is an interval in $\mathbb{R}$ if and only if $s = v$ or $t = u$. The following lemma, which is equivalent to [73, 1.5 Lemma], gives a necessary and sufficient condition for a union of two intervals in $\mathbb{R}^m$ to be an interval in $\mathbb{R}^m$.

Lemma 2.1.2. *Let* $[\boldsymbol{s}, \boldsymbol{t}]$ *and* $[\boldsymbol{u}, \boldsymbol{v}]$ *be non-overlapping intervals in* $\mathbb{R}^m$. *Then* $[\boldsymbol{s}, \boldsymbol{t}] \cup [\boldsymbol{u}, \boldsymbol{v}]$ *is an interval in* $\mathbb{R}^m$ *if and only if there exists* $j \in \{1, \ldots, m\}$ *with the following properties:*

(i) $(s_j, t_j) \cap (u_j, v_j)$ *is empty;*
(ii) $[s_j, t_j] \cap [u_j, v_j]$ *is non-empty;*
(iii) $[s_i, t_i] = [u_i, v_i]$ *for all* $i \in \{1, \ldots, m\} \backslash \{j\}$.

Proof. ($\Longrightarrow$) Suppose that $[\boldsymbol{s}, \boldsymbol{t}] \cup [\boldsymbol{u}, \boldsymbol{v}]$ is an interval in $\mathbb{R}^m$. Since $[\boldsymbol{s}, \boldsymbol{t}]$ and $[\boldsymbol{u}, \boldsymbol{v}]$ are assumed to be non-overlapping intervals in $\mathbb{R}^m$, there exists $j \in \{1, \ldots, m\}$ such that $(s_j, t_j) \cap (u_j, v_j)$ is empty; that is, (i) holds.

To prove (ii), it suffices to prove that

$$[s_1, t_1] \cup [u_1, v_1], \ldots, [s_m, t_m] \cup [u_m, v_m]$$

are intervals in $\mathbb{R}$, which is equivalent to the condition that $\prod_{i=1}^m([s_i, t_i] \cup [u_i, v_i])$ is an interval in $\mathbb{R}^m$. Since $[\boldsymbol{s}, \boldsymbol{t}] \cup [\boldsymbol{u}, \boldsymbol{v}]$ is assumed to be an interval in $\mathbb{R}^m$, we get

$$[\boldsymbol{s}, \boldsymbol{t}] \cup [\boldsymbol{u}, \boldsymbol{v}] = \prod_{i=1}^m([s_i, t_i] \cup [u_i, v_i]). \tag{2.1.1}$$

It remains to prove that (iii) holds. To do this, we use (i) and (ii) to fix $x_j \in [u_j, v_j] \backslash [s_j, t_j]$. If $x_i \in [s_i, t_i]$ for all $i \in \{1, \ldots, m\} \backslash \{j\}$, then it follows from our choice of x_j and (2.1.1) that $\boldsymbol{x} \in [\boldsymbol{u}, \boldsymbol{v}]$. Hence $[s_i, t_i] \subseteq [u_i, v_i]$ for all $i \in \{1, \ldots, m\} \backslash \{j\}$. Since a similar argument shows that $[u_i, v_i] \subseteq [s_i, t_i]$ for all $i \in \{1, \ldots, m\} \backslash \{j\}$, we get (iii).

($\Longleftarrow$) Conversely, suppose that (i), (ii) and (iii) are satisfied. In this case, it is not difficult to check that (2.1.1) holds. □

In the proof of Theorem 1.3.5, $\mu_1(\emptyset) := 0$, and $\mu_1([u, v]) := v - u$ whenever $-\infty < u \leq v < \infty$. Similarly, we let $\mu_m(\emptyset) := 0$, and

$$\mu_m\Big(\prod_{k=1}^m [u_k, v_k]\Big) := \prod_{i=1}^m (v_i - u_i)$$

whenever $-\infty < u_i \leq v_i < \infty$ for $i = 1, \ldots, m$. The following theorem is obvious.

Theorem 2.1.3. *If* I *and* J *are intervals in* $\mathbb{R}^m$ *and* $J \subseteq I$, *then* $\mu_m(J) \leq \mu_m(I)$.

For the rest of this section, we fix an interval $[\boldsymbol{a}, \boldsymbol{b}]$ in $\mathbb{R}^m$. A division of $[\boldsymbol{a}, \boldsymbol{b}]$ is a finite collection of pairwise non-overlapping intervals in $\mathbb{R}^m$ whose union is $[\boldsymbol{a}, \boldsymbol{b}]$. Our goal is to show that if D is a *division* of $[\boldsymbol{a}, \boldsymbol{b}]$, then

$$\mu_m([\boldsymbol{a}, \boldsymbol{b}]) = \sum_{I \in D} \mu_m(I).$$

Example 2.1.4. The collection

$$\{[0,1]\times[0,2],[1,2]\times[0,1],[1,2]\times[1,2]\}$$

is a division of $[0,2]\times[0,2]$.

We begin a special kind of division. A division D of $[\boldsymbol{a},\boldsymbol{b}]$ is a *net* if for each $k\in\{1,\ldots,m\}$ there exists a division D_k of $[a_k,b_k]$ such that

$$D=\left\{\prod_{k=1}^{m}[s_k,t_k]:[s_k,t_k]\in D_k \text{ for } k=1,\ldots,m\right\}.$$

Example 2.1.5. The collection

$$\{[0,1]\times[0,1],[0,1]\times[1,2],[1,2]\times[0,1],[1,2]\times[1,2]\}$$

is a net of $[0,2]\times[0,2]$.

Let $\mathcal{I}_m([\boldsymbol{a},\boldsymbol{b}])$ be the collection of all subintervals of $[\boldsymbol{a},\boldsymbol{b}]$. The following lemma is obvious.

Lemma 2.1.6. *If $I\in\mathcal{I}_m([\boldsymbol{a},\boldsymbol{b}])$, then there exists a net D of $[\boldsymbol{a},\boldsymbol{b}]$ such that $I\in D$ and the cardinality of D is not more than 3^m.*

The following lemma is a consequence of Lemma 2.1.6.

Lemma 2.1.7. *If $\{I_1,\ldots,I_p\}\subset\mathcal{I}_m([\boldsymbol{a},\boldsymbol{b}])$ is a finite collection of non-overlapping intervals in $\mathbb{R}^m$, then there exists a net D_0 of $[\boldsymbol{a},\boldsymbol{b}]$ with the following property: if $J\in D_0$ and $J\cap I_r\in\mathcal{I}_m([\boldsymbol{a},\boldsymbol{b}])$ for some $r\in\{1,\ldots,p\}$, then $J\subseteq I_r$.*

Proof. For each $k\in\{1,\ldots,p\}$ we apply Lemma 2.1.6 to select a net D_k of $[\boldsymbol{a},\boldsymbol{b}]$ such that $I_k\in D_k$. Then

$$D_0:=\left\{\bigcap_{k=1}^{p}J_k\in\mathcal{I}_m([\boldsymbol{a},\boldsymbol{b}]):J_k\in D_k \text{ for } k=1,\ldots,p\right\}$$

is a net of $[\boldsymbol{a},\boldsymbol{b}]$ with the desired property. □

The following lemma is a consequence of Lemma 2.1.2.

Lemma 2.1.8. *If* $[\boldsymbol{s},\boldsymbol{t}]$ *and* $[\boldsymbol{u},\boldsymbol{v}]$ *are non-overlapping subintervals of* $[\boldsymbol{a},\boldsymbol{b}]$ *such that* $[\boldsymbol{s},\boldsymbol{t}]\cup[\boldsymbol{u},\boldsymbol{v}]\in\mathcal{I}_m([\boldsymbol{a},\boldsymbol{b}])$, *then*

$$\mu_m([\boldsymbol{s},\boldsymbol{t}]\cup[\boldsymbol{u},\boldsymbol{v}])=\mu_m([\boldsymbol{s},\boldsymbol{t}])+\mu_m([\boldsymbol{u},\boldsymbol{v}]).$$

Proof. Let $j\in\{1,\ldots,m\}$ be given as in Lemma 2.1.2. Then

$$\begin{aligned}
&\mu_m([\boldsymbol{s},\boldsymbol{t}]\cup[\boldsymbol{u},\boldsymbol{v}])\\
&=\mu_m\Big(\prod_{i=1}^{m}([s_i,t_i]\cup[u_i,v_i])\Big)\\
&=(t_j-s_j+v_j-u_j)\prod_{\substack{i=1\\ i\neq j}}^{m}(t_i-s_i)\\
&=\mu_m([\boldsymbol{s},\boldsymbol{t}])+\mu_m([\boldsymbol{u},\boldsymbol{v}]).
\end{aligned}$$

□

The following theorem is a consequence of Lemmas 2.1.7 and 2.1.8.

Theorem 2.1.9. *If* $\{I_1,\ldots,I_p\}$ *is a division of* $[\boldsymbol{a},\boldsymbol{b}]$, *then*

$$\mu_m([\boldsymbol{a},\boldsymbol{b}])=\sum_{k=1}^{p}\mu_m(I_k).$$

Proof. If $\{I_1,\ldots,I_p\}$ is a net of $[\boldsymbol{a},\boldsymbol{b}]$, then the result is a consequence of Lemma 2.1.8.

On the other hand, suppose that $\{I_1,\ldots,I_p\}$ is not a net of $[\boldsymbol{a},\boldsymbol{b}]$. In this case, we let D_0 be given as in Lemma 2.1.7. Thus

$$\begin{aligned}
&\sum_{k=1}^{p}\mu_m(I_k)\\
&=\sum_{k=1}^{p}\sum_{J\in D_0}\mu_m(J\cap I_k)\\
&=\sum_{J\in D_0}\sum_{k=1}^{p}\mu_m(J\cap I_k)\\
&=\sum_{J\in D_0}\mu_m(J)\\
&=\mu_m(I).
\end{aligned}$$

□

Some generalizations of μ_m will be given in Chapters 3 and 5.

2.2 The Henstock-Kurzweil integral

Unless specified otherwise, for the rest of this book the space $\mathbb{R}^m$ will be equipped with the maximum norm $|||\cdot|||$, where

$$|||\boldsymbol{x}||| = \max\{|x_i| : i = 1, \ldots, m\},$$

and $[\boldsymbol{a}, \boldsymbol{b}]$ denotes a fixed interval in $\mathbb{R}^m$. Given $\boldsymbol{x} \in \mathbb{R}^m$ and $r > 0$, we set

$$B(\boldsymbol{x}, r) := \{\boldsymbol{y} \in \mathbb{R}^m : |||\boldsymbol{x} - \boldsymbol{y}||| < r\},$$

where $\boldsymbol{x} - \boldsymbol{y} := (x_1 - y_1, \ldots, x_m - y_m)$ $(\boldsymbol{y} \in \mathbb{R}^m)$.

Definition 2.2.1.

(i) A *point-interval pair* $(\boldsymbol{t}, I)$ consists of a point $\boldsymbol{t} \in \mathbb{R}^m$ and an interval I in $\mathbb{R}^m$. Here $\boldsymbol{t}$ is known as the *tag* of I.

(ii) A *Perron partition* of $[\boldsymbol{a}, \boldsymbol{b}]$ is a finite collection $\{(\boldsymbol{t}_1, [\boldsymbol{u}_1, \boldsymbol{v}_1]), \ldots, (\boldsymbol{t}_p, [\boldsymbol{u}_p, \boldsymbol{v}_p])\}$ of point-interval pairs, where $\{[\boldsymbol{u}_1, \boldsymbol{v}_1], \ldots, [\boldsymbol{u}_p, \boldsymbol{v}_p]\}$ is a division of $[\boldsymbol{a}, \boldsymbol{b}]$, and $\boldsymbol{t}_k \in [\boldsymbol{u}_k, \boldsymbol{v}_k]$ for $k = 1, \ldots, p$.

(iii) Let $P = \{(\boldsymbol{t}_1, [\boldsymbol{u}_1, \boldsymbol{v}_1]), \ldots, (\boldsymbol{t}_p, [\boldsymbol{u}_p, \boldsymbol{v}_p])\}$ be a Perron partition of $[\boldsymbol{a}, \boldsymbol{b}]$ and let δ be a gauge (i.e. positive function) defined on $\{\boldsymbol{t}_1, \ldots, \boldsymbol{t}_p\}$. The Perron partition P is said to be *δ-fine* if $[\boldsymbol{u}_k, \boldsymbol{v}_k] \subset B(\boldsymbol{t}_k, \delta(\boldsymbol{t}_k))$ for $k = 1, \ldots, p$.

We are now ready to state and prove the following crucial result.

Theorem 2.2.2 (Cousin's Lemma). *If δ is a gauge on $[\boldsymbol{a}, \boldsymbol{b}]$, then there exists a δ-fine Perron partition of $[\boldsymbol{a}, \boldsymbol{b}]$.*

Proof. Proceeding towards a contradiction, suppose that $[\boldsymbol{a}, \boldsymbol{b}]$ does not have a δ-fine Perron partition. For each $k \in \{1, \ldots, m\}$, we divide $[a_k, b_k]$ into $[a_k, \frac{a_k+b_k}{2}]$ and $[\frac{a_k+b_k}{2}, b_k]$ so that $[\boldsymbol{a}, \boldsymbol{b}]$ is a union of 2^m pairwise non-overlapping intervals in $\mathbb{R}^m$. Since $[\boldsymbol{a}, \boldsymbol{b}]$ does not have a δ-fine Perron partition, we can select an interval I_1 from the above-mentioned 2^m intervals so that there is no δ-fine Perron partition of I_1. Using induction, we construct a sequence $(I_n)_{n=1}^{\infty}$ of subintervals of $[\boldsymbol{a}, \boldsymbol{b}]$ so that the following conditions hold for every $n \in \mathbb{N}$:

(i) $I_n \supset I_{n+1}$;

(ii) no δ-fine Perron partition of I_n exists;

(iii) $\max\{|||\boldsymbol{t} - \boldsymbol{s}||| : \boldsymbol{s}, \boldsymbol{t} \in I_n\} = \frac{1}{2^n} |||\boldsymbol{b} - \boldsymbol{a}|||$.

Since properties (i) and (iii) hold for every $n \in \mathbb{N}$, it follows from the Nested Interval Theorem that

$$\bigcap_{k=1}^{\infty} I_k = \{\boldsymbol{t}_0\}$$

for some $\boldsymbol{t}_0 \in \mathbb{R}^m$. On the other hand, since $\bigcap_{k=1}^{\infty} I_k = \{\boldsymbol{t}_0\}$ and $\delta(\boldsymbol{x}_0) > 0$, it follows from property (iii) that there exists $N \in \mathbb{N}$ such that $\{(\boldsymbol{t}_0, I_N)\}$ is a δ-fine Perron partition of I_N. But this contradicts property (ii). Therefore a δ-fine Perron partition of $[\boldsymbol{a}, \boldsymbol{b}]$ exists. □

Let $C[\boldsymbol{a}, \boldsymbol{b}]$ be the space of continuous functions on $[\boldsymbol{a}, \boldsymbol{b}]$. A simple application of Cousin's Lemma gives the following result.

Theorem 2.2.3. *If $f \in C[\boldsymbol{a}, \boldsymbol{b}]$, then f is uniformly continuous on $[\boldsymbol{a}, \boldsymbol{b}]$.*

Proof. Exercise. □

Following the proof of Theorem 2.2.3, we get the following corollary.

Corollary 2.2.4. *If $f \in C[\boldsymbol{a}, \boldsymbol{b}]$, then f is bounded on $[\boldsymbol{a}, \boldsymbol{b}]$.*

Let $\{(\boldsymbol{t}_1, [\boldsymbol{u}_1, \boldsymbol{v}_1]), \ldots, (\boldsymbol{t}_p, [\boldsymbol{u}_p, \boldsymbol{v}_p])\}$ be a Perron partition of $[\boldsymbol{a}, \boldsymbol{b}]$. For any function f defined on $\{\boldsymbol{t}_1, \ldots, \boldsymbol{t}_p\}$, we set

$$S(f, P) = \sum_{k=1}^{p} f(\boldsymbol{t}_k)\mu_m([\boldsymbol{u}_k, \boldsymbol{v}_k]).$$

We begin with the definition of the Riemann integral.

Definition 2.2.5. A function $f : [\boldsymbol{a}, \boldsymbol{b}] \longrightarrow \mathbb{R}$ is said to be *Riemann integrable* on $[\boldsymbol{a}, \boldsymbol{b}]$ if there exists $A_0 \in \mathbb{R}$ with the following property: given $\varepsilon > 0$ there exists a <u>constant</u> gauge δ on $[\boldsymbol{a}, \boldsymbol{b}]$ such that

$$|S(f, P) - A_0| < \varepsilon \tag{2.2.1}$$

for each δ-fine Perron partition P of $[\boldsymbol{a}, \boldsymbol{b}]$.

The collection of all functions that are Riemann integrable on $[\boldsymbol{a}, \boldsymbol{b}]$ will be denoted by $R[\boldsymbol{a}, \boldsymbol{b}]$.

Once we omit the word "constant" from Definition 2.2.5, we obtain the following modification of Definition 2.2.5.

Definition 2.2.6. A function $f : [\boldsymbol{a}, \boldsymbol{b}] \longrightarrow \mathbb{R}$ is said to be *Henstock-Kurzweil integrable* on $[\boldsymbol{a}, \boldsymbol{b}]$ if there exists $A \in \mathbb{R}$ with the following property: given $\varepsilon > 0$ there exists a gauge δ on $[\boldsymbol{a}, \boldsymbol{b}]$ such that

$$|S(f, P) - A| < \varepsilon \tag{2.2.2}$$

for each δ-fine Perron partition P of $[\boldsymbol{a}, \boldsymbol{b}]$.

The collection of all functions that are Henstock-Kurzweil integrable on $[\boldsymbol{a}, \boldsymbol{b}]$ will be denoted by $HK[\boldsymbol{a}, \boldsymbol{b}]$.. It is easy to deduce from Definitions 2.2.5 and 2.2.6 that if $f \in R[\boldsymbol{a}, \boldsymbol{b}]$, then $f \in HK[\boldsymbol{a}, \boldsymbol{b}]$. In this case, Cousin's Lemma shows that there is a unique number satisfying Definitions 2.2.5 and 2.2.6.

Theorem 2.2.7. *There is at most one number A satisfying Definition 2.2.6.*

Proof. Suppose that both numbers A_1 and A_2 satisfy Definition 2.2.6. We claim that $A_1 = A_2$.

Let $\varepsilon > 0$ be given. Since A_1 satisfies Definition 2.2.6, there exists a gauge δ_1 on $[\boldsymbol{a}, \boldsymbol{b}]$ such that

$$|S(f, P_1) - A_1| < \frac{\varepsilon}{2}$$

for each δ_1-fine Perron partition P_1 of $[\boldsymbol{a}, \boldsymbol{b}]$. Similarly, there exists a gauge δ_2 on $[\boldsymbol{a}, \boldsymbol{b}]$ such that

$$|S(f, P_2) - A_2| < \frac{\varepsilon}{2}$$

for each δ_2-fine Perron partition P_2 of $[\boldsymbol{a}, \boldsymbol{b}]$.

Define a gauge δ on $[\boldsymbol{a}, \boldsymbol{b}]$ by setting

$$\delta(\boldsymbol{x}) = \min\{\delta_1(\boldsymbol{x}), \delta_2(\boldsymbol{x})\}.$$

In view of Cousin's Lemma, we may fix a δ-fine Perron partition P of $[\boldsymbol{a}, \boldsymbol{b}]$. Using our definition of δ, it is not difficult to check that P is both δ_1-fine and δ_2-fine. Thus

$$|A_1 - A_2| \leq |S(f, P) - A_1| + |S(f, P) - A_2| < \varepsilon.$$

Since $\varepsilon > 0$ is arbitrary, we conclude that $A_1 = A_2$. □

The above theorem shows that if $f \in HK[\boldsymbol{a}, \boldsymbol{b}]$, then there is a unique number A satisfying Definition 2.2.6. In this case, we write A as $(HK)\int_{[\boldsymbol{a},\boldsymbol{b}]} f$, $(HK)\int_{[\boldsymbol{a},\boldsymbol{b}]} f(\boldsymbol{x})\, d\boldsymbol{x}$ or $(HK)\int_{[\boldsymbol{a},\boldsymbol{b}]} f(\boldsymbol{t})\, d\boldsymbol{t}$, which is known as the *Henstock-Kurzweil integral* of f over $[\boldsymbol{a}, \boldsymbol{b}]$. If, in addition, $f \in R[\boldsymbol{a}, \boldsymbol{b}]$, the Riemann integral of f over $[\boldsymbol{a}, \boldsymbol{b}]$ will be denoted by $\int_{[\boldsymbol{a},\boldsymbol{b}]} f$, $\int_{[\boldsymbol{a},\boldsymbol{b}]} f(\boldsymbol{x})\, d\boldsymbol{x}$ or $\int_{[\boldsymbol{a},\boldsymbol{b}]} f(\boldsymbol{t})\, d\boldsymbol{t}$.

Example 2.2.8. Let f be a constant real-valued function defined on $[\boldsymbol{a}, \boldsymbol{b}]$. Then f is Riemann and hence Henstock-Kurzweil integrable on $[\boldsymbol{a}, \boldsymbol{b}]$.

Further examples of Henstock-Kurzweil integrable functions will be given in Section 2.3.

2.3 Simple properties

The aim of this section is to prove some basic properties of the Henstock-Kurzweil integral via Definition 2.2.6.

Theorem 2.3.1. *If $f, g \in HK[\boldsymbol{a}, \boldsymbol{b}]$, then $f + g \in HK[\boldsymbol{a}, \boldsymbol{b}]$ and*

$$(HK)\int_{[\boldsymbol{a},\boldsymbol{b}]} (f+g) = (HK)\int_{[\boldsymbol{a},\boldsymbol{b}]} f + (HK)\int_{[\boldsymbol{a},\boldsymbol{b}]} g.$$

Proof. Let $\varepsilon > 0$ be given. Since $f \in HK[\boldsymbol{a}, \boldsymbol{b}]$, there exists a gauge δ_1 on $[\boldsymbol{a}, \boldsymbol{b}]$ such that

$$\left| S(f, P_1) - (HK)\int_{[\boldsymbol{a},\boldsymbol{b}]} f \right| < \frac{\varepsilon}{2}$$

for each δ_1-fine Perron partition P_1 of $[\boldsymbol{a}, \boldsymbol{b}]$. Similarly, there exists a gauge δ_2 on $[\boldsymbol{a}, \boldsymbol{b}]$ such that

$$\left| S(g, P_2) - (HK)\int_{[\boldsymbol{a},\boldsymbol{b}]} g \right| < \frac{\varepsilon}{2}$$

for each δ_2-fine Perron partition P_2 of $[\boldsymbol{a}, \boldsymbol{b}]$.

Define a gauge δ on $[\boldsymbol{a}, \boldsymbol{b}]$ by setting

$$\delta(\boldsymbol{x}) = \min\{\delta_1(\boldsymbol{x}), \delta_2(\boldsymbol{x})\}, \tag{2.3.1}$$

and let P be a δ-fine Perron partition of $[\boldsymbol{a}, \boldsymbol{b}]$. Since (2.3.1) implies that the δ-fine Perron partition P is both δ_1-fine and δ_2-fine, the identity

$$S(f+g, P) = S(f, P) + S(g, P)$$

and the triangle inequality yield

$$\begin{aligned} &\left| S(f+g, P) - \left\{ (HK)\int_{[\boldsymbol{a},\boldsymbol{b}]} f + (HK)\int_{[\boldsymbol{a},\boldsymbol{b}]} g \right\} \right| \\ &\leq \left| S(f, P) - (HK)\int_{[\boldsymbol{a},\boldsymbol{b}]} f \right| + \left| S(g, P) - (HK)\int_{[\boldsymbol{a},\boldsymbol{b}]} g \right| \\ &< \varepsilon. \end{aligned}$$

Since $\varepsilon > 0$ is arbitrary, we conclude that $f + g \in HK[\boldsymbol{a}, \boldsymbol{b}]$ and

$$(HK)\int_{[\boldsymbol{a},\boldsymbol{b}]} (f+g) = (HK)\int_{[\boldsymbol{a},\boldsymbol{b}]} f + (HK)\int_{[\boldsymbol{a},\boldsymbol{b}]} g.$$

□

Theorem 2.3.2. *If $f \in HK[\boldsymbol{a}, \boldsymbol{b}]$ and $c \in \mathbb{R}$, then $cf \in HK[\boldsymbol{a}, \boldsymbol{b}]$ and*

$$(HK)\int_{[\boldsymbol{a},\boldsymbol{b}]} cf = c\left\{(HK)\int_{[\boldsymbol{a},\boldsymbol{b}]} f\right\}.$$

Proof. Let $\varepsilon > 0$ be given. Since $f \in HK[\boldsymbol{a}, \boldsymbol{b}]$, there exists a gauge δ on $[\boldsymbol{a}, \boldsymbol{b}]$ such that

$$\left|S(f, P_1) - (HK)\int_{[\boldsymbol{a},\boldsymbol{b}]} f\right| < \frac{\varepsilon}{|c|+1}$$

for each δ-fine Perron partition P_1 of $[\boldsymbol{a}, \boldsymbol{b}]$. If P is a δ-fine Perron partition of $[\boldsymbol{a}, \boldsymbol{b}]$, then

$$\left|S(cf, P) - c\left\{(HK)\int_{[\boldsymbol{a},\boldsymbol{b}]} f\right\}\right| = |c|\left|S(f, P) - (HK)\int_{[\boldsymbol{a},\boldsymbol{b}]} f\right| < \varepsilon.$$

Since $\varepsilon > 0$ is arbitrary, the theorem is proved. □

Theorem 2.3.3. *If $f, g \in HK[\boldsymbol{a}, \boldsymbol{b}]$ and $f(\boldsymbol{x}) \leq g(\boldsymbol{x})$ for all $\boldsymbol{x} \in [\boldsymbol{a}, \boldsymbol{b}]$, then*

$$(HK)\int_{[\boldsymbol{a},\boldsymbol{b}]} f \leq (HK)\int_{[\boldsymbol{a},\boldsymbol{b}]} g.$$

Proof. Let $\varepsilon > 0$ be given. Since $f \in HK[\boldsymbol{a}, \boldsymbol{b}]$, there exists a gauge δ_1 on $[\boldsymbol{a}, \boldsymbol{b}]$ such that

$$\left|S(f, P_1) - (HK)\int_{[\boldsymbol{a},\boldsymbol{b}]} f\right| < \frac{\varepsilon}{2}$$

for each δ_1-fine Perron partition P_1 of $[\boldsymbol{a}, \boldsymbol{b}]$. Similarly, there exists a gauge δ_2 on $[\boldsymbol{a}, \boldsymbol{b}]$ such that

$$\left|S(g, P_2) - (HK)\int_{[\boldsymbol{a},\boldsymbol{b}]} g\right| < \frac{\varepsilon}{2}$$

for each δ_2-fine Perron partition P_2 of $[\boldsymbol{a}, \boldsymbol{b}]$.

Define a gauge δ on $[\boldsymbol{a}, \boldsymbol{b}]$ by setting

$$\delta(\boldsymbol{x}) = \min\{\delta_1(\boldsymbol{x}), \delta_2(\boldsymbol{x})\}, \tag{2.3.2}$$

and we apply Cousin's Lemma to fix a δ-fine Perron partition P_0 of $[\boldsymbol{a}, \boldsymbol{b}]$. Since (2.3.2) implies that P_0 is both δ_1-fine and δ_2-fine, we conclude that

$$(HK)\int_{[\boldsymbol{a},\boldsymbol{b}]} f < S(f, P_0) + \frac{\varepsilon}{2} \leq S(g, P_0) + \frac{\varepsilon}{2} < (HK)\int_{[\boldsymbol{a},\boldsymbol{b}]} g + \varepsilon,$$

and the desired inequality follows from the arbitrariness of ε. □

The following theorem gives a necessary and sufficient condition for a function f to be Henstock-Kurzweil integrable on $[\boldsymbol{a}, \boldsymbol{b}]$.

Theorem 2.3.4 (Cauchy Criterion). *A function* $f : [\boldsymbol{a}, \boldsymbol{b}] \longrightarrow \mathbb{R}$ *belongs to* $HK[\boldsymbol{a}, \boldsymbol{b}]$ *if and only if given* $\varepsilon > 0$ *there exists a gauge* δ *on* $[\boldsymbol{a}, \boldsymbol{b}]$ *such that*

$$|S(f, P) - S(f, Q)| < \varepsilon$$

for each pair of δ*-fine Perron partitions* P *and* Q *of* $[\boldsymbol{a}, \boldsymbol{b}]$.

Proof. ($\Longrightarrow$) Let $\varepsilon > 0$ be given. Since $f \in HK[\boldsymbol{a}, \boldsymbol{b}]$, there exists a gauge δ on $[\boldsymbol{a}, \boldsymbol{b}]$ such that

$$\left| S(f, P_0) - (HK) \int_{[\boldsymbol{a}, \boldsymbol{b}]} f \right| < \frac{\varepsilon}{2}$$

for each δ-fine Perron partition P_0 of $[\boldsymbol{a}, \boldsymbol{b}]$. Hence, for each pair of δ-fine Perron partitions P and Q of $[\boldsymbol{a}, \boldsymbol{b}]$, we have

$$\begin{aligned} |S(f, P) - S(f, Q)| &\leq \left| S(f, P) - (HK) \int_{[\boldsymbol{a}, \boldsymbol{b}]} f \right| + \left| S(f, Q) - (HK) \int_{[\boldsymbol{a}, \boldsymbol{b}]} f \right| \\ &< \varepsilon. \end{aligned}$$

($\Longleftarrow$) For each $n \in \mathbb{N}$ we let δ_n be a gauge on $[\boldsymbol{a}, \boldsymbol{b}]$ such that

$$|S(f, Q_n) - S(f, R_n)| < \frac{1}{n}$$

for each pair of δ_n-fine Perron partitions Q_n and R_n of $[\boldsymbol{a}, \boldsymbol{b}]$. Next we define a gauge Δ_n on $[\boldsymbol{a}, \boldsymbol{b}]$ by setting

$$\Delta_n(\boldsymbol{x}) = \min\{\delta_1(\boldsymbol{x}), \ldots, \delta_n(\boldsymbol{x})\},$$

and apply Cousin's Lemma to fix a Δ_n-fine Perron partition of $[\boldsymbol{a}, \boldsymbol{b}]$.

We claim that $(S(f, P_n))_{n=1}^{\infty}$ is a Cauchy sequence of real numbers. Let $\varepsilon > 0$ be given and choose a positive integer N so that $\frac{1}{N} < \varepsilon$. If n_1 and n_2 are positive integers satisfying $\min\{n_1, n_2\} \geq N$, then P_{n_1} and P_{n_2} are both $\Delta_{\min\{n_1, n_2\}}$-fine Perron partitions of $[\boldsymbol{a}, \boldsymbol{b}]$ and so

$$|S(f, P_{n_1}) - S(f, P_{n_2})| < \frac{1}{\min\{n_1, n_2\}} \leq \frac{1}{N} < \varepsilon.$$

Consequently, $(S(f, P_n))_{n=1}^{\infty}$ is a Cauchy sequence of real numbers. Since $\mathbb{R}$ is complete, the sequence $(S(f, P_n))_{n=1}^{\infty}$ of real numbers converges to some real number A.

It remains to prove that $f \in HK[\boldsymbol{a}, \boldsymbol{b}]$ and $A = (HK)\int_{[\boldsymbol{a},\boldsymbol{b}]} f$. Let P be a Δ_N-fine Perron partition P of $[\boldsymbol{a}, \boldsymbol{b}]$. Since the sequence $(\Delta_n)_{n=1}^{\infty}$ of gauges is non-increasing, we see that the Perron partition P_n is Δ_N-fine for every integers $n \geq N$. Thus

$$|S(f,P) - A| = \lim_{n\to\infty} |S(f,P) - S(f,P_n)| \leq \frac{1}{N} < \varepsilon.$$

Since $\varepsilon > 0$ is arbitrary, we conclude that $f \in HK[\boldsymbol{a}, \boldsymbol{b}]$ and $A = (HK)\int_{[\boldsymbol{a},\boldsymbol{b}]} f$. □

We are now ready to give an important class of Henstock-Kurzweil integrable functions.

Theorem 2.3.5. *If $f \in C[\boldsymbol{a}, \boldsymbol{b}]$, then $f \in HK[\boldsymbol{a}, \boldsymbol{b}]$.*

Proof. Let $\varepsilon > 0$ be given. Since f is continuous on $[\boldsymbol{a}, \boldsymbol{b}]$, for each $\boldsymbol{x} \in [\boldsymbol{a}, \boldsymbol{b}]$ there exists $\delta(\boldsymbol{x}) > 0$ such that

$$|f(\boldsymbol{y}) - f(\boldsymbol{x})| < \frac{\varepsilon}{2\prod_{i=1}^{m}(b_i - a_i)}$$

whenever $\boldsymbol{y} \in [a,b] \cap (\boldsymbol{x} - \delta(\boldsymbol{x}), \boldsymbol{x} + \delta(\boldsymbol{x}))$.

Clearly, the function $\boldsymbol{x} \mapsto \delta(\boldsymbol{x})$ is a gauge on $[\boldsymbol{a}, \boldsymbol{b}]$. To this end we let $P = \{(\boldsymbol{t}_1, I_1), \ldots, (\boldsymbol{t}_p, I_p)\}$ and $Q = \{(\boldsymbol{w}_1, J_1), \ldots, (\boldsymbol{w}_q, J_q)\}$ be two δ-fine Perron partitions of $[\boldsymbol{a}, \boldsymbol{b}]$. If $I_r \cap J_k$ is non-empty for some $r \in \{1, \ldots, p\}$ and $k \in \{1, \ldots, q\}$, we select and fix a point $\boldsymbol{z}_{r,k} \in I_r \cap J_k$. On the other hand, if $I_r \cap J_s$ is empty for some $r \in \{1, \ldots, p\}$ and $s \in \{1, \ldots, q\}$, we set $\boldsymbol{z}_{r,s} = \boldsymbol{a}$. Thus

$$\begin{aligned}
&|S(f,P) - S(f,Q)| \\
&= \left|\sum_{j=1}^{p} f(\boldsymbol{t}_j)\mu_m(I_j) - \sum_{k=1}^{q} f(\boldsymbol{w}_k)\mu_m(J_k)\right| \\
&= \left|\sum_{j=1}^{p}\sum_{k=1}^{q} f(\boldsymbol{t}_j)\mu_m(I_j \cap J_k) - \sum_{k=1}^{q}\sum_{j=1}^{p} f(\boldsymbol{w}_k)\mu_m(I_j \cap J_k)\right| \\
&\leq \left|\sum_{j=1}^{p}\sum_{k=1}^{q} f(\boldsymbol{t}_j)\mu_m(I_j \cap J_k) - \sum_{j=1}^{p}\sum_{k=1}^{q} f(\boldsymbol{z}_{j,k})\mu_m(I_j \cap J_k)\right| \\
&\quad + \left|\sum_{k=1}^{q}\sum_{j=1}^{p} f(\boldsymbol{w}_k)\mu_m(I_j \cap J_k) - \sum_{j=1}^{p}\sum_{k=1}^{q} f(\boldsymbol{z}_{j,k})\mu_m(I_j \cap J_k)\right| \\
&< \varepsilon.
\end{aligned}$$

An application of Theorem 2.3.4 completes the proof. □

The proof of Theorem 2.3.6 is slightly more complicated than that of Theorem 1.3.6.

Theorem 2.3.6. *If* $f \in HK[\boldsymbol{a},\boldsymbol{b}]$, *then* $f \in HK(I)$ *for each* $I \in \mathcal{I}_m([\boldsymbol{a},\boldsymbol{b}])$.

Proof. Let $\varepsilon > 0$ be given. According to Theorem 2.3.4, there is a gauge δ on $[\boldsymbol{a},\boldsymbol{b}]$ such that

$$|S(f,P) - S(f,Q)| < \varepsilon$$

for each pair of δ-fine Perron partitions P and Q of $[\boldsymbol{a},\boldsymbol{b}]$.

Suppose that $I \in \mathcal{I}_m([\boldsymbol{a},\boldsymbol{b}])$. If $I = [\boldsymbol{a},\boldsymbol{b}]$, then the theorem is true. Henceforth we assume that $I \neq [\boldsymbol{a},\boldsymbol{b}]$; in this case, we apply Lemma 2.1.6 to select a finite collection $\mathcal{C}$ of intervals so that $I \notin \mathcal{C}$ and $\mathcal{C} \cup \{I\}$ is a net of $[\boldsymbol{a},\boldsymbol{b}]$. For each $J \in \mathcal{C} \cup \{I\}$, we apply Cousin's Lemma to fix a δ-fine Perron partition P_J of J. If P_I and Q_I are δ-fine Perron partitions of I, then

$$P = P_I \cup \bigcup_{J\in\mathcal{C}} P_J \text{ and } Q = Q_I \cup \bigcup_{J\in\mathcal{C}} P_J$$

are δ-fine Perron partitions of $[\boldsymbol{a},\boldsymbol{b}]$ such that

$$S(f,P_I) = S(f,P) - \sum_{J\in\mathcal{C}} S(f,P_J)$$

and

$$S(f,Q_I) = S(f,Q) - \sum_{J\in\mathcal{C}} S(f,P_J).$$

Thus

$$|S(f,P_I) - S(f,Q_I)| = |S(f,P) - S(f,Q)| < \varepsilon,$$

and the theorem follows from Theorem 2.3.4. □

Remark 2.3.7. If $f \in HK[\boldsymbol{a},\boldsymbol{b}]$ and I is a *degenerate* subinterval of $[\boldsymbol{a},\boldsymbol{b}]$, we define the Henstock-Kurzweil integral of f over I to be zero.

Before we state and prove the next theorem, we need the following

Definition 2.3.8. For any $\boldsymbol{x} \in \mathbb{R}^m$ and non-empty set $X \subseteq \mathbb{R}^m$, we define

$$\operatorname{dist}(\boldsymbol{x}, X) := \inf\{\, |||\boldsymbol{x} - \boldsymbol{s}||| : \boldsymbol{s} \in X\}.$$

The proof of Theorem 2.3.9 is more involved than that of Theorem 1.3.8. Indeed, if $m \geq 2$ and $\{[\boldsymbol{s},\boldsymbol{t}],[\boldsymbol{u},\boldsymbol{v}]\}$ is a division of $[\boldsymbol{a},\boldsymbol{b}]$, then the intersection $[\boldsymbol{s},\boldsymbol{t}] \cap [\boldsymbol{u},\boldsymbol{v}]$ contains more than one point.

Theorem 2.3.9. *Let* $\{[\boldsymbol{s},\boldsymbol{t}],[\boldsymbol{u},\boldsymbol{v}]\}$ *be a division of* $[\boldsymbol{a},\boldsymbol{b}]$. *If* $f \in HK[\boldsymbol{s},\boldsymbol{t}] \cap HK[\boldsymbol{u},\boldsymbol{v}]$, *then* $f \in HK[\boldsymbol{a},\boldsymbol{b}]$ *and*

$$(HK)\int_{[\boldsymbol{a},\boldsymbol{b}]} f = (HK)\int_{[\boldsymbol{s},\boldsymbol{t}]} f + (HK)\int_{[\boldsymbol{u},\boldsymbol{v}]} f.$$

Proof. Given $\varepsilon > 0$ we select a gauge δ_1 on $[\boldsymbol{s},\boldsymbol{t}]$ such that

$$\left|S(f,P_1) - (HK)\int_{[\boldsymbol{s},\boldsymbol{t}]} f\right| < \frac{\varepsilon}{2}$$

for each δ_1-fine Perron partition P_1 of $[\boldsymbol{s},\boldsymbol{t}]$. Similarly, there exists a gauge δ_2 on $[\boldsymbol{u},\boldsymbol{v}]$ such that

$$\left|S(f,P_2) - (HK)\int_{[\boldsymbol{u},\boldsymbol{v}]} f\right| < \frac{\varepsilon}{2}$$

for each δ_2-fine Perron partition P_2 of $[\boldsymbol{u},\boldsymbol{v}]$.

Define a gauge δ on $[\boldsymbol{a},\boldsymbol{b}]$ by setting

$$\delta(\boldsymbol{x}) = \begin{cases} \min\{\delta_1(\boldsymbol{x}),\delta_2(\boldsymbol{x})\} & \text{if } \boldsymbol{x} \in [\boldsymbol{s},\boldsymbol{t}] \cap [\boldsymbol{u},\boldsymbol{v}], \\ \min\{\operatorname{dist}(\boldsymbol{x},[\boldsymbol{u},\boldsymbol{v}]),\delta_1(\boldsymbol{x})\} & \text{if } \boldsymbol{x} \in [\boldsymbol{s},\boldsymbol{t}]\backslash[\boldsymbol{u},\boldsymbol{v}], \\ \min\{\operatorname{dist}(\boldsymbol{x},[\boldsymbol{u},\boldsymbol{v}]),\delta_2(\boldsymbol{x})\} & \text{if } \boldsymbol{x} \in [\boldsymbol{u},\boldsymbol{v}]\backslash[\boldsymbol{s},\boldsymbol{t}], \end{cases}$$

and let P be a δ-fine Perron partition of $[\boldsymbol{a},\boldsymbol{b}]$. Then

$$\{(\boldsymbol{x},I) \in P : \boldsymbol{x} \in [\boldsymbol{s},\boldsymbol{t}] \text{ and } \mu_m(I \cap [\boldsymbol{s},\boldsymbol{t}]) > 0\}$$

is a δ-fine Perron partition of $[\boldsymbol{s},\boldsymbol{t}]$, and

$$\{(\boldsymbol{x},I) \in P : \boldsymbol{x} \in [\boldsymbol{u},\boldsymbol{v}] \text{ and } \mu_m(I \cap [\boldsymbol{u},\boldsymbol{v}]) > 0\}$$

is a δ-fine Perron partition of $[\boldsymbol{u},\boldsymbol{v}]$. Thus

$$\begin{aligned} &\left|S(f,P) - \left\{(HK)\int_{[\boldsymbol{s},\boldsymbol{t}]} f + (HK)\int_{[\boldsymbol{u},\boldsymbol{v}]} f\right\}\right| \\ &\leq \left|\sum\{f(\boldsymbol{x})\mu_m(I \cap [\boldsymbol{s},\boldsymbol{t}]) : (\boldsymbol{x},I) \in P \text{ and } \boldsymbol{x} \in [\boldsymbol{s},\boldsymbol{t}]\} - (HK)\int_{[\boldsymbol{s},\boldsymbol{t}]} f\right| \\ &+ \left|\sum\{f(\boldsymbol{x})\mu_m(I \cap [\boldsymbol{u},\boldsymbol{v}]) : (\boldsymbol{x},I) \in P \text{ and } \boldsymbol{x} \in [\boldsymbol{u},\boldsymbol{v}]\} - (HK)\int_{[\boldsymbol{u},\boldsymbol{v}]} f\right| \\ &< \varepsilon. \end{aligned}$$

Since $\varepsilon > 0$ is arbitrary, the theorem is proved. □

Theorem 2.3.10. *Let $f : [\boldsymbol{a}, \boldsymbol{b}] \longrightarrow \mathbb{R}$ and let D be a division of $[\boldsymbol{a}, \boldsymbol{b}]$. If $f \in HK(J)$ for all $J \in D$, then $f \in HK[\boldsymbol{a}, \boldsymbol{b}]$ and*

$$(HK)\int_{[\boldsymbol{a},\boldsymbol{b}]} f = \sum_{J\in D} (HK)\int_J f.$$

Proof. In view of Theorem 2.3.6 and Lemma 2.1.7 we may assume that D is a net of $[\boldsymbol{a}, \boldsymbol{b}]$. In this case, we apply Theorem 2.3.9 repeatedly to get the result. □

For each $f \in HK[\boldsymbol{a}, \boldsymbol{b}]$, it follows from Theorem 2.3.6 that $f \in HK(I)$ for each $I \in \mathcal{I}_m([\boldsymbol{a}, \boldsymbol{b}])$. In this case, we can define an interval function $F : \mathcal{I}_m([\boldsymbol{a}, \boldsymbol{b}]) \longrightarrow \mathbb{R}$ by setting

$$F(I) = (HK)\int_I f. \tag{2.3.3}$$

Here F is known as the *indefinite Henstock-Kurzweil integral*, or in short the indefinite HK-integral, of f. Theorem 2.3.10 tells us that F is an additive interval function; that is, if D is a division of some $I \in \mathcal{I}_m([\boldsymbol{a}, \boldsymbol{b}])$, then

$$F(I) = \sum_{J\in D} F(J).$$

Definition 2.3.11. Let $\boldsymbol{x} \in [\boldsymbol{a}, \boldsymbol{b}]$ and let $F : \mathcal{I}_m([\boldsymbol{a}, \boldsymbol{b}]) \longrightarrow \mathbb{R}$. The interval function F is said to be strongly derivable at $\boldsymbol{x}$ if the following limit

$$D_s F(\boldsymbol{x}) := \lim_{\substack{|||\boldsymbol{v}-\boldsymbol{u}|||\to 0 \\ \boldsymbol{x}\in[\boldsymbol{u},\boldsymbol{v}]\in\mathcal{I}_m([\boldsymbol{a},\boldsymbol{b}])}} \frac{F([\boldsymbol{u}, \boldsymbol{v}])}{|||\boldsymbol{v} - \boldsymbol{u}|||} \text{ exists.}$$

The following theorem is an m-dimensional analogue of Theorem 1.2.8.

Theorem 2.3.12. *Let $F : \mathcal{I}_m([\boldsymbol{a}, \boldsymbol{b}]) \longrightarrow \mathbb{R}$ be an additive interval function. If F is strongly derivable at every point $\boldsymbol{x} \in [\boldsymbol{a}, \boldsymbol{b}]$, then D_sF is Henstock-Kurzweil integrable on $[\boldsymbol{a}, \boldsymbol{b}]$ and*

$$(HK)\int_{[\boldsymbol{a},\boldsymbol{b}]} D_s F = F([\boldsymbol{a}, \boldsymbol{b}]).$$

Proof. Let $\varepsilon > 0$ be given. Since F is strongly derivable at every point of $[\boldsymbol{a}, \boldsymbol{b}]$, for each $\boldsymbol{x} \in [\boldsymbol{a}, \boldsymbol{b}]$ there exists $\delta(\boldsymbol{x}) > 0$ such that

$$|D_s F(\boldsymbol{x})\mu_m(I) - F(I)| < \frac{\varepsilon \mu_m(I)}{1 + \prod_{i=1}^m (b_i - a_i)}$$

for each $I \in \mathcal{I}_m([\boldsymbol{a}, \boldsymbol{b}])$ satisfying $\boldsymbol{x} \in I \subset B(\boldsymbol{x}, \delta(\boldsymbol{x}))$. It follows that δ is a gauge on $[\boldsymbol{a}, \boldsymbol{b}]$. If P is any δ-fine Perron partition of $[\boldsymbol{a}, \boldsymbol{b}]$, then

$$\begin{aligned}
&|S(D_sF, P) - F([\boldsymbol{a}, \boldsymbol{b}])| \\
&= \left| \sum_{(\boldsymbol{t}, I) \in P} D_sF(\boldsymbol{t})\mu_m(I) - F([\boldsymbol{a}, \boldsymbol{b}]) \right| \\
&= \left| \sum_{(\boldsymbol{t}, I) \in P} \{D_sF(\boldsymbol{t})\mu_m(I) - F(I)\} \right| \\
&\leq \sum_{(\boldsymbol{t}, I) \in P} |D_sF(\boldsymbol{t})\mu_m(I) - F(I)| \\
&< \sum_{(\boldsymbol{t}, I) \in P} \frac{\varepsilon \mu_m(I)}{1 + \prod_{i=1}^{m}(b_i - a_i)} \\
&< \varepsilon.
\end{aligned}$$

Since $\varepsilon > 0$ is arbitrary, we conclude that $D_sF \in HK[\boldsymbol{a}, \boldsymbol{b}]$ and

$$(HK) \int_{[\boldsymbol{a}, \boldsymbol{b}]} D_sF = F([\boldsymbol{a}, \boldsymbol{b}]).$$

□

2.4 Saks-Henstock Lemma

The aim of this section is to establish the important Saks-Henstock Lemma for the Henstock-Kurzweil integral (Theorem 2.4.4). We begin with some definitions.

Definition 2.4.1. A finite collection $\{(\boldsymbol{t}_1, [\boldsymbol{u}_1, \boldsymbol{v}_1]), \ldots, (\boldsymbol{t}_p, [\boldsymbol{u}_p, \boldsymbol{v}_p])\}$ of point-interval pair is said to be a *Perron subpartition* of $[\boldsymbol{a}, \boldsymbol{b}]$ if $\boldsymbol{t}_k \in [\boldsymbol{u}_k, \boldsymbol{v}_k]$ for $k = 1, \ldots, p$, and $\{[\boldsymbol{u}_1, \boldsymbol{v}_1], \ldots, [\boldsymbol{u}_p, \boldsymbol{v}_p]\}$ is a finite collection of non-overlapping intervals in $[\boldsymbol{a}, \boldsymbol{b}]$.

Definition 2.4.2. Let $\{(\boldsymbol{t}_1, [\boldsymbol{u}_1, \boldsymbol{v}_1]), \ldots, (\boldsymbol{t}_p, [\boldsymbol{u}_p, \boldsymbol{v}_p])\}$ be a Perron subpartition of $[\boldsymbol{a}, \boldsymbol{b}]$ and let δ be a gauge on $\{\boldsymbol{t}_1, \ldots, \boldsymbol{t}_p\}$. The Perron subpartition $\{(\boldsymbol{t}_1, [\boldsymbol{u}_1, \boldsymbol{v}_1]), \ldots, (\boldsymbol{t}_p, [\boldsymbol{u}_p, \boldsymbol{v}_p])\}$ is said to be *δ-fine* if $[\boldsymbol{u}_k, \boldsymbol{v}_k] \subset B(\boldsymbol{t}_k, \delta(\boldsymbol{t}_k))$ for $k = 1, \ldots, p$.

By replacing δ-fine Perron partitions by δ-fine Perron subpartitions in Definition 2.2.6, we obtain the following lemma.

Lemma 2.4.3. *Let $f \in HK[\boldsymbol{a},\boldsymbol{b}]$ and let $\varepsilon > 0$. If δ is a gauge on $[\boldsymbol{a},\boldsymbol{b}]$ such that*

$$\left| \sum_{(\boldsymbol{x},J)\in Q} f(\boldsymbol{x})\mu_m(J) - (HK)\int_{[\boldsymbol{a},\boldsymbol{b}]} f \right| < \varepsilon$$

for each δ-fine Perron partition Q of $[\boldsymbol{a},\boldsymbol{b}]$, then

$$\left| \sum_{(\boldsymbol{t},I)\in P} \left\{ f(\boldsymbol{t})\mu_m(I) - (HK)\int_I f \right\} \right| \le \varepsilon \tag{2.4.1}$$

for each δ-fine Perron subpartition P of $[\boldsymbol{a},\boldsymbol{b}]$.

Proof. Let P be a δ-fine Perron subpartition of $[\boldsymbol{a},\boldsymbol{b}]$. If $\bigcup_{(\boldsymbol{t},I)\in P} I = [\boldsymbol{a},\boldsymbol{b}]$, then (2.4.1) follows from Theorem 2.3.10. Henceforth we assume that $\bigcup_{(\boldsymbol{t},I)\in P} I \subset [\boldsymbol{a},\boldsymbol{b}]$; in this case, we choose non-overlapping intervals $J_1, \dots, J_q$ such that $\{J_k : k = 1, \dots, q\} \cup \bigcup_{(\boldsymbol{t},I)\in P}\{I\}$ is a division of $[\boldsymbol{a},\boldsymbol{b}]$. For each $k \in \{1, \dots, q\}$ we infer from Theorem 2.3.6 that $f \in HK(J_k)$. Thus for each $\eta > 0$ there exists a gauge δ_k on J_k such that

$$\left| S(f, P_k) - (HK)\int_{J_k} f \right| < \frac{\eta}{q}$$

for each δ_k-fine Perron partition P_k of J_k. Since δ and δ_k are gauges on J_k, we can apply Cousin's Lemma (Theorem 2.2.2) to select and fix a $\min\{\delta, \delta_k\}$-fine Perron partition Q_k of J_k. Then $Q := P \cup \bigcup_{k=1}^q Q_k$ is a δ-fine Perron partition of $[\boldsymbol{a},\boldsymbol{b}]$ such that

$$S(f,Q) = S(f,P) + \sum_{k=1}^q S(f,Q_k)$$

and

$$(HK)\int_{[\boldsymbol{a},\boldsymbol{b}]} f = \sum_{(\boldsymbol{t},I)\in P} (HK)\int_I f + \sum_{k=1}^q (HK)\int_{J_k} f.$$

Therefore

$$\begin{aligned}
&\left| \sum_{(\boldsymbol{t},I)\in P} \left\{ f(\boldsymbol{t})\mu_m(I) - (HK)\int_I f \right\} \right| \\
&= \left| \left\{ S(f,Q) - \sum_{k=1}^q S(f,Q_k) \right\} - \left\{ (HK)\int_{[\boldsymbol{a},\boldsymbol{b}]} f - \sum_{k=1}^q (HK)\int_{J_k} f \right\} \right| \\
&\le \left| S(f,Q) - (HK)\int_{[\boldsymbol{a},\boldsymbol{b}]} f \right| + \sum_{k=1}^q \left| S(f,Q_k) - (HK)\int_{J_k} f \right| \\
&< \varepsilon + \eta.
\end{aligned}$$

Since $\eta > 0$ is arbitrary, the lemma is proved. □

We are now ready to state and prove the following crucial result, which plays an important role for the rest of this monograph.

Theorem 2.4.4 (Saks-Henstock). *If $f \in HK[\boldsymbol{a}, \boldsymbol{b}]$, then for each $\varepsilon > 0$ there exists a gauge δ on $[\boldsymbol{a}, \boldsymbol{b}]$ such that*

$$\sum_{(\boldsymbol{t},[\boldsymbol{u},\boldsymbol{v}])\in P} \left| f(\boldsymbol{t})\mu_m([\boldsymbol{u},\boldsymbol{v}]) - (HK)\int_{[\boldsymbol{u},\boldsymbol{v}]} f \right| < \varepsilon \tag{2.4.2}$$

for each δ-fine Perron subpartition P of $[\boldsymbol{a}, \boldsymbol{b}]$.

Proof. Since f is Henstock-Kurzweil integrable on $[\boldsymbol{a}, \boldsymbol{b}]$, it follows from Lemma 2.4.3 that there exists a gauge δ on $[\boldsymbol{a}, \boldsymbol{b}]$ such that

$$\left| \sum_{(\boldsymbol{x},J)\in Q} \left\{ f(\boldsymbol{x})\mu_m(J) - (HK)\int_J f \right\} \right| < \frac{\varepsilon}{2} \tag{2.4.3}$$

for each δ-fine Perron subpartition Q of $[\boldsymbol{a}, \boldsymbol{b}]$. Let P be a δ-fine Perron subpartition of $[\boldsymbol{a}, \boldsymbol{b}]$, let

$$P^+ = \left\{ (\boldsymbol{t}, I) \in P : f(\boldsymbol{t})\mu_m(I) - (HK)\int_I f \geq 0 \right\}$$

and let $P^- = P \backslash P^+$. By (2.4.3),

$$\begin{aligned}
&\sum_{(\boldsymbol{t},I)\in P} \left| f(\boldsymbol{t})\mu_m(I) - (HK)\int_I f \right| \\
&= \sum_{(\boldsymbol{t},I)\in P^+} \left\{ f(\boldsymbol{t})\mu_m(I) - (HK)\int_I f \right\} - \sum_{(\boldsymbol{t},I)\in P^-} \left\{ f(\boldsymbol{t})\mu_m(I) - (HK)\int_I f \right\} \\
&< \frac{\varepsilon}{2} + \frac{\varepsilon}{2} \\
&= \varepsilon.
\end{aligned}$$

□

Our next goal is to establish a higher-dimensional analogue of Theorem 1.4.5; see Theorem 2.4.10. However, this result does not appear to be an immediate consequence of the Saks-Henstock Lemma. We will show that it is a corollary of the Strong Saks-Henstock Lemma (Theorem 2.4.7). We begin with the following observation.

Observation 2.4.5. Let $F : \mathcal{I}_1([a,b]) \longrightarrow \mathbb{R}$ be an additive interval function, and let $t \in [a,b]$. Define the function $\widetilde{F}_t : [a,b] \longrightarrow \mathbb{R}$ by

$$\widetilde{F}_t(x) = \begin{cases} F([t,x]) & \text{if } t \leq x \leq b, \\ -F([x,t]) & \text{if } a \leq x < t. \end{cases}$$

Then $F([u,v]) = \widetilde{F}(v) - \widetilde{F}(u)$ for all $[u,v] \in \mathcal{I}_1([a,b])$.

In order to formulate an m-dimensional analogue of Observation 2.4.5 (cf. Lemma 2.4.6), we need to fix some notations.

For each $\boldsymbol{t}, \boldsymbol{x} \in [\boldsymbol{a}, \boldsymbol{b}]$, let $< \boldsymbol{t}, \boldsymbol{x} >= \prod_{k=1}^m [\alpha_k, \beta_k]$, where

$$[\alpha_k, \beta_k] = \begin{cases} [t_k, x_k] \text{ if } t_k \leq x_k \leq \beta_k, \\ [x_k, t_k] \text{ if } \alpha_k \leq x_k < t_k \end{cases}$$

for $k = 1, \ldots, m$.

For each $[\boldsymbol{u}, \boldsymbol{v}] \in \mathcal{I}_m([\boldsymbol{a}, \boldsymbol{b}])$ we let $\mathcal{V}[\boldsymbol{u}, \boldsymbol{v}]$ be the collection all vertices of $[\boldsymbol{u}, \boldsymbol{v}]$:

$$\begin{aligned} &\mathcal{V}[\boldsymbol{u}, \boldsymbol{v}] \\ &= \{(u_1 + \varepsilon_1(v_1 - u_1), \ldots, u_m + \varepsilon_m(v_m - u_m)) : \varepsilon_i \in \{0,1\} \\ &\quad \text{for } \ i = 1, \ldots, m\}. \end{aligned}$$

Recall the the signum function on $\mathbb{R}$ is defined by $\text{sgn}(0) := 0$ and $\text{sgn}(x) := \frac{x}{|x|}$ $(x \in \mathbb{R}\backslash\{0\})$. The following lemma is an m-dimensional analogue of Observation 2.4.5.

Lemma 2.4.6. *Let $H : \mathcal{I}_m([\boldsymbol{a}, \boldsymbol{b}]) \longrightarrow \mathbb{R}$ be an additive interval function, and let $\boldsymbol{t} \in [\boldsymbol{a}, \boldsymbol{b}]$. Define the function $\widetilde{H}_{\boldsymbol{t}} : [\boldsymbol{a}, \boldsymbol{b}] \longrightarrow \mathbb{R}$ by*

$$\widetilde{H}_{\boldsymbol{t}}(\boldsymbol{x}) = H(< \boldsymbol{t}, \boldsymbol{x} >) \prod_{i=1}^m sgn(x_i - t_i).$$

Then the inequality

$$|H([\boldsymbol{u}, \boldsymbol{v}])| \leq \sum_{\boldsymbol{x} \in \mathcal{V}[\boldsymbol{u}, \boldsymbol{v}]} \left| \widetilde{H}_{\boldsymbol{t}}(\boldsymbol{x}) \right|$$

holds for every $[\boldsymbol{u}, \boldsymbol{v}] \in \mathcal{I}_m([\boldsymbol{a}, \boldsymbol{b}])$.

Proof. Using Observation 2.4.5 repeatedly, we get

$$\begin{aligned} &H([\boldsymbol{u}, \boldsymbol{v}]) \\ &= \sum_{\varepsilon_1=0}^1 \cdots \sum_{\varepsilon_m=0}^1 (-1)^{\sum_{i=1}^m (1-\varepsilon_i)} \widetilde{H}_{\boldsymbol{t}}(u_1 + \varepsilon_1(v_1 - u_1), \ldots, u_m + \varepsilon_m(v_m - u_m)) \end{aligned}$$

and the lemma follows. □

The following theorem is a consequence of Lemma 2.4.6 and Theorem 2.4.4.

Theorem 2.4.7 (Strong version of Saks-Henstock Lemma). *If $f \in HK[\boldsymbol{a},\boldsymbol{b}]$, then for each $\varepsilon > 0$ there exists a gauge δ on $[\boldsymbol{a},\boldsymbol{b}]$ such that*

$$\sum_{(\boldsymbol{t},I)\in P} \sup_{J\in\mathcal{I}_m(I)} \left| f(\boldsymbol{t})\mu_m(J) - (HK)\int_J f \right| < \varepsilon$$

for each δ-fine Perron subpartition P of $[\boldsymbol{a},\boldsymbol{b}]$.

Proof. For each $\varepsilon > 0$ we use the Saks-Henstock Lemma to select a gauge δ on $[\boldsymbol{a},\boldsymbol{b}]$ so that

$$\sum_{(\boldsymbol{x},I')\in Q} \left| f(\boldsymbol{x})\mu_m(I') - (HK)\int_{I'} f \right| < \frac{\varepsilon}{2^{m+1}} \tag{2.4.4}$$

for each δ-fine Perron subpartition Q of $[\boldsymbol{a},\boldsymbol{b}]$.

Let $\{(\boldsymbol{t}_1, I_1), \dots, (\boldsymbol{t}_p, I_p)\}$ be a δ-fine Perron subpartition of $[\boldsymbol{a},\boldsymbol{b}]$. Without loss of generality, we may assume that $\boldsymbol{t}$ is a vertex of I whenever the point-interval pair $(\boldsymbol{t}, I)$ belongs to P. For $k = 1, \dots, m$, we let $J_k \in \mathcal{I}_m(I_k)$ and write

$$\mathcal{V}(I_k) = \{\boldsymbol{w}_{k,1}, \dots, \boldsymbol{w}_{k,2^m}\}.$$

According Lemma 2.4.6 and Remark 2.3.7,

$$\begin{aligned}
&\sum_{k=1}^{p} \left| f(\boldsymbol{t}_k)\mu_m(J_k) - (HK)\int_{J_k} f \right| \\
&\leq \sum_{k=1}^{p}\sum_{j=1}^{2^m} \left| f(\boldsymbol{t}_k)\mu_m(<\boldsymbol{t}_k, \boldsymbol{w}_{k,j}>) - (HK)\int_{<\boldsymbol{t}_k,\boldsymbol{w}_{k,j}>} f \right| \\
&= \sum_{j=1}^{2^m}\sum_{k=1}^{p} \left| f(\boldsymbol{t}_k)\mu_m(<\boldsymbol{t}_k, \boldsymbol{w}_{k,j}>) - (HK)\int_{<\boldsymbol{t}_k,\boldsymbol{w}_{k,j}>} f \right| \\
&< \frac{2^m\varepsilon}{2^{m+1}} \\
&< \varepsilon.
\end{aligned}$$

□

Theorem 2.4.8. *If $f \in HK[\boldsymbol{a},\boldsymbol{b}]$, then given $\varepsilon > 0$ there exists $\eta > 0$ such that*

$$\left| (HK)\int_{[\boldsymbol{u},\boldsymbol{v}]} f \right| < \varepsilon$$

whenever $[\boldsymbol{u},\boldsymbol{v}] \in \mathcal{I}_m([\boldsymbol{a},\boldsymbol{b}])$ with $\mu_m([\boldsymbol{u},\boldsymbol{v}]) < \eta$.

Proof. For each $\varepsilon > 0$ we use Theorem 2.4.7 to select a gauge δ on $[a,b]$ such that

$$\sum_{k=1}^{p} \sup_{J_k \in \mathcal{I}_m(I_k)} \left| f(t_k)\mu_m(J_k) - (HK)\int_{J_k} f \right| < \frac{\varepsilon}{2}$$

for each δ-fine Perron subpartition $\{(t_1, I_1), \ldots, (t_p, I_p)\}$ of $[a,b]$. According to Cousin's Lemma (Theorem 2.2.2), we may fix a δ-fine Perron partition P of $[a,b]$, and let $\eta = \frac{\varepsilon}{2\max\{|f(t)|:(t,I)\in P\}+1}$. If $[u,v] \in \mathcal{I}_m([a,b])$ with $\mu_m([u,v]) < \eta$, then

$$\left| (HK) \int_{[u,v]} f \right| \leq \sum_{(t,I)\in P} \left| (HK) \int_{I\cap[u,v]} f \right|$$

$$\leq \sum_{(t,I)\in P} |f(t)\mu_m(I \cap [u,v])|$$

$$+ \sum_{(t,I)\in P} \left| f(t)\mu_m(I \cap [u,v]) - (HK) \int_{[u,v]\cap I} f \right|$$

$$< \varepsilon. \qquad \square$$

For any subsets X and Y of $[a,b]$, we write $X \Delta Y = (X \backslash Y) \cup (Y \backslash X)$. For any $I, J \in \mathcal{I}_m([a,b])$, we put $\mu_m(I \Delta J) := \mu_m(I) + \mu_m(J) - \mu_m(I \cap J)$. We are now ready to state and prove a refinement of Theorem 2.4.8.

Theorem 2.4.9. *Let $f \in HK[a,b]$ and let F be the indefinite Henstock-Kurzweil integral of f. Then for each $\varepsilon > 0$ there exists $\eta_0 > 0$ such that*

$$|F(I) - F(J)| < \varepsilon$$

for each $I, J \in \mathcal{I}_m([a,b])$ with $\mu_m(I \Delta J) < \eta_0$.

Proof. For each $\varepsilon > 0$ we choose $\eta_0 > 0$ corresponding to $\frac{\varepsilon}{3^{m+1}}$ in Theorem 2.4.8.

Suppose that $I, J \in \mathcal{I}_m([a,b])$ with $\mu_m(I \Delta J) < \eta_0$. If $\mu_m(I \cap J) = 0$, then I and J are non-overlapping intervals with $\mu_m(I) + \mu_m(J) < \eta_0$. In this case, our choice of η_0 yields

$$|F(I) - F(J)| < \frac{2\varepsilon}{3^{m+1}} < \varepsilon.$$

On the other hand, suppose that $\mu_m(I \cap J) > 0$. Since F is additive (Theorem 2.3.10), it follows from the triangle inequality, Lemma 2.1.6 and our choice of η_0 that

$$|F(I) - F(J)| \leq |F(I) - F(I \cap J)| + |F(J) - F(I \cap J)| < \frac{2(3^m)\varepsilon}{3^{m+1}} < \varepsilon. \qquad \square$$

Let $(\boldsymbol{a}, \boldsymbol{b}] := \prod_{i=1}^{m}(a_i, b_i]$. The following theorem is an m-dimensional analogue of Theorem 1.4.5.

Theorem 2.4.10. *Let $f \in HK[\boldsymbol{a}, \boldsymbol{b}]$ and let*

$$\widetilde{F}(\boldsymbol{x}) = \begin{cases} (HK)\int_{[\boldsymbol{a},\boldsymbol{x}]} f & \text{if } \boldsymbol{x} \in (\boldsymbol{a}, \boldsymbol{b}], \\ 0 & \text{if } \boldsymbol{x} \in [\boldsymbol{a}, \boldsymbol{b}]\backslash(\boldsymbol{a}, \boldsymbol{b}]. \end{cases}$$

Then $\widetilde{F}$ is continuous on $[\boldsymbol{a}, \boldsymbol{b}]$.

Proof. For each $\varepsilon > 0$ we select $\eta_0 > 0$ associated to ε according to Theorem 2.4.9. Next we pick $\delta > 0$ so that $\mu_m([\boldsymbol{a}, \boldsymbol{x}]\Delta[\boldsymbol{a}, \boldsymbol{y}]) < \eta_0$ whenever $\boldsymbol{x}, \boldsymbol{y} \in [\boldsymbol{a}, \boldsymbol{b}]$ with $|||\boldsymbol{x} - \boldsymbol{y}||| < \delta$. An application of Theorem 2.4.9 yields the result. □

The following theorem is an m-dimensional analogue of Theorem 1.4.6.

Theorem 2.4.11. *Let $f : [\boldsymbol{a}, \boldsymbol{b}] \longrightarrow \mathbb{R}$. If $f \in HK[\boldsymbol{c}, \boldsymbol{d}]$ for every $[\boldsymbol{c}, \boldsymbol{d}] \in \mathcal{I}_m([\boldsymbol{a}, \boldsymbol{b}])$ disjoint from $\{\boldsymbol{b}\}$, then $f \in HK[\boldsymbol{a}, \boldsymbol{b}]$ if and only if*

$$\lim_{\substack{\boldsymbol{x}\to\boldsymbol{b}\\ \boldsymbol{x}\in[\boldsymbol{a},\boldsymbol{b}]}} (HK)\int_{[\boldsymbol{a},\boldsymbol{x}]} f \text{ exists.} \tag{2.4.5}$$

In this case,

$$\lim_{\substack{\boldsymbol{x}\to\boldsymbol{b}\\ \boldsymbol{x}\in[\boldsymbol{a},\boldsymbol{b}]}} (HK)\int_{[\boldsymbol{a},\boldsymbol{x}]} f = (HK)\int_{[\boldsymbol{a},\boldsymbol{b}]} f.$$

Proof. ($\Longrightarrow$) This follows from Theorem 2.4.10.

($\Longleftarrow$) Conversely, suppose that (2.4.5) holds. In this case, it is easy to check that the limit

$$\lim_{\substack{\boldsymbol{x}\to\boldsymbol{b}\\ \boldsymbol{x}\in[\boldsymbol{u},\boldsymbol{b}]}} (HK)\int_{[\boldsymbol{u},\boldsymbol{x}]} f$$

exists whenever $[\boldsymbol{u}, \boldsymbol{b}] \in \mathcal{I}_m([\boldsymbol{a}, \boldsymbol{b}])$. Now we define the interval function $F : \mathcal{I}_m([\boldsymbol{a}, \boldsymbol{b}]) \longrightarrow \mathbb{R}$ by setting

$$F([\boldsymbol{u}, \boldsymbol{v}]) = \begin{cases} (HK)\int_{[\boldsymbol{u},\boldsymbol{v}]} f & \text{if } [\boldsymbol{u}, \boldsymbol{v}] \in \mathcal{I}_m([\boldsymbol{a}, \boldsymbol{b}]) \text{ and } \boldsymbol{b} \notin [\boldsymbol{u}, \boldsymbol{v}], \\ \\ \lim_{\substack{\boldsymbol{x}\to\boldsymbol{b}\\ \boldsymbol{x}\in[\boldsymbol{u},\boldsymbol{b}]}} (HK)\int_{[\boldsymbol{u},\boldsymbol{x}]} f & \text{if } [\boldsymbol{u}, \boldsymbol{v}] \in \mathcal{I}_m([\boldsymbol{a}, \boldsymbol{b}]) \text{ and } \boldsymbol{v} = \boldsymbol{b}. \end{cases}$$

Then F is an additive interval function on $\mathcal{I}_m([\boldsymbol{a}, \boldsymbol{b}])$. Next we let

$$\widetilde{F}(\boldsymbol{x}) = \begin{cases} (HK)\int_{[\boldsymbol{a},\boldsymbol{x}]} f & \text{if } \boldsymbol{x} \in [\boldsymbol{a}, \boldsymbol{b}]\backslash\{\boldsymbol{a}\}, \\ \\ \lim_{\substack{\boldsymbol{x}\to\boldsymbol{b}\\ \boldsymbol{x}\in[\boldsymbol{a},\boldsymbol{b}]}} (HK)\int_{[\boldsymbol{a},\boldsymbol{x}]} f & \text{if } \boldsymbol{x} = \boldsymbol{b} \end{cases}$$

so that we can apply Lemma 2.4.6 with $H = F$ and $\boldsymbol{t} = \boldsymbol{a}$ to conclude that

$$|F([\boldsymbol{u},\boldsymbol{v}])| \leq \sum_{\boldsymbol{x}\in\mathcal{V}([\boldsymbol{u},\boldsymbol{v}])} \left|\widetilde{F}(\boldsymbol{x})\right| \quad \text{whenever } [\boldsymbol{u},\boldsymbol{v}] \in \mathcal{I}_m([\boldsymbol{a},\boldsymbol{b}]). \tag{2.4.6}$$

It remains to prove that $f \in HK[\boldsymbol{a},\boldsymbol{b}]$. For each $\varepsilon > 0$ we use (2.4.6) and (2.4.5) to choose $\eta \in (0, \frac{1}{2}\min_{k=1,\ldots,m}(b_k - a_k))$ so that

$$|F([\boldsymbol{x},\boldsymbol{b}])| < \frac{\varepsilon}{3} \text{ whenever } \boldsymbol{x} \in \prod_{i=1}^{m}(b_i - \eta, b_i]. \tag{2.4.7}$$

For any given $n \in \mathbb{N} \cup \{0\}$, we observe that the set

$$U_n := \left(\prod_{i=1}^{m}\left[b_i - \frac{b_i - a_i}{n+1}, b_i\right]\right) \Big\backslash \left(\prod_{i=1}^{m}\left(b_i - \frac{b_i - a_i}{n+2}, b_i\right]\right)$$

can be written as a union of $2^m - 1$ pairwise non-overlapping intervals in $\mathbb{R}^m$. Hence we infer from the definition of F and Saks-Henstock Lemma that there exists a gauge δ_n on $(\prod_{i=1}^m[b_i - \frac{b_i-a_i}{n+1}, b_i])\backslash(\prod_{i=1}^m[b_i - \frac{b_i-a_i}{n+2}, b_i])$ such that

$$\sum_{(\boldsymbol{w},J)\in Q_n} |f(\boldsymbol{w})\mu_m(J) - F(J)| < \frac{\varepsilon}{3(2^{n+1})} \tag{2.4.8}$$

for each δ_n-fine Perron subpartition Q_n of U_n.

Define a gauge δ on $[\boldsymbol{a},\boldsymbol{b}]$ by setting

$$\delta(\boldsymbol{x}) = \begin{cases} \min\{\delta_n(\boldsymbol{x}), \operatorname{dist}(\boldsymbol{x}, \prod_{i=1}^m[b_i - \frac{b_i-a_i}{n+2}, b_i])\} \\ \qquad \text{if } \boldsymbol{x} \in (\prod_{i=1}^m[b_i - \frac{b_i-a_i}{n+1}, b_i])\backslash(\prod_{i=1}^m[b_i - \frac{b_i-a_i}{n+2}, b_i]) \\ \qquad \text{for some } n \in \mathbb{N}\cup\{0\}, \\ \min\{\eta, \frac{\varepsilon}{3(1+|f(\boldsymbol{b})|)}\} \qquad \text{if } \boldsymbol{x} = \boldsymbol{b} \end{cases}$$

and choose a δ-fine Perron partition P of $[\boldsymbol{a}, \boldsymbol{b}]$. According to our choice of $\delta(\boldsymbol{b})$, (2.4.7) and (2.4.8), we have

$$\begin{aligned}
&|S(f, P) - F([\boldsymbol{a}, \boldsymbol{b}])| \\
&\leq \left| \sum_{\substack{(\boldsymbol{t}, I) \in P \\ \boldsymbol{t} \neq \boldsymbol{b}}} \left\{ f(\boldsymbol{t})\mu_m(I) - F(I) \right\} \right| + \left| \sum_{\substack{(\boldsymbol{t}, I) \in P \\ \boldsymbol{t} = \boldsymbol{b}}} \left\{ f(\boldsymbol{t})\mu_m(I) - F(I) \right\} \right| \\
&< \left| \sum_{\substack{(\boldsymbol{t}, I) \in P \\ \boldsymbol{t} \neq \boldsymbol{b}}} \left\{ f(\boldsymbol{t})\mu_m(I) - F(I) \right\} \right| + \frac{\varepsilon}{3} + \frac{\varepsilon}{3} \\
&\leq \sum_{n=0}^{\infty} \sum_{\substack{(\boldsymbol{t}, I) \in P \\ \boldsymbol{t} \in (\prod_{i=1}^m [b_i - \frac{b_i - a_i}{n+1}, b_i]) \setminus (\prod_{i=1}^m [b_i - \frac{b_i - a_i}{n+2}, b_i])}} |f(\boldsymbol{t})\mu_m(I) - F(I)| + \frac{2\varepsilon}{3} \\
&< \varepsilon.
\end{aligned}$$

Since $\varepsilon > 0$ is arbitrary, $f \in HK[\boldsymbol{a}, \boldsymbol{b}]$. □

A similar reasoning yields the following result.

Theorem 2.4.12. *Let $f : [\boldsymbol{a}, \boldsymbol{b}] \longrightarrow \mathbb{R}$ and suppose that $f \in HK[\boldsymbol{c}, \boldsymbol{d}]$ for every $[\boldsymbol{c}, \boldsymbol{d}] \in \mathcal{I}_m([\boldsymbol{a}, \boldsymbol{b}])$ disjoint from $\{\boldsymbol{a}\}$. Then $f \in HK[\boldsymbol{a}, \boldsymbol{b}]$ if and only if*

$$\lim_{\substack{\boldsymbol{x} \to \boldsymbol{a} \\ \boldsymbol{x} \in [\boldsymbol{a}, \boldsymbol{b}]}} (HK) \int_{[\boldsymbol{x}, \boldsymbol{b}]} f \text{ exists.}$$

In this case,

$$\lim_{\substack{\boldsymbol{x} \to \boldsymbol{a} \\ \boldsymbol{x} \in [\boldsymbol{a}, \boldsymbol{b}]}} (HK) \int_{[\boldsymbol{x}, \boldsymbol{b}]} f = (HK) \int_{[\boldsymbol{a}, \boldsymbol{b}]} f.$$

Further extensions of Theorems 2.4.11 and 2.4.12 will be given in Chapters 4 and 5.

2.5 Fubini's Theorem

The aim of this section is to show that Fubini's Theorem holds for the multiple Henstock-Kurzweil integral (Theorem 2.5.5). We begin with the following definition.

Definition 2.5.1. If $X \subseteq [\boldsymbol{a}, \boldsymbol{b}]$ is any set, the characteristic function χ_X of X is defined by

$$\chi_X(\boldsymbol{x}) = \begin{cases} 1 & \text{if } \boldsymbol{x} \in X, \\ 0 & \text{otherwise.} \end{cases}$$

We need four lemmas in order to prove Theorem 2.5.5 below.

Lemma 2.5.2. *Let $f : [\boldsymbol{a}, \boldsymbol{b}] \longrightarrow \mathbb{R}$ and suppose that $Z \subset [\boldsymbol{a}, \boldsymbol{b}]$. Then the following assertions hold.*

(i) *If $\chi_Z \in HK[\boldsymbol{a}, \boldsymbol{b}]$ and $(HK)\int_{[\boldsymbol{a},\boldsymbol{b}]} \chi_Z = 0$, then $f\chi_Z$ and $|f\chi_Z|$ belong to $HK[\boldsymbol{a}, \boldsymbol{b}]$ and*

$$(HK)\int_{[\boldsymbol{a},\boldsymbol{b}]} f\chi_Z = (HK)\int_{[\boldsymbol{a},\boldsymbol{b}]} |f\chi_Z| = 0.$$

(ii) *If $f(\boldsymbol{x}) > 0$ for all $\boldsymbol{x} \in [\boldsymbol{a}, \boldsymbol{b}]$, then the converse of (i) holds.*

Proof. (i) Let $\varepsilon > 0$ be given. For each $k \in \mathbb{N} \cup \{0\}$ there exists a gauge δ_k on $[\boldsymbol{a}, \boldsymbol{b}]$ such that

$$\sum_{(\boldsymbol{x},J)\in P_k} \chi_Z(\boldsymbol{x})\mu_m(J) < \frac{\varepsilon}{(k+1)2^{k+1}}$$

for each δ_k-fine Perron subpartition P_k of $[\boldsymbol{a}, \boldsymbol{b}]$.

Define a gauge δ on $[\boldsymbol{a}, \boldsymbol{b}]$ so that

$$\delta(\boldsymbol{x}) = \delta_k(\boldsymbol{x})$$

whenever $k \in \mathbb{N} \cup \{0\}$ and $\boldsymbol{x} \in [\boldsymbol{a}, \boldsymbol{b}]$ with $k \leq |f(\boldsymbol{x})| < k+1$. If P is a δ-fine Perron partition of $[\boldsymbol{a}, \boldsymbol{b}]$, then

$$|S(f\chi_Z, P)| \leq \sum_{k=0}^{\infty} \sum_{\substack{(\boldsymbol{t},I)\in P \\ k\leq|f(\boldsymbol{t})|<k+1}} |f(\boldsymbol{t})\chi_Z(\boldsymbol{t})|\, \mu_m(I) < \varepsilon.$$

Since $\varepsilon > 0$ is arbitrary, we conclude that $f\chi_Z \in HK[\boldsymbol{a}, \boldsymbol{b}]$ and $(HK)\int_{[\boldsymbol{a},\boldsymbol{b}]} f\chi_Z = 0$. Similarly, we have $|f\chi_Z| \in HK[\boldsymbol{a}, \boldsymbol{b}]$ and $(HK)\int_{[\boldsymbol{a},\boldsymbol{b}]} |f\chi_Z| = 0$. This completes the proof of (i).

Since the proof of (ii) is similar to that of (i), the lemma is proved. □

For the rest of this section, we let r and s be positive integers so that $r + s = m$. We also fix an interval E_r (resp. E_s) in $\mathbb{R}^r$ (resp. $\mathbb{R}^s$).

Lemma 2.5.3. *Let $f \in HK(E_r \times E_s)$, let $Z_r = \{\xi \in E_r : f(\xi, \cdot) \notin HK(E_s)\}$, and let $Z_s = \{\eta \in E_s : f(\cdot, \eta) \notin HK(E_r)\}$. Then the following assertions hold.*

(i) $(HK)\int_{E_r} \chi_{Z_r}(\xi)\, d\xi = 0$.
(ii) $(HK)\int_{E_s} \chi_{Z_s}(\eta)\, d\eta = 0$.

Proof. (i) Using the definition of Z_r together with Theorem 2.3.4, for each $\xi \in Z_r$ there exists $g(\xi) > 0$ with the following property: for each gauge δ_s on E_s there exist δ_s-fine Perron partitions $Q_{1,\xi}$, $Q_{2,\xi}$ of E_s such that

$$\sum_{(\eta_1,J_1)\in Q_{1,\xi}} f(\xi,\eta_1)\mu_s(J_1) - \sum_{(\eta_2,J_2)\in Q_{2,\xi}} f(\xi,\eta_2)\mu_s(J_2) \geq g(\xi). \quad (2.5.1)$$

We will first prove that $(HK)\int_{E_r} g(\xi)\, d\xi = 0$, where $g(\xi) := 0$ whenever $\xi \in E_r\backslash Z_r$.

Let $\varepsilon > 0$ and select a gauge Δ on $E_r \times E_s$ so that

$$|S(f,Q_1) - S(f,Q_2)| < \varepsilon \quad (2.5.2)$$

for each pair of Δ-fine Perron partitions Q_1, Q_2 of $E_r \times E_s$. In particular, for each $\xi \in Z_r$ the function $\Delta(\xi,\cdot)$ is a gauge on E_s; hence (2.5.1) implies that there exist two $\Delta(\xi,\cdot)$-fine Perron partitions $T_{1,\xi}$, $T_{2,\xi}$ of E_s such that

$$\sum_{(\eta_3,J_3)\in T_{1,\xi}} f(\xi,\eta_3)\mu_s(J_3) - \sum_{(\eta_4,J_4)\in T_{2,\xi}} f(\xi,\eta_4)\mu_s(J_4) \geq g(\xi). \quad (2.5.3)$$

On the other hand, for each $\xi \in E_r\backslash Z_r$ we fix a $\Delta(\xi,\cdot)$-fine Perron partition T_ξ of E_s; in this case, we set $T_{1,\xi} = T_{2,\xi} = T_\xi$.

Define a gauge δ_r on E_r by setting

$$\delta_r(\xi) = \min\{\Delta(\xi,\eta) : (\eta, J) \in T_{1,\xi} \cup T_{2,\xi}\},$$

and let P be a δ_r-fine Perron partition of E_r. Since

$$P_1 = \{((\xi,\eta), I \times J) : (\xi, I) \in P \text{ and } (\eta, J) \in T_{1,\xi}\}$$

and

$$P_2 = \{((\xi,\eta), I \times J) : (\xi, I) \in P \text{ and } (\eta, J) \in T_{2,\xi}\}$$

are Δ-fine Perron partitions of $E_r \times E_s$, we infer from (2.5.3) and (2.5.2) that

$$\begin{aligned}
&\sum_{(\xi,I)\in P} g(\xi)\ \mu_r(I) \\
&\leq \sum_{(\xi,I)\in P} \Bigg(\sum_{(\eta_5,J_5)\in T_{1,\xi}} f(\xi,\eta_5)\mu_s(J_5) - \sum_{(\eta_6,J_6)\in T_{2,\xi}} f(\xi,\eta_6)\mu_s(J_6) \Bigg) \mu_r(I) \\
&= S(f,P_1) - S(f,P_2) \\
&< \varepsilon.
\end{aligned}$$

Since $\varepsilon > 0$ is arbitrary, we conclude that $(HK)\int_{E_r} g(\xi)\, d\xi = 0$. Now we apply Lemma 2.5.2(ii) with $m = r$ and $f = g\chi_{Z_r} + \chi_{E_r \setminus Z_r}$ to conclude that $(HK)\int_{E_r} \chi_{Z_r}(\xi)\, d\xi = 0$.

Since the proof of (ii) is similar to that of (i), the lemma follows. □

If f and g are real functions defined on E_r and E_s respectively, we define the function $f \otimes g : E_r \times E_s \longrightarrow \mathbb{R}$ by setting

$$(f \otimes g)(\xi, \eta) = f(\xi) g(\eta).$$

Lemma 2.5.4. *Let f, Z_r and Z_s be given as in Lemma 2.5.3. Then the following assertions hold.*

(i) $\chi_{Z_r} \otimes \chi_{Z_s} \in HK(E_r \times E_s)$ *and*

$$(HK)\int_{E_r \times E_s} (\chi_{Z_r} \otimes \chi_{Z_s})(\xi, \eta)\, d(\xi, \eta) = 0.$$

(ii) $f\chi_{Z_r \times Z_s} \in HK(E_r \times E_s)$ *and*

$$(HK)\int_{E_r \times E_s} f(\xi, \eta)\chi_{Z_r \times Z_s}(\xi, \eta)\, d(\xi, \eta) = 0.$$

Proof. (i) Let $\varepsilon > 0$ and select a gauge Δ_r on E_r so that

$$\sum_{(\xi', I') \in P'} \chi_{Z_r}(\xi')\mu_r(I') < \frac{\varepsilon}{1 + \mu_s(E_s)}$$

for each Δ_r-fine Perron subpartition P' of E_r.

Define a gauge δ on $E_r \times E_s$ by setting $\delta(\xi, \eta) = \Delta_r(\xi)$. For each δ-fine Perron partition P of $E_r \times E_s$, we fix a net $D_s = \{J_1, \dots, J_N\}$ of E_s so that the following condition is met: if $J \in D_s$, then $J \subseteq J'$ for some $((\xi', \eta'), I' \times J') \in P$. Then

$$\{(\xi, I) : ((\xi, \eta), I \times J) \in P \text{ and } J_k \subseteq J\} \quad (k = 1, \dots, N)$$

are Δ_r-fine Perron subpartitions P' of E_r; hence

$$\begin{aligned} S(\chi_{Z_r} \otimes \chi_{Z_s}, P) &\leq \sum_{k=1}^{N} \sum_{((\xi,\eta), I \times J) \in P} \chi_{Z_r}(\xi)\mu_r(I)\mu_s(J \cap J_k) \\ &\leq \sum_{k=1}^{N} \sum_{\substack{((\xi,\eta), I \times J) \in P \\ J_k \subseteq J}} \chi_{Z_r}(\xi)\mu_r(I)\mu_s(J \cap J_k) \\ &< \sum_{k=1}^{N} \frac{\varepsilon}{1 + \mu_s(E_s)} \mu_s(J_k) \\ &< \varepsilon. \end{aligned}$$

Since $\varepsilon > 0$ is arbitrary, assertion (i) holds.

Using (i) and Lemma 2.5.2(i) with $m = r+s$, we get (ii). This completes the proof of the lemma. □

Theorem 2.5.5 (Fubini). *Let $f \in HK(E_r \times E_s)$, let $Z_r = \{\xi \in E_r : f(\xi, \cdot) \notin HK(E_s)\}$, and let $Z_s = \{\eta \in E_s : f(\cdot, \eta) \notin HK(E_r)\}$. Then the following statements are true.*

(i) $(HK) \int_{E_r} \chi_{Z_r}(\xi)\, d\xi = 0$.
(ii) $(HK) \int_{E_s} \chi_{Z_s}(\eta)\, d\eta = 0$.
(iii) *There exists $f_0 \in HK(E_r \times E_s)$ with the following properties:*
 (a) *$f_0(\xi,\eta) = f(\xi,\eta)$ whenever $(\xi,\eta) \in (E_r \times E_s)\backslash(Z_r \times Z_s)$;*
 (b) *the Henstock-Kurzweil integrals*

$$(HK)\int_{E_r}\left\{(HK)\int_{E_s} f_0(\xi,\eta)\, d\eta\right\} d\xi,$$

$$(HK)\int_{E_s}\left\{(HK)\int_{E_r} f_0(\xi,\eta)\, d\xi\right\} d\eta$$

 exist;
 (c)

$$(HK)\int_{E_r\times E_s} f = (HK)\int_{E_r}\left\{(HK)\int_{E_s} f_0(\xi,\eta)\, d\eta\right\} d\xi$$
$$= (HK)\int_{E_s}\left\{(HK)\int_{E_r} f_0(\xi,\eta)\, d\xi\right\} d\eta.$$

Proof. Statements (i) and (ii) follow from Lemma 2.5.3. To prove part (a) of (iii), we define the function $f_0 : E_r \times E_s \longrightarrow \mathbb{R}$ by setting $f_0 = f\chi_{(E_r\times E_s)\backslash(Z_r\times Z_s)}$. By Lemma 2.5.4 and Theorem 2.3.1, $f_0 \in HK(E_r \times E_s)$ and

$$(HK)\int_{E_r\times E_s} f_0 = (HK)\int_{E_r\times E_s} f. \tag{2.5.4}$$

Now we prove part (b) of (iii). For each $\varepsilon > 0$ we use (2.5.4) to select a gauge Δ on $E_r \times E_s$ such that

$$\left|S(f_0, P) - (HK)\int_{E_r\times E_s} f\right| < \frac{\varepsilon}{2} \tag{2.5.5}$$

for each Δ-fine Perron partition P of $E_r \times E_s$. Since the definition of f_0 implies that the function $\eta \mapsto f_0(\xi,\eta)$ belongs to $HK(E_s)$ whenever $\xi \in E_r$, for each $\xi \in E_r$ we select a $\Delta(\xi,\cdot)$-fine Perron partition T_ξ of E_s so that

$$\left|S(f_0(\xi,\cdot)) - (HK)\int_{E_s} f_0(\xi,\eta)\, d\eta\right| < \frac{\varepsilon}{2(1+\mu_r(E_r))}. \tag{2.5.6}$$

Define a gauge δ_r on E_r by setting

$$\delta_r(\xi) = \min\{\Delta(\xi, \eta) : (\eta, J) \in T_\xi\},$$

and let Q_r be any δ_r-fine Perron partition of E_r. Then

$$P_0 = \{((\xi, \eta), I \times J) : (\xi, I) \in Q_r \text{ and } (\eta, J) \in T_\xi\}$$

is a Δ-fine Perron partition of $E_r \times E_s$. Therefore the triangle inequality, (2.5.5) and (2.5.6) yield

$$\left| \sum_{(\xi,I)\in Q_r} \left((HK) \int_{E_s} f_0(\xi, \eta)\, d\eta \right) \mu_r(I) - (HK) \int_{E_r \times E_s} f \right|$$

$$\leq \left| \sum_{(\xi,I)\in Q_r} \left((HK) \int_{E_s} f_0(\xi, \eta)\, d\eta \right) \mu_r(I) - S(f_0, P_0) \right|$$

$$+ \left| S(f_0, P_0) - (HK) \int_{E_r \times E_s} f \right|$$

$$< \left| \sum_{(\xi,I)\in Q_r} \left\{ \left((HK) \int_{E_s} f_0(\xi, \eta)\, d\eta \right) \mu_r(I) - \sum_{(\eta,J)\in T_\xi} f_0(\xi, \eta) \mu_r(I) \mu_s(J) \right\} \right|$$

$$+ \frac{\varepsilon}{2}$$

$$\leq \sum_{(\xi,I)\in Q_r} \left| \left((HK) \int_{E_s} f_0(\xi, \eta)\, d\eta \right) \mu_r(I) - \sum_{(\eta,J)\in T_\xi} f_0(\xi, \eta) \mu_r(I) \mu_s(J) \right| + \frac{\varepsilon}{2}$$

$$< \varepsilon.$$

Since $\varepsilon > 0$ is arbitrary, we conclude that the function $\xi \mapsto (HK) \int_{E_s} f_0(\xi, \eta)\, d\eta$ belongs to $HK(E_r)$ and

$$(HK) \int_{E_r} \left\{ (HK) \int_{E_s} f_0(\xi, \eta)\, d\eta \right\} d\xi = (HK) \int_{E_r \times E_s} f.$$

Similarly, we get

$$(HK) \int_{E_s} \left\{ (HK) \int_{E_r} f_0(\xi, \eta)\, d\xi \right\} d\eta = (HK) \int_{E_r \times E_s} f.$$

This completes the proof of the theorem. □

Example 2.5.6. Let $m = 2$ and define the function $f : [0, 1]^2 \longrightarrow \mathbb{R}$ by

$$f(\xi, \eta) = \sum_{k=0}^{\infty} \frac{2^{4k}}{k+1} \chi_{[2^{-k-1}, 2^{-k}]^2}(\xi, \eta) \sin(2^{2k}\pi\xi) \sin(2^{2k}\pi\eta).$$

Then $f \in HK([0, 1]^2)$ and $(HK) \int_{[0,1]^2} f = 0$.

Proof. First, we observe that f is continuous on $[0,1]^2\setminus\{(0,0)\}$. Hence, by Theorem 2.3.5, $f \in HK(I)$ for every interval $I \in \mathcal{I}_2([0,1]^2)$ satisfying $I \cap \{(0,0)\} = \emptyset$. In view of Theorem 2.4.12, it remains to prove that

$$\sup_{(x_1,x_2)\in[0,1)^2\setminus[0,2^{-n-1}]^2} \left|(HK)\int_{[x_1,1]\times[x_2,1]} f\right| \leq \frac{4}{n+1} \tag{2.5.7}$$

for all $n \in \mathbb{N}$.

A direct computation shows that

$$(HK)\int_{[x_1,1]\times[x_2,1]} f = 0$$

whenever $(x_1,x_2) \in [0,1]^2\setminus(\{(0,0)\} \cup \bigcup_{k=1}^{\infty}[2^{-k-1},2^{-k}]^2)$. On the other hand, suppose that $(x_1,x_2) \in [2^{-n-1},2^{-n}] \times [2^{-n-1},2^{-n}]$ for some $n \in \mathbb{N}$. By Fubini's Theorem,

$$\begin{aligned}\left|(HK)\int_{[x_1,2^{-n}]\times[x_2,2^{-n}]} f\right| &= \left|\frac{(-1+\cos(2^{2n}\pi x_1))(-1+\cos(2^{2n}\pi x_2))}{n+1}\right| \\ &\leq \frac{4}{n+1}.\end{aligned}$$

Combining the above cases yields (2.5.7) to be proved. $\square$

The following example shows that the implication

$$f \in HK[\boldsymbol{a},\boldsymbol{b}] \Longrightarrow |f| \in HK[\boldsymbol{a},\boldsymbol{b}]$$

is false.

Example 2.5.7. If f is given as in Example 2.5.6, then $|f| \notin HK([0,1]^2)$.

Proof. For each $n \in \mathbb{N}$, we have

$$\begin{aligned}&(HK)\int_{[2^{-n-1},1]^2} |f| \\ &= \sum_{k=0}^{n}(HK)\int_{[2^{-k-1},2^{-k}]^2} \frac{2^{4k}}{k+1}\left|\sin(2^{2k}\pi\xi)\sin(2^{2k}\pi\eta)\right| \, d(\xi,\eta) \\ &= \sum_{k=0}^{n}\frac{2^{4k}}{k+1}\left(\int_{2^{-k-1}}^{2^{-k}}\left|\sin(2^{2k}\pi x)\right| \, dx\right)^2 \\ &= \frac{1}{\pi^2}\sum_{k=0}^{n}\frac{2^{2k}}{k+1}.\end{aligned}$$

Since the series $\sum_{k=0}^{\infty}\frac{2^{2k}}{k+1}$ diverges, it follows from Theorem 2.4.12 that $|f| \notin HK([0,1]^2)$. $\square$

The next example shows that the implication

$$f \in HK[\boldsymbol{a}, \boldsymbol{b}] \text{ and } g \in C[\boldsymbol{a}, \boldsymbol{b}] \Longrightarrow fg \in HK[\boldsymbol{a}, \boldsymbol{b}]$$

is false.

Example 2.5.8. Let $m = 2$ and define the function $g : [0,1]^2 \longrightarrow \mathbb{R}$ by

$$g(\xi, \eta) = \sum_{k=0}^{\infty} \frac{1}{2^{2k}} \sin^3(2^{2k}\pi\xi) \sin^3(2^{2k}\pi\eta).$$

If f is given as in Example 2.5.6, $fg \notin HK([0,1]^2)$.

Proof. Proceeding towards a contradiction, suppose that $fg \in HK([0,1]^2)$. Then for each $n \in \mathbb{N}$ we have

$$\begin{aligned}
&(HK)\int_{[0,\frac{1}{2^n}]^2} fg - (HK)\int_{[0,\frac{1}{2^{2n}}]^2} fg \\
&= \sum_{k=n}^{2n-1} \left\{ (HK)\int_{[0,\frac{1}{2^k}]^2} fg - (HK)\int_{[0,\frac{1}{2^{k+1}}]^2} fg \right\} \\
&= \sum_{k=n}^{2n-1} \frac{2^{2k}}{k+1} \left(\int_{2^{-k-1}}^{2^{-k}} \sin^4(2^{2k}\pi x)\, dx \right)^2 \\
&\geq \frac{9}{128}.
\end{aligned}$$

This is a contradiction, since Theorem 2.4.9 implies

$$\lim_{n\to\infty} \left\{ (HK)\int_{[0,\frac{1}{2^n}]^2} fg - (HK)\int_{[0,\frac{1}{2^{2n}}]^2} fg \right\} = 0.$$

This contradiction completes the proof. □

The following example shows that if $m \geq 2$, the hypotheses of Theorems 2.4.11 and 2.4.12 cannot be relaxed.

Example 2.5.9. Let $m = 2$ and let

$$h(x_1, x_2) = \begin{cases} \frac{1}{x_1^2} & \text{if } (x_1, x_2) \in (0,1] \times (\frac{2}{3}, 1], \\ -\frac{1}{x_1^2} & \text{if } (x_1, x_2) \in (0,1] \times (\frac{1}{3}, \frac{2}{3}], \\ 0 & \text{otherwise.} \end{cases}$$

Then

$$\lim_{\substack{(x_1,x_2)\to(0,0) \\ (x_1,x_2)\in(0,1)^2}} (HK)\int_{[x_1,1]\times[x_2,1]} h \tag{2.5.8}$$

exists but $h \notin HK([0,1]^2)$.

Proof. It is easy to check that the limit (2.5.8) exists. If $h \in HK([0,1]^2)$, then $h|_{[0,1]\times[\frac{2}{3},1]} \in HK([0,1]\times[\frac{2}{3},1])$, a contradiction to Fubini's Theorem. The proof is complete. □

The following example shows that the converse of Theorem 2.5.5 is false.

Example 2.5.10. Define $f : [0,\pi]^2 \longrightarrow \mathbb{R}$ by

$$f(x_1,x_2) = \begin{cases} 2^{2n+3}\sin(2^{n+1}(x_1 - \frac{\pi}{2^n}))\sin(2^{n+2}x_2) \\ \qquad\qquad \text{if } (x_1,x_2) \in [\frac{\pi}{2^n}, \frac{\pi}{2^{n-1}}] \times [0, \frac{\pi}{2^{n+1}}] \text{ for some } n \in \mathbb{N}, \\ 0 \qquad\qquad \text{otherwise.} \end{cases}$$

Direct computations show that the following iterated integrals

$$\int_a^b \left(\int_c^d f(x_1,x_2)\,dx_2\right)dx_1, \int_c^d \left(\int_a^b f(x_1,x_2)\,dx_1\right)dx_2$$

exist and coincide for every interval $[a,b]\times[c,d] \subseteq [0,\pi]^2$. On the other hand, since $\lim_{n\to\infty} \mu_2([\frac{\pi}{2^n}, \frac{3\pi}{2^{n+1}}] \times [0, \frac{\pi}{2^{n+2}}]) = 0$ and

$$\sup_{n\in\mathbb{N}} (HK)\int_{[\frac{\pi}{2^n}, \frac{3\pi}{2^{n+1}}]\times[0,\frac{\pi}{2^{n+2}}]} f = 4,$$

it follows from Theorem 2.4.8 that $f \notin HK([0,\pi]^2)$.

2.6 Notes and Remarks

It is known that $R[\boldsymbol{a},\boldsymbol{b}]$ is a linear space. Further properties of multiple Riemann integrals can be found in [1, Chapter 14].

The multiple Henstock-Kurzweil integral is equivalent to the multiple Perron integral in the following sense: a function which is integrable in one sense is integrable in the other sense and both integrals coincide; see also Schwabík [147] or Ostaszewski [134].

The proof of Theorem 2.4.7 is similar to that of [93, Theorem 3.4]. For other results concerning interval functions, see Aversa and Laczkovich [3], and Laczkovich [83, 84].

K. Karták [68] proved Theorems 2.4.11 and 2.4.12 in the context of Perron integration. For a more general version of such results, see Faure and J. Mawhin [37].

A Fubini's Theorem holds for multiple Perron integrals (cf. [118]). For further results concerning Fubini's Theorem, see Kurzweil [72] or Ostaszewski [134].

Examples 2.5.6 and 2.5.8 are due to Kurzweil [73].

Chapter 3

Lebesgue integrable functions

3.1 Introduction

It is known that if $f \in HK[\boldsymbol{a}, \boldsymbol{b}]$, then $|f|$ need not belong to $HK[\boldsymbol{a}, \boldsymbol{b}]$; see, for instance, Example 2.5.7. Moreover, Example 2.5.8 shows that if $f \in HK([0,1]^2)$ and $g \in C([0,1]^2)$, then fg need not belong to $HK([0,1]^2)$. In this chapter, we study a subclass of Henstock-Kurzweil integrable functions that removes the above "undesirable" properties.

Definition 3.1.1. A function $f : [\boldsymbol{a}, \boldsymbol{b}] \longrightarrow \mathbb{R}$ is said to be *Lebesgue integrable* on $[\boldsymbol{a}, \boldsymbol{b}]$ if both f and $|f|$ belong to $HK[\boldsymbol{a}, \boldsymbol{b}]$.

The collection of all functions that are Lebesgue integrable on $[\boldsymbol{a}, \boldsymbol{b}]$ will be denoted by $L^1[\boldsymbol{a}, \boldsymbol{b}]$. If $f \in L^1[\boldsymbol{a}, \boldsymbol{b}]$, we write $(HK)\int_{[\boldsymbol{a},\boldsymbol{b}]} f$ as $\int_{[\boldsymbol{a},\boldsymbol{b}]} f \, d\mu_m$, $\int_{[\boldsymbol{a},\boldsymbol{b}]} f(\boldsymbol{x}) \, d\mu_m(\boldsymbol{x})$ or $\int_{[\boldsymbol{a},\boldsymbol{b}]} f(\boldsymbol{t}) \, d\mu_m(\boldsymbol{t})$. According to Definition 3.1.1 and Example 2.5.7, $L^1[\boldsymbol{a}, \boldsymbol{b}] \subset HK[\boldsymbol{a}, \boldsymbol{b}]$.

The following simple lemma gives a necessary condition for a function to be Lebesgue integrable on $[\boldsymbol{a}, \boldsymbol{b}]$.

Lemma 3.1.2. *If $f \in L^1[\boldsymbol{a}, \boldsymbol{b}]$, then*

$$\sup\left\{\sum_{I \in D}\left|\int_I f \, d\mu_m\right| : D \text{ is a division of } [\boldsymbol{a}, \boldsymbol{b}]\right\} \leq \int_{[\boldsymbol{a},\boldsymbol{b}]} |f| \, d\mu_m.$$

Proof. Let D be a division of $[\boldsymbol{a}, \boldsymbol{b}]$. According to Theorems 2.3.3 and 2.3.10, we have

$$\sum_{I \in D}\left|\int_I f \, d\mu_m\right| \leq \sum_{I \in D}\int_I |f| \, d\mu_m = \int_{[\boldsymbol{a},\boldsymbol{b}]} |f| \, d\mu_m.$$

Since D is an arbitrary division of $[\boldsymbol{a}, \boldsymbol{b}]$, the lemma follows. □

The above necessary condition turns out to be a sufficient condition for f to be Lebesgue integrable there.

Theorem 3.1.3. *If $f \in HK[\boldsymbol{a}, \boldsymbol{b}]$, then $f \in L^1[\boldsymbol{a}, \boldsymbol{b}]$ if and only if*

$$\sup\left\{\sum_{I\in D}\left|(HK)\int_I f\right| : D \text{ is a division of } [\boldsymbol{a}, \boldsymbol{b}]\right\} < \infty. \tag{3.1.1}$$

In this case,

$$\int_{[\boldsymbol{a},\boldsymbol{b}]} |f|\, d\mu_m = \sup\left\{\sum_{I\in D}\left|(HK)\int_I f\right| : D \text{ is a division of } [\boldsymbol{a}, \boldsymbol{b}]\right\}. \tag{3.1.2}$$

Proof. ($\Longrightarrow$) This follows from Lemma 3.1.2.

($\Longleftarrow$) Conversely, suppose that (3.1.1) holds. Let V denote the left-hand side of (3.1.1). Since $f \in HK[\boldsymbol{a}, \boldsymbol{b}]$, it suffices to prove that $|f| \in HK[\boldsymbol{a}, \boldsymbol{b}]$ and

$$(HK)\int_{[\boldsymbol{a},\boldsymbol{b}]} |f| = V. \tag{3.1.3}$$

For each $\varepsilon > 0$ we fix a division D_0 of $[\boldsymbol{a}, \boldsymbol{b}]$ so that

$$V < \sum_{J\in D_0}\left|(HK)\int_J f\right| + \frac{\varepsilon}{2}. \tag{3.1.4}$$

By the Saks-Henstock Lemma there exists a gauge δ on $[\boldsymbol{a}, \boldsymbol{b}]$ such that

$$\sum_{(\boldsymbol{w},I')\in Q}\left|f(\boldsymbol{w})\mu_m(I') - (HK)\int_{I'} f\right| < \frac{\varepsilon}{2}$$

for each δ-fine Perron subpartition Q of $[\boldsymbol{a}, \boldsymbol{b}]$. By making δ smaller, if necessary, we may assume that if $(\boldsymbol{w}, I') \in Q$ and $\mu_m(I' \cap K) > 0$ for some $K \in D_0$, then $I' \subseteq K$. Thus

$$(HK)\int_J f = \sum_{\substack{(\boldsymbol{w},I')\in Q\\ I'\subseteq J}} (HK)\int_{I'} f \text{ for all } J \in D_0. \tag{3.1.5}$$

To this end, we consider any δ-fine Perron partition P of $[\boldsymbol{a}, \boldsymbol{b}]$. According to our choice of δ, (3.1.4) and (3.1.5), we have

$$\begin{aligned}
&\left| \sum_{(\boldsymbol{t},I)\in P} |f(\boldsymbol{t})|\, \mu_m(I) - V \right| \\
&\leq \sum_{(\boldsymbol{t},I)\in P} \left| \left\{ |f(\boldsymbol{t})|\, \mu_m(I) - \left| (HK)\int_I f \right| \right\} \right| + V - \sum_{(\boldsymbol{t},I)\in P} \left| (HK)\int_I f \right| \\
&\leq \sum_{(\boldsymbol{t},I)\in P} \left| f(\boldsymbol{t})\mu_m(I) - (HK)\int_I f \right| + V - \sum_{J\in D_0} \left| (HK)\int_J f \right| \\
&< \varepsilon.
\end{aligned}$$

Since $\varepsilon > 0$ is arbitrary, we conclude that $|f| \in HK[\boldsymbol{a}, \boldsymbol{b}]$ and (3.1.3) holds. □

The following theorem is a consequence of Theorems 2.3.3 and 3.1.3.

Theorem 3.1.4. *If $f, g \in HK[\boldsymbol{a}, \boldsymbol{b}]$ and $|f(\boldsymbol{x})| \leq g(\boldsymbol{x})$ for all $\boldsymbol{x} \in [\boldsymbol{a}, \boldsymbol{b}]$, then $f \in L^1[\boldsymbol{a}, \boldsymbol{b}]$.*

The proof of the following theorem is left to the reader.

Theorem 3.1.5. *$L^1[\boldsymbol{a}, \boldsymbol{b}]$ is a linear space.*

Definition 3.1.6. Let f and g be real-valued functions defined on $[\boldsymbol{a}, \boldsymbol{b}]$.

(i) We define the **maximum** of f and g, denoted by $\max\{f, g\}$ or $f \vee g$, by

$$(\max\{f, g\})(\boldsymbol{x}) := \max\{f(\boldsymbol{x}), g(\boldsymbol{x})\} \text{ for all } \boldsymbol{x} \in [\boldsymbol{a}, \boldsymbol{b}].$$

(ii) We define the **minimum** of f and g, denoted by $\min\{f, g\}$ or $f \wedge g$, by

$$(\min\{f, g\})(\boldsymbol{x}) := \min\{f(\boldsymbol{x}), g(\boldsymbol{x})\} \text{ for all } \boldsymbol{x} \in [\boldsymbol{a}, \boldsymbol{b}].$$

The proofs of the following theorems are left to the reader.

Theorem 3.1.7. *If $f, g \in L^1[\boldsymbol{a}, \boldsymbol{b}]$, then $\max\{f, g\}$ and $\min\{f, g\}$ also belong to $L^1[\boldsymbol{a}, \boldsymbol{b}]$.*

Theorem 3.1.8. *If $f \in L^1[\boldsymbol{a}, \boldsymbol{b}]$, then $|f| \in L^1[\boldsymbol{a}, \boldsymbol{b}]$ and*

$$\left| \int_{[\boldsymbol{a},\boldsymbol{b}]} f \, d\mu_m \right| \leq \int_{[\boldsymbol{a},\boldsymbol{b}]} |f| \, d\mu_m.$$

Exercise 3.1.9. Let $m = 1$ and let F be given as in Example 1.2.6. Show that $F' \in HK[0,1] \backslash L^1[0,1]$.

3.2 Some convergence theorems for Lebesgue integrals

In this section we use the Saks Henstock Lemma to prove some convergence theorems for the Lebesgue integral. The first convergence theorem is the following version of the Monotone Convergence Theorem.

Theorem 3.2.1. *Let $f : [\boldsymbol{a}, \boldsymbol{b}] \longrightarrow \mathbb{R}$ and let $(f_n)_{n=1}^{\infty}$ be a sequence in $L^1[\boldsymbol{a}, \boldsymbol{b}]$. If $(f_n(\boldsymbol{x}))_{n=1}^{\infty}$ is monotone and $f(\boldsymbol{x}) = \lim_{n\to\infty} f_n(\boldsymbol{x})$ for all $\boldsymbol{x} \in [\boldsymbol{a}, \boldsymbol{b}]$, then $f \in L^1[\boldsymbol{a}, \boldsymbol{b}]$ if and only if the sequence $\left(\int_{[\boldsymbol{a},\boldsymbol{b}]} f_n \, d\mu_m\right)_{n=1}^{\infty}$ is bounded. In this case,*

$$\lim_{n\to\infty} \int_{[\boldsymbol{a},\boldsymbol{b}]} f_n \, d\mu_m = \int_{[\boldsymbol{a},\boldsymbol{b}]} f \, d\mu_m.$$

Proof. We may assume that $0 \leq f_n(\boldsymbol{x}) \leq f_{n+1}(\boldsymbol{x}) \leq f(\boldsymbol{x})$ for all $\boldsymbol{x} \in [\boldsymbol{a}, \boldsymbol{b}]$ and $n \in \mathbb{N}$.

($\Longrightarrow$) Suppose that $f \in L^1[\boldsymbol{a}, \boldsymbol{b}]$. Since $f \in L^1[\boldsymbol{a}, \boldsymbol{b}]$ and $f_n(\boldsymbol{x}) \leq f_{n+1}(\boldsymbol{x}) \leq f(\boldsymbol{x})$ for all $\boldsymbol{x} \in [\boldsymbol{a}, \boldsymbol{b}]$ and $n \in \mathbb{N}$, it follows from Theorem 2.3.3 that the sequence $\left(\int_{[\boldsymbol{a},\boldsymbol{b}]} f_n \, d\mu_m\right)_{n=1}^{\infty}$ is bounded.

($\Longleftarrow$) Suppose that the sequence $\left(\int_{[\boldsymbol{a},\boldsymbol{b}]} f_n \, d\mu_m\right)_{n=1}^{\infty}$ is increasing and bounded. In this case, this sequence converges to a real number A, where

$$A = \sup_{n\in\mathbb{N}} \int_{[\boldsymbol{a},\boldsymbol{b}]} f_n \, d\mu_m.$$

We claim that $f \in HK[\boldsymbol{a}, \boldsymbol{b}]$ and $(HK) \int_{[\boldsymbol{a},\boldsymbol{b}]} f = A$. Let $\varepsilon > 0$ be given and select a positive integer N so that

$$A - \int_{[\boldsymbol{a},\boldsymbol{b}]} f_N \, d\mu_m < \frac{\varepsilon}{3}.$$

For each $n \in \mathbb{N}$ we use the Saks-Henstock Lemma to select a gauge δ_n on $[\boldsymbol{a}, \boldsymbol{b}]$ such that

$$\sum_{(\boldsymbol{x},J)\in Q_n} \left| f_n(\boldsymbol{x})\mu_m(J) - \int_J f_n \, d\mu_m \right| < \frac{\varepsilon}{3(2^n)}$$

for each δ_n-fine Perron subpartition Q_n of $[\boldsymbol{a}, \boldsymbol{b}]$.

For each $\boldsymbol{x} \in [\boldsymbol{a}, \boldsymbol{b}]$ the increasing sequence $(f_n(\boldsymbol{x}))_{n=1}^{\infty}$ converges to $f(\boldsymbol{x})$; hence there exists a positive integer $r(\boldsymbol{x}) > N$ such that

$$f(\boldsymbol{x}) - f_n(\boldsymbol{x}) < \frac{\varepsilon}{3\mu_m([\boldsymbol{a}, \boldsymbol{b}])} \tag{3.2.1}$$

for all integers $n \geq r(\boldsymbol{x})$.

Define a gauge δ on $[\boldsymbol{a}, \boldsymbol{b}]$ by setting

$$\delta(\boldsymbol{x}) = \delta_{r(\boldsymbol{x})}(\boldsymbol{x}),$$

and let P be a δ-fine Perron partition of $[\boldsymbol{a}, \boldsymbol{b}]$. Next for each $k \in \mathbb{N}$ we let

$$P_k = \{(\boldsymbol{t}, I) \in P : r(\boldsymbol{t}) = k\}$$

and use Cousin's Lemma to choose positive integers $k_1, \ldots, k_s$ so that $k_1 < \cdots < k_s$, $P = \bigcup_{i=1}^{s} P_{k_i}$ and $P_{k_i} \neq \emptyset$ for $i = 1, \ldots, s$. It follows that

$$\begin{aligned}
&|S(f, P) - A| \\
&= \left| \sum_{(\boldsymbol{t}, I) \in P} f(\boldsymbol{t}) \mu_m(I) - A \right| \\
&= \left| \sum_{i=1}^{s} \sum_{(\boldsymbol{t}, I) \in P_{k_i}} f(\boldsymbol{t}) \mu_m(I) - A \right| \qquad \text{(by our choice of } P_{k_1}, \ldots, P_{k_s}) \\
&\leq \left| \sum_{i=1}^{s} \sum_{(\boldsymbol{t}, I) \in P_{k_i}} (f(\boldsymbol{t}) - f_{r(\boldsymbol{t})}(\boldsymbol{t})) \mu_m(I) \right| \\
&+ \left| \sum_{i=1}^{s} \sum_{(\boldsymbol{t}, I) \in P_{k_i}} \left\{ f_{r(\boldsymbol{t})}(\boldsymbol{t}) \mu_m(I) - \int_I f_{r(\boldsymbol{t})} \, d\mu_m \right\} \right| \\
&+ \left| A - \sum_{i=1}^{s} \sum_{(\boldsymbol{t}, I) \in P_{k_i}} \int_I f_{r(\boldsymbol{t})} \, d\mu_m \right| \\
&= S_1 + S_2 + S_3, \qquad (3.2.2)
\end{aligned}$$

say.

We claim that $\max\{S_1, S_2, S_3\} < \frac{\varepsilon}{3}$.

(i) Using (3.2.1), we get $S_1 < \frac{\varepsilon}{3}$:

$$S_1 \leq \sum_{i=1}^{s} \sum_{(\boldsymbol{t}, I) \in P_{k_i}} \left| f(\boldsymbol{t}) - f_{r(\boldsymbol{t})}(\boldsymbol{t}) \right| \mu_m(I) < \frac{\varepsilon}{3}.$$

(ii) For each $i = 1, \ldots, s$ we note that P_{k_i} is a δ_{k_i}-fine Perron subpartition of $[\boldsymbol{a}, \boldsymbol{b}]$. Hence, by the triangle inequality and our choice of gauge δ, we have $S_2 < \frac{\varepsilon}{3}$:

$$S_2 \leq \sum_{i=1}^{s} \sum_{(\boldsymbol{t}, I) \in P_{k_i}} \left| f_{r(\boldsymbol{t})}(\boldsymbol{t}) \mu_m(I) - \int_I f_{r(\boldsymbol{t})} \, d\mu_m \right| < \sum_{i=1}^{s} \frac{\varepsilon}{3(2^{k_i})} < \frac{\varepsilon}{3}.$$

(iii) We have $S_3 < \frac{\varepsilon}{3}$:

$$\begin{aligned}
&\int_{[\boldsymbol{a},\boldsymbol{b}]} f_N \, d\mu_m \\
&= \sum_{(\boldsymbol{t},I)\in P} \int_I f_N \, d\mu_m && \text{(since } P \text{ is a partition of } [\boldsymbol{a},\boldsymbol{b}]) \\
&= \sum_{i=1}^{s} \sum_{(\boldsymbol{t},I)\in P_{k_i}} \int_I f_N \, d\mu_m && \text{(by our choice of } P_{k_1},\ldots,P_{k_s}) \\
&\leq \sum_{i=1}^{s} \sum_{(\boldsymbol{t},I)\in P_{k_i}} \int_I f_{r(\boldsymbol{t})} \, d\mu_m && \text{(since } r(\boldsymbol{t}) \geq N) \\
&\leq A && \text{(by the definition of } A) \\
&< \int_{[\boldsymbol{a},\boldsymbol{b}]} f_N \, d\mu_m + \frac{\varepsilon}{3} && \text{(by our choice of } N).
\end{aligned}$$

Combining the above inequalities, we get

$$|S(f,P) - A| < \varepsilon.$$

Since $\varepsilon > 0$ is arbitrary and $f(\boldsymbol{x}) = \lim_{n\to\infty} f_n(\boldsymbol{x}) \geq 0$ for all $\boldsymbol{x} \in [\boldsymbol{a},\boldsymbol{b}]$, we conclude that $f \in L^1[\boldsymbol{a},\boldsymbol{b}]$ and

$$\int_{[\boldsymbol{a},\boldsymbol{b}]} f \, d\mu_m = \lim_{n\to\infty} \int_{[\boldsymbol{a},\boldsymbol{b}]} f_n \, d\mu_m.$$

□

Example 3.2.2. Let $m = 2$ and let

$$f(x_1,x_2) = \begin{cases} \frac{2^{2n}}{n^2} \text{ if } (x_1,x_2) \in (\frac{1}{2^n}, \frac{1}{2^{n-1}})^2 \text{ for some } n \in \mathbb{N}, \\ 0 \quad \text{otherwise.} \end{cases}$$

Then $f \in L^1([0,1]^2)$ and

$$\int_{[0,1]^2} f \, d\mu_2 = \sum_{k=1}^{\infty} \frac{1}{k^2}.$$

Proof. For each $n \in \mathbb{N}$ we let

$$f_n(x_1,x_2) = \sum_{k=1}^{n} \frac{2^{2k}}{k^2} \chi_{(\frac{1}{2^k},\frac{1}{2^{k-1}})^2}(x_1,x_2).$$

Then $(f_n)_{n=1}^{\infty}$ is a sequence in $L^1([0,1]^2)$. For each $(x_1, x_2) \in [0,1]^2$, the sequence $(f_n(x_1,x_2))_{n=1}^{\infty}$ is increasing and $\lim_{n\to\infty} f_n(x_1,x_2) = f(x_1,x_2)$. Since we also have

$$\int_{[0,1]^2} f_n \, d\mu_2 = \sum_{k=1}^{n} \left(\frac{2^{2k}}{k^2}\right)\left(\frac{1}{2^{2k}}\right) = \sum_{k=1}^{n} \frac{1}{k^2} \le 2$$

for all $n \in \mathbb{N}$, an application of Theorem 3.2.1 shows that $f \in L^1([0,1]^2)$ and

$$\int_{[0,1]^2} f \, d\mu_2 = \lim_{n\to\infty} \int_{[0,1]^2} f_n \, d\mu_2 = \sum_{k=1}^{\infty} \frac{1}{k^2}.$$

□

Our next aim is to prove a slight improvement of Theorem 3.2.1. We need the following crucial lemma.

Lemma 3.2.3. *Let $(f_n)_{n=1}^{\infty}$ be an increasing sequence in $L^1[\boldsymbol{a},\boldsymbol{b}]$ and suppose that $\sup_{n\in\mathbb{N}}\{\int_{[\boldsymbol{a},\boldsymbol{b}]} f_n \, d\mu_m\}$ is finite. If $Z = \{\boldsymbol{x} \in [\boldsymbol{a},\boldsymbol{b}] : \lim_{n\to\infty} f_n(\boldsymbol{x}) = \infty\}$, then $\chi_Z \in L^1[\boldsymbol{a},\boldsymbol{b}]$ and $\int_{[\boldsymbol{a},\boldsymbol{b}]} \chi_Z \, d\mu_m = 0$.*

Proof. Without loss of generality, we may assume that $f_n \ge 0$ for all $n \in \mathbb{N}$.

We will first construct a decreasing sequence $(g_n)_{n=1}^{\infty}$ in $L^1[\boldsymbol{a},\boldsymbol{b}]$ such that $\lim_{n\to\infty} g_n(\boldsymbol{x}) = \chi_Z(\boldsymbol{x})$ for all $\boldsymbol{x} \in [\boldsymbol{a},\boldsymbol{b}]$ and

$$0 \le \int_{[\boldsymbol{a},\boldsymbol{b}]} g_n \, d\mu_m \le \frac{1}{n} \sup_{j\in\mathbb{N}} \left\{ \int_{[\boldsymbol{a},\boldsymbol{b}]} f_j \, d\mu_m \right\} \text{ for all } n \in \mathbb{N}. \quad (3.2.3)$$

For each $j, n \in \mathbb{N}$, we define a function $g_{j,n}$ on $[\boldsymbol{a},\boldsymbol{b}]$ by setting

$$g_{j,n} := \min\left\{1, \frac{f_j}{n}\right\}.$$

By Theorems 3.1.7 and 2.3.3, $(g_{j,n})_{(j,n)\in\mathbb{N}^2}$ is a double sequence in $L^1[\boldsymbol{a},\boldsymbol{b}]$ and

$$0 \le \int_{[\boldsymbol{a},\boldsymbol{b}]} g_{j,n} \, d\mu_m \le \frac{1}{n} \sup_{j\in\mathbb{N}} \left\{ \int_{[\boldsymbol{a},\boldsymbol{b}]} f_j \, d\mu_m \right\} \text{ for all } (j,n) \in \mathbb{N}^2. \quad (3.2.4)$$

Since $(f_n)_{n=1}^{\infty}$ is increasing, for each $n \in \mathbb{N}$ the sequence $(g_{j,n})_{j=1}^{\infty}$ is increasing and

$$g_n(\boldsymbol{x}) := \lim_{j\to\infty} g_{j,n}(\boldsymbol{x}) = \begin{cases} 1 & \text{if } \boldsymbol{x} \in Z, \\ \min\{1, \frac{f(\boldsymbol{x})}{n}\} & \text{if } \boldsymbol{x} \in [\boldsymbol{a},\boldsymbol{b}] \setminus Z. \end{cases} \quad (3.2.5)$$

Hence (3.2.4), (3.2.5) and Theorem 3.2.1 imply that $g_n \in L^1[\boldsymbol{a}, \boldsymbol{b}]$. It is now clear that the sequence $(g_n)_{n=1}^{\infty}$ has the desired properties.

Finally, since $(g_n)_{n=1}^{\infty}$ is a decreasing sequence of non-negative Lebesgue integrable functions on $[\boldsymbol{a}, \boldsymbol{b}]$ and $\lim_{n\to\infty} g_n(\boldsymbol{x}) = \chi_Z(\boldsymbol{x})$ for all $\boldsymbol{x} \in [\boldsymbol{a}, \boldsymbol{b}]$, the conclusion follows from Theorem 3.2.1 and (3.2.4). □

In order to formulate an improvement of Theorem 3.2.1, we need the following definition.

Definition 3.2.4. Let $Q(\boldsymbol{x})$ be a statement concerning the point $\boldsymbol{x} \in [\boldsymbol{a}, \boldsymbol{b}]$. If the set $E \subset [\boldsymbol{a}, \boldsymbol{b}]$ satisfies $\int_{[\boldsymbol{a},\boldsymbol{b}]} \chi_E \, d\mu_m = 0$, we say that E is *μ_m-negligible* and $Q(\boldsymbol{x})$ *holds for μ_m-almost all $\boldsymbol{x} \in [\boldsymbol{a}, \boldsymbol{b}]$*.

The following result is a refinement of Theorem 3.2.1.

Theorem 3.2.5 (Monotone Convergence Theorem). *Let $f:[\boldsymbol{a}, \boldsymbol{b}] \longrightarrow \mathbb{R}$ and let $(f_n)_{n=1}^{\infty}$ be a sequence in $L^1[\boldsymbol{a}, \boldsymbol{b}]$. If the sequence $(f_n(\boldsymbol{x}))_{n=1}^{\infty}$ is monotone and $f(\boldsymbol{x}) = \lim_{n\to\infty} f_n(\boldsymbol{x})$ for μ_m-almost all $\boldsymbol{x} \in [\boldsymbol{a}, \boldsymbol{b}]$, then $f \in L^1[\boldsymbol{a}, \boldsymbol{b}]$ if and only if the sequence $\left(\int_{[\boldsymbol{a},\boldsymbol{b}]} f_n \, d\mu_m\right)_{n=1}^{\infty}$ is bounded. In this case,*

$$\lim_{n\to\infty} \int_{[\boldsymbol{a},\boldsymbol{b}]} f_n \, d\mu_m = \int_{[\boldsymbol{a},\boldsymbol{b}]} f \, d\mu_m.$$

Proof. Let Z be given as in Lemma 3.2.3. Then Z is μ_m-negligible and the following conditions are satisfied for every $\boldsymbol{x} \in [\boldsymbol{a}, \boldsymbol{b}]$:

(i) $\left(f_n(\boldsymbol{x})\chi_{[\boldsymbol{a},\boldsymbol{b}]\setminus Z}(\boldsymbol{x})\right)_{n=1}^{\infty}$ is monotone;
(ii) $f(\boldsymbol{x})\chi_{[\boldsymbol{a},\boldsymbol{b}]\setminus Z}(\boldsymbol{x}) = \lim_{n\to\infty} f_n(\boldsymbol{x})\chi_{[\boldsymbol{a},\boldsymbol{b}]\setminus Z}(\boldsymbol{x})$.

Since Z is μ_m-negligible, it follows from our assumptions and Lemma 2.5.2(i) that $\left(f_n\chi_{[\boldsymbol{a},\boldsymbol{b}]\setminus Z}\right)_{n=1}^{\infty}$ is a sequence in $L^1[\boldsymbol{a}, \boldsymbol{b}]$. Hence, by Theorem 3.2.1, $f\chi_{[\boldsymbol{a},\boldsymbol{b}]\setminus Z} \in L^1[\boldsymbol{a}, \boldsymbol{b}]$ if and only if the sequence $\left(\int_{[\boldsymbol{a},\boldsymbol{b}]} f_n\chi_{[\boldsymbol{a},\boldsymbol{b}]\setminus Z} \, d\mu_m\right)_{n=1}^{\infty}$ is bounded. As Z is μ_m-negligble and f is real-valued, the theorem follows from Lemma 2.5.2(i).

□

Exercise 3.2.6. Prove that $\sum_{k=0}^{\infty} \frac{1}{(4k+1)(4k+2)} = \frac{1}{4}\ln 2 + \frac{\pi}{8}$.

The following result is a consequence of the Monotone Convergence Theorem.

Theorem 3.2.7 (Fatou's Lemma). *Let $f : [\boldsymbol{a}, \boldsymbol{b}] \longrightarrow \mathbb{R}$, let $(f_n)_{n=1}^{\infty}$ be a sequence in $L^1[\boldsymbol{a}, \boldsymbol{b}]$, and suppose that $f_n(\boldsymbol{x}) \geq 0$ for μ_m-almost all $\boldsymbol{x} \in [\boldsymbol{a}, \boldsymbol{b}]$ and $n \in \mathbb{N}$. If $\liminf_{n\to\infty} \int_{[\boldsymbol{a},\boldsymbol{b}]} f_n \, d\mu_m$ is finite and $\lim_{n\to\infty} f_n(\boldsymbol{x}) = f(\boldsymbol{x})$ for μ_m-almost all $\boldsymbol{x} \in [\boldsymbol{a}, \boldsymbol{b}]$, then $f \in L^1[\boldsymbol{a}, \boldsymbol{b}]$ and*

$$\int_{[\boldsymbol{a},\boldsymbol{b}]} f \, d\mu_m \leq \liminf_{n\to\infty} \int_{[\boldsymbol{a},\boldsymbol{b}]} f_n \, d\mu_m.$$

Proof. Define a sequence $(\phi_n)_{n=1}^{\infty}$ of non-negative functions on $[\boldsymbol{a}, \boldsymbol{b}]$ by setting

$$\phi_n := \inf_{k\geq n} f_k$$

for $n = 1, 2, \ldots$. Then the sequence $(\phi_n(\boldsymbol{x}))_{n=1}^{\infty}$ is increasing and

$$\lim_{n\to\infty} \phi_n(\boldsymbol{x}) = f(\boldsymbol{x})$$

for μ_m-almost all $\boldsymbol{x} \in [\boldsymbol{a}, \boldsymbol{b}]$. In view of the Monotone Convergence Theorem, it suffices to prove that $(\phi_n)_{n=1}^{\infty}$ is a sequence in $L^1[\boldsymbol{a}, \boldsymbol{b}]$ satisfying

$$\sup_{n\in\mathbb{N}} \left\{ \int_{[\boldsymbol{a},\boldsymbol{b}]} \phi_n \, d\mu_m \right\} < \infty.$$

For each $n \in \mathbb{N}$ we consider the decreasing sequence $\left(\min_{i=n,\ldots,k} f_i(\boldsymbol{x}) \right)_{k=n}^{\infty}$. By Theorem 3.1.7, this particular sequence is in $L^1[\boldsymbol{a}, \boldsymbol{b}]$. Since we also have

$$\lim_{k\to\infty} \min_{i=n,\ldots,k} f_i(\boldsymbol{x}) = \phi_n(\boldsymbol{x})$$

for μ_m-almost all $\boldsymbol{x} \in [\boldsymbol{a}, \boldsymbol{b}]$, the Monotone Convergence Theorem implies that $\phi_n \in L^1[\boldsymbol{a}, \boldsymbol{b}]$. Finally, the sequence $\left(\int_{[\boldsymbol{a},\boldsymbol{b}]} \phi_n \, d\mu_m \right)_{n=1}^{\infty}$ is bounded because the sequence $(\phi_n(\boldsymbol{x}))_{n=1}^{\infty}$ of non-negative functions is increasing for μ_m-almost all $\boldsymbol{x} \in [\boldsymbol{a}, \boldsymbol{b}]$ and

$$0 \leq \int_{[\boldsymbol{a},\boldsymbol{b}]} \phi_n \, d\mu_m \leq \liminf_{n\to\infty} \int_{[\boldsymbol{a},\boldsymbol{b}]} f_n \, d\mu_m < \infty \text{ for } n = 1, 2, \ldots.$$ □

Theorem 3.2.8 (Lebesgue's Dominated Convergence Theorem). *Let $f : [\boldsymbol{a}, \boldsymbol{b}] \longrightarrow \mathbb{R}$ and let $(f_n)_{n=1}^{\infty}$ be a sequence in $L^1[\boldsymbol{a}, \boldsymbol{b}]$ such that $\lim_{n\to\infty} f_n(\boldsymbol{x}) = f(\boldsymbol{x})$ for μ_m-almost all $\boldsymbol{x} \in [\boldsymbol{a}, \boldsymbol{b}]$. Suppose that there exists $g \in L^1[\boldsymbol{a}, \boldsymbol{b}]$ such that*

$$|f_n(\boldsymbol{x})| \leq g(\boldsymbol{x}) \quad \textit{for } \mu_m\textit{-almost all } \boldsymbol{x} \in [\boldsymbol{a}, \boldsymbol{b}] \textit{ and } n \in \mathbb{N}. \tag{3.2.6}$$

Then $f \in L^1[\boldsymbol{a}, \boldsymbol{b}]$ and

$$\lim_{n\to\infty} \int_{[\boldsymbol{a},\boldsymbol{b}]} f_n \, d\mu_m = \int_{[\boldsymbol{a},\boldsymbol{b}]} f \, d\mu_m. \tag{3.2.7}$$

Proof. According to our assumptions and Theorem 2.3.3, $(f_n + g)_{n=1}^{\infty}$ is a sequence of non-negative functions in $L^1[\boldsymbol{a}, \boldsymbol{b}]$ satisfying

$$\liminf_{n\to\infty} \int_{[\boldsymbol{a},\boldsymbol{b}]} (f_n + g)\, d\mu_m < \infty.$$

Hence, by Fatou's Lemma, $f + g \in L^1[\boldsymbol{a}, \boldsymbol{b}]$ and

$$\int_{[\boldsymbol{a},\boldsymbol{b}]} (f + g)\, d\mu_m \le \liminf_{n\to\infty} \int_{[\boldsymbol{a},\boldsymbol{b}]} (f_n + g)\, d\mu_m. \tag{3.2.8}$$

Since $g \in L^1[\boldsymbol{a}, \boldsymbol{b}]$, we infer from (3.2.8) that $f \in L^1[\boldsymbol{a}, \boldsymbol{b}]$ and

$$\int_{[\boldsymbol{a},\boldsymbol{b}]} f\, d\mu_m \le \liminf_{n\to\infty} \int_{[\boldsymbol{a},\boldsymbol{b}]} f_n\, d\mu_m. \tag{3.2.9}$$

We will next prove that the equality (3.2.7) holds. Following the proof of (3.2.8), we apply Fatou's Lemma to the sequence $(g - f_n)_{n=1}^{\infty}$ of non-negative functions to conclude that $g - f \in L^1[\boldsymbol{a}, \boldsymbol{b}]$ and

$$\int_{[\boldsymbol{a},\boldsymbol{b}]} (g - f)\, d\mu_m \le \liminf_{n\to\infty} \int_{[\boldsymbol{a},\boldsymbol{b}]} (g - f_n)\, d\mu_m \tag{3.2.10}$$

Since $g \in L^1[\boldsymbol{a}, \boldsymbol{b}]$, we infer from (3.2.10) that

$$\int_{[\boldsymbol{a},\boldsymbol{b}]} g\, d\mu_m - \int_{[\boldsymbol{a},\boldsymbol{b}]} f\, d\mu_m \le \int_{[\boldsymbol{a},\boldsymbol{b}]} g\, d\mu_m - \limsup_{n\to\infty} \int_{[\boldsymbol{a},\boldsymbol{b}]} f_n\, d\mu_m;$$

that is,

$$\limsup_{n\to\infty} \int_{[\boldsymbol{a},\boldsymbol{b}]} f_n\, d\mu_m \le \int_{[\boldsymbol{a},\boldsymbol{b}]} f\, d\mu_m. \tag{3.2.11}$$

The proof is now complete since (3.2.7) follows from (3.2.9), (3.2.11) and the following inequality

$$\liminf_{n\to\infty} \int_{[\boldsymbol{a},\boldsymbol{b}]} f_n\, d\mu_m \le \limsup_{n\to\infty} \int_{[\boldsymbol{a},\boldsymbol{b}]} f_n\, d\mu_m.$$

□

Exercise 3.2.9. Show that we can have strict inequality in Fatou's Lemma.

Exercise 3.2.10. Show that $\pi = 4 \sum_{k=1}^{\infty} \frac{(-1)^{k-1}}{2k-1}$.

3.3 μ_m-measurable sets

The aim of this section is to give some examples of bounded Lebesgue integrable functions. We begin with the study of closed sets.

Definition 3.3.1. A set $Y \subseteq [\boldsymbol{a}, \boldsymbol{b}]$ is said to be *closed* if every convergent sequence in Y has its limit in Y.

Example 3.3.2. $[0, 1]^m$ is a closed set.

We need the following lemma in order to prove Theorem 3.3.4 below.

Lemma 3.3.3. *If a set $Y \subseteq [\boldsymbol{a}, \boldsymbol{b}]$ is closed, then*

$$Y = \bigcap_{n=1}^{\infty} \bigcup_{\boldsymbol{y} \in Y} B\Big(\boldsymbol{y}, \frac{1}{n}\Big).$$

Proof. We may assume that Y is non-empty. Since the obvious inclusion $Y \subseteq \bigcap_{n=1}^{\infty} \bigcup_{\boldsymbol{y}\in Y} B(\boldsymbol{y}, \frac{1}{n})$ implies that the set $\bigcap_{n=1}^{\infty} \bigcup_{\boldsymbol{y}\in Y} B(\boldsymbol{y}, \frac{1}{n})$ is also non-empty, it remains to prove that

$$Y \supseteq \bigcap_{n=1}^{\infty} \bigcup_{\boldsymbol{y} \in Y} B\Big(\boldsymbol{y}, \frac{1}{n}\Big).$$

Let $\boldsymbol{x} \in \bigcap_{n=1}^{\infty} \bigcup_{\boldsymbol{y}\in Y} B(\boldsymbol{y}, \frac{1}{n})$. For each $n \in \mathbb{N}$ there exists $\boldsymbol{x}_n \in Y$ such that $|||\boldsymbol{x}_n - \boldsymbol{x}||| < \frac{1}{n}$. It follows that the sequence $(\boldsymbol{x}_n)_{n=1}^{\infty}$ in Y converges to $\boldsymbol{x}$. Since Y is closed, $\boldsymbol{x} \in Y$. □

The following theorem gives a useful characterization of non-empty closed sets.

Theorem 3.3.4. *A non-empty set $Y \subseteq [\boldsymbol{a}, \boldsymbol{b}]$ is closed if and only if*

$$Y = \{\boldsymbol{x} \in [\boldsymbol{a}, \boldsymbol{b}] : \mathit{dist}(\boldsymbol{x}, Y) = 0\}.$$

Proof. Since it is obvious that

$$Y \subseteq \{\boldsymbol{x} \in [\boldsymbol{a}, \boldsymbol{b}] : \operatorname{dist}(\boldsymbol{x}, Y) = 0\},$$

it suffices to prove that the reverse inclusion holds. To this end, we let $\boldsymbol{y} \in [\boldsymbol{a}, \boldsymbol{b}]$ with $\operatorname{dist}(\boldsymbol{y}, Y) = 0$. Then there exists a sequence $(\boldsymbol{y}_n)_{n=1}^{\infty}$ in Y such that $\lim_{n\to\infty} |||\boldsymbol{y}_n - \boldsymbol{y}||| = 0$. Since Y is closed, $\boldsymbol{y} \in Y$. □

Exercise 3.3.5. Prove the following properties of closed sets:

(i) The union of any finite collection of closed sets is closed.

(ii) The intersection of any collection of closed sets is closed.

Exercise 3.3.6. Show that a countable union of closed sets need not be closed.

Theorem 3.3.7. *Suppose that the set $W \subseteq [\boldsymbol{a}, \boldsymbol{b}]$ is non-empty. Then the function $\boldsymbol{x} \mapsto dist(\boldsymbol{x}, W)$ is continuous on $[\boldsymbol{a}, \boldsymbol{b}]$.*

Proof. Since

$$|\operatorname{dist}(\boldsymbol{s}, W) - \operatorname{dist}(\boldsymbol{t}, W)| \leq |||\boldsymbol{s} - \boldsymbol{t}|||$$

for all $\boldsymbol{s}, \boldsymbol{t} \in [\boldsymbol{a}, \boldsymbol{b}]$, the result follows. □

Corollary 3.3.8. *Suppose that the set $W \subseteq [\boldsymbol{a}, \boldsymbol{b}]$ is non-empty. Then the function $\boldsymbol{x} \mapsto dist(\boldsymbol{x}, W)$ is bounded on $[\boldsymbol{a}, \boldsymbol{b}]$.*

Proof. This follows from Theorem 3.3.7 and Corollary 2.2.4. □

For any non-empty subsets X and Y of $\mathbb{R}^m$, we write

$$\operatorname{dist}(X, Y) := \inf\left\{ |||\boldsymbol{t} - \boldsymbol{s}||| : \boldsymbol{s} \in X, \boldsymbol{t} \in Y \right\}.$$

Theorem 3.3.9. *Suppose that X and Y are non-empty closed subsets of $[\boldsymbol{a}, \boldsymbol{b}]$. Then there exists $\boldsymbol{z} \in X$ such that*

$$dist(\boldsymbol{z}, Y) = dist(X, Y).$$

Proof. If X and Y are not disjoint, then the result follows from Theorem 3.3.4. Henceforth, we will assume that X and Y are disjoint.

Proceeding towards a contradiction, suppose that

$$\operatorname{dist}(\boldsymbol{z}, Y) > \operatorname{dist}(X, Y) \text{ for all } \boldsymbol{z} \in X. \tag{3.3.1}$$

According to Theorem 3.3.7, for each $\boldsymbol{t} \in [\boldsymbol{a}, \boldsymbol{b}]$ there exists $\delta(\boldsymbol{t}) > 0$ such that

$$\operatorname{dist}(\boldsymbol{x}, Y) > \operatorname{dist}(\boldsymbol{t}, Y) - \frac{1}{2}(\operatorname{dist}(\boldsymbol{t}, Y) - \operatorname{dist}(X, Y)) \tag{3.3.2}$$

whenever $\boldsymbol{x} \in B(\boldsymbol{t}, \delta(\boldsymbol{t})) \cap [\boldsymbol{a}, \boldsymbol{b}]$. Thanks to Theorem 3.3.4, we may assume that $\delta(\boldsymbol{t}) < \operatorname{dist}(\boldsymbol{t}, X)$ for all $\boldsymbol{t} \in [\boldsymbol{a}, \boldsymbol{b}] \backslash X$. In view of our construction of δ and Cousin's Lemma, there exists a δ-fine Perron partition P of $[\boldsymbol{a}, \boldsymbol{b}]$ such that

$$X \subseteq \bigcup_{\substack{(\boldsymbol{t}, I) \in P \\ \boldsymbol{t} \in X}} I;$$

hence (3.3.2) and (3.3.1) imply

$$\operatorname{dist}(\boldsymbol{z}, Y) \geq \rho > \operatorname{dist}(X, Y) \text{ for all } \boldsymbol{z} \in X,$$

where

$$\rho = \frac{1}{2} \min_{\substack{(\boldsymbol{t}, I) \in P \\ \boldsymbol{t} \in X}} \left\{ \operatorname{dist}(X, Y) + \operatorname{dist}(\boldsymbol{t}, Y) \right\}.$$

Since $\boldsymbol{z} \in X$ is arbitrary, we conclude that $\operatorname{dist}(X, Y)$ is not the greatest lower bound of the function $\boldsymbol{x} \mapsto \operatorname{dist}(\boldsymbol{x}, Y)$ on X. This contradiction proves the theorem. □

We can now state and prove the following useful result.

Theorem 3.3.10. *Let X and Y be non-empty closed subsets of $[\boldsymbol{a}, \boldsymbol{b}]$. Then X and Y are disjoint if and only if $\operatorname{dist}(X, Y) > 0$.*

Proof. ($\Longrightarrow$) This follows from Theorems 3.3.9 and 3.3.4.

($\Longleftarrow$) Exercise. □

Definition 3.3.11. A set $X \subseteq [\boldsymbol{a}, \boldsymbol{b}]$ is said to be *μ_m-measurable* if $\chi_X \in L^1[\boldsymbol{a}, \boldsymbol{b}]$. In this case, we write $\int_{[\boldsymbol{a}, \boldsymbol{b}]} \chi_X \, d\mu_m$ as $\mu_m(X)$.

The following theorem is a consequence of Theorem 3.3.4 and Lebesgue's Dominated Convergence Theorem.

Theorem 3.3.12. *If a set $Y \subseteq [\boldsymbol{a}, \boldsymbol{b}]$ is closed, then Y is μ_m-measurable.*

Proof. We may assume that Y is non-empty. By Lemma 3.3.3,

$$Y = \bigcap_{k=1}^{\infty} G_k,$$

where

$$G_n = \bigcup_{\boldsymbol{y} \in Y} B(\boldsymbol{y}, \frac{1}{n}) \quad \text{for } n = 1, 2, \ldots.$$

We will next construct an appropriate sequence $(f_n)_{n=1}^{\infty}$ of uniformly bounded continuous functions on $[\boldsymbol{a}, \boldsymbol{b}]$ such that $\lim_{n \to \infty} f_n(\boldsymbol{x}) = \chi_Y(\boldsymbol{x})$ for all $\boldsymbol{x} \in [\boldsymbol{a}, \boldsymbol{b}]$. Let $n \in \mathbb{N}$ be given. Since both sets Y and $[\boldsymbol{a}, \boldsymbol{b}] \setminus G_n$ are closed and disjoint, Theorem 3.3.4 implies that $\operatorname{dist}(\boldsymbol{x}, Y) + \operatorname{dist}(\boldsymbol{x}, [\boldsymbol{a}, \boldsymbol{b}] \setminus G_n) > 0$ for all $\boldsymbol{x} \in [\boldsymbol{a}, \boldsymbol{b}]$. Now we define the function $f_n : [\boldsymbol{a}, \boldsymbol{b}] \longrightarrow \mathbb{R}$ by setting

$$f_n(\boldsymbol{x}) = \frac{\operatorname{dist}(\boldsymbol{x}, [\boldsymbol{a}, \boldsymbol{b}] \setminus G_n)}{\operatorname{dist}(\boldsymbol{x}, Y) + \operatorname{dist}(\boldsymbol{x}, [\boldsymbol{a}, \boldsymbol{b}] \setminus G_n)}.$$

According to Theorems 3.3.7 and 2.3.5, $(f_n)_{n=1}^{\infty}$ is a sequence in $L^1[\boldsymbol{a}, \boldsymbol{b}]$. Since we also have $\sup_{n\in\mathbb{N}} |f_n(\boldsymbol{x})| \leq 1$ and $\lim_{n\to\infty} f_n(\boldsymbol{x}) = \chi_Y(\boldsymbol{x})$ for all $\boldsymbol{x} \in [\boldsymbol{a}, \boldsymbol{b}]$, the result follows from Lebesgue's Dominated Convergence Theorem. □

Our next goal is to obtain a modification of Theorem 3.3.12. We need the following definition.

Definition 3.3.13. A set $O \subseteq \mathbb{R}^m$ is open if for each $\boldsymbol{x} \in O$ there exists $\delta(\boldsymbol{x}) > 0$ such that $B(\boldsymbol{x}, \delta(\boldsymbol{x})) \subset O$.

Example 3.3.14. The set $(0,1)^m$ is open.

Exercise 3.3.15. Prove the following properties of open sets:

(i) The intersection of any finite collection of open sets is open.
(ii) The union of any collection of open sets is open.

Exercise 3.3.16. Show that a countable intersection of open sets need not be open.

Theorem 3.3.17. *A set $Y \subseteq [\boldsymbol{a}, \boldsymbol{b}]$ is closed if and only if $\mathbb{R}^m \backslash Y$ is open.*

Proof. We may suppose that Y is non-empty and $Y \neq \mathbb{R}^m$.

($\Longrightarrow$) Suppose that Y is closed. To show that $\mathbb{R}^m \backslash Y$ is open, we let $\boldsymbol{z} \in \mathbb{R}^m \backslash Y$. Since Theorem 3.3.4 implies that $\text{dist}(\boldsymbol{z}, Y) > 0$, we use the continuity of the function $\boldsymbol{x} \mapsto \text{dist}(\boldsymbol{x}, Y)$ to choose $\delta(\boldsymbol{z}) > 0$ so that $\text{dist}(\boldsymbol{x}, Y) > 0$ whenever $\boldsymbol{x} \in B(\boldsymbol{z}, \delta(\boldsymbol{z}))$. By Theorem 3.3.4 again, $B(\boldsymbol{z}, \delta(\boldsymbol{z})) \subset \mathbb{R}^m \backslash Y$. Since $\boldsymbol{z} \in \mathbb{R}^m \backslash Y$ is arbitrary, $\mathbb{R}^m \backslash Y$ is open.

Conversely, suppose that $\mathbb{R}^m \backslash Y$ is open. To prove that Y is closed, we let $(\boldsymbol{x}_n)_{n=1}^{\infty}$ be a sequence in Y converging to some $\boldsymbol{x} \in \mathbb{R}^m$. Proceeding towards a contradiction, suppose that $\boldsymbol{x} \in \mathbb{R}^m \backslash Y$. Since $\mathbb{R}^m \backslash Y$ is open, there exists $\delta(\boldsymbol{x}) > 0$ such that $B(\boldsymbol{x}, \delta(\boldsymbol{x})) \subset \mathbb{R}^m \backslash Y$; hence our choice of $(\boldsymbol{x}_n)_{n=1}^{\infty}$ implies that $\boldsymbol{x}_n \in \mathbb{R}^m \backslash Y$ for all sufficiently large integers n, a contradiction. This contradiction shows that $\boldsymbol{x} \in Y$. □

The technique used in the proof of Theorem 3.3.18 is important.

Theorem 3.3.18. *If a set $O \subset [\boldsymbol{a}, \boldsymbol{b}]$ is open and non-empty, then O can be written as a countable union of pairwise non-overlapping intervals in $\mathbb{R}^m$.*

Proof. We will use Cousin's Lemma to construct a desired sequence of pairwise non-overlapping intervals in $\mathbb{R}^m$. Using Theorems 3.3.17 and 3.3.4

with $Y = [\boldsymbol{a},\boldsymbol{b}]\backslash O$, we define a gauge δ_1 on $[\boldsymbol{a},\boldsymbol{b}]$ by setting

$$\delta_1(\boldsymbol{x}) = \begin{cases} 1 & \text{if } \boldsymbol{x} \in [\boldsymbol{a},\boldsymbol{b}]\backslash O, \\ \operatorname{dist}(\boldsymbol{x}, [\boldsymbol{a},\boldsymbol{b}]\backslash O) & \text{if } \boldsymbol{x} \in O. \end{cases}$$

Hence, by Cousin's Lemma, there exists δ_1-fine Perron partition P_1 of $[\boldsymbol{a},\boldsymbol{b}]$. Since our choice of δ_1 implies that the following closed sets

$$\bigcup_{\substack{(t,I)\in P_1 \\ t\in O}} I \text{ and } [\boldsymbol{a},\boldsymbol{b}]\backslash O$$

are disjoint, we infer from Theorem 3.3.10 that

$$\operatorname{dist}\Bigg(\bigcup_{\substack{(t,I)\in P_1 \\ t\in O}} I, [\boldsymbol{a},\boldsymbol{b}]\backslash O\Bigg) > 0.$$

Set $P_1 := P$, $O_1 := O$, $O_2 := O \setminus \bigcup_{\substack{(t,I)\in P_1 \\ t\in O_1}} I$, and define a gauge δ_2 on $[\boldsymbol{a},\boldsymbol{b}]$ by setting

$$\delta_2(\boldsymbol{x}) = \begin{cases} \min\Bigg\{\frac{1}{2}, \operatorname{dist}\Bigg([\boldsymbol{a},\boldsymbol{b}]\backslash O, \bigcup_{\substack{(t,I)\in P_1 \\ t\in O_1}} I\Bigg)\Bigg\} & \text{if } \boldsymbol{x} \in [\boldsymbol{a},\boldsymbol{b}]\backslash O_1, \\ \operatorname{dist}(\boldsymbol{x}, [\boldsymbol{a},\boldsymbol{b}]\backslash O_1) & \text{if } \boldsymbol{x} \in O_1. \end{cases}$$

Again, Cousin's Lemma enables us to fix a δ_2-fine Perron partition P_2 of $[\boldsymbol{a},\boldsymbol{b}]$.

Proceeding inductively, we construct a decreasing sequence $(O_n)_{n=1}^{\infty}$ of open sets, a sequence $(\delta_n)_{n=1}^{\infty}$ of gauges on $[\boldsymbol{a},\boldsymbol{b}]$, and a sequence $(P_k)_{k=1}^{\infty}$ of Perron partitions of $[\boldsymbol{a},\boldsymbol{b}]$ such that the following conditions hold for every $n \in \mathbb{N}\backslash\{1\}$:

(a) If $k \in \{1, \ldots, n-1\}$, then P_k is a δ_k-fine Perron partition of $[\boldsymbol{a},\boldsymbol{b}]$;

(b) $O_n = O \setminus \bigcup_{k=1}^{n-1} \bigcup_{\substack{(t,I)\in P_k \\ t\in O_k}} I$;

(c)

$$\delta_n(\boldsymbol{x}) = \begin{cases} \min\Bigg\{\frac{1}{n}, \operatorname{dist}\Bigg([\boldsymbol{a},\boldsymbol{b}]\backslash O, \bigcup_{k=1}^{n-1} \bigcup_{\substack{(t,I)\in P_k \\ t\in O_k}} I\Bigg)\Bigg\} & \text{if } \boldsymbol{x} \in [\boldsymbol{a},\boldsymbol{b}]\backslash O_n, \\ \operatorname{dist}(\boldsymbol{x}, [\boldsymbol{a},\boldsymbol{b}]\backslash O_n) & \text{if } \boldsymbol{x} \in O_n. \end{cases}$$

According to our construction of $(P_n)_{n=1}^{\infty}$, we have

$$[\boldsymbol{a},\boldsymbol{b}]\backslash O \subseteq \bigcap_{n=1}^{\infty} \bigcup_{\substack{(\boldsymbol{t},I)\in P_n \\ \boldsymbol{t}\in[\boldsymbol{a},\boldsymbol{b}]\backslash O}} I.$$

Next we follow the proof of Lemma 3.3.3 to conclude that

$$[\boldsymbol{a},\boldsymbol{b}]\backslash O = \bigcap_{n=1}^{\infty} \bigcup_{\substack{(\boldsymbol{t},I)\in P_n \\ \boldsymbol{t}\in[\boldsymbol{a},\boldsymbol{b}]\backslash O}} I.$$

Finally, since

$$\bigcup_{k=1}^{n} \bigcup_{\substack{(\boldsymbol{t},I)\in P_k \\ \boldsymbol{t}\in O_k}} I \cup \bigcup_{\substack{(\boldsymbol{t},I)\in P_n \\ \boldsymbol{t}\in[\boldsymbol{a},\boldsymbol{b}]\backslash O}} I = [\boldsymbol{a},\boldsymbol{b}]$$

for all $n \in \mathbb{N}$, we get

$$O = \bigcup_{n=1}^{\infty} \bigcup_{k=1}^{n} \bigcup_{\substack{(\boldsymbol{t},I)\in P_k \\ \boldsymbol{t}\in O_k}} I = \bigcup_{n=1}^{\infty} \bigcup_{\substack{(\boldsymbol{t},I)\in P_n \\ \boldsymbol{t}\in O_n}} I.$$

□

The following theorem is a consequence of Theorem 3.3.18 and the Monotone Convergence Theorem.

Theorem 3.3.19. *If $O \subseteq [\boldsymbol{a},\boldsymbol{b}]$ is open and non-empty, then O is μ_m-measurable.*

There are μ_m-measurable sets that are neither open nor closed sets.

Example 3.3.20. The μ_1-measurable set $\mathbb{Q} \cap [0,1]$ is neither open nor closed.

The following result is a consequence of the Monotone Convergence Theorem.

Theorem 3.3.21. *If $(X_n)_{n=1}^{\infty}$ is an increasing sequence of μ_m-measurable subsets of $[\boldsymbol{a},\boldsymbol{b}]$, then $\bigcup_{k=1}^{\infty} X_k$ is μ_m-measurable and*

$$\mu_m\left(\bigcup_{k=1}^{\infty} X_k\right) = \lim_{n\to\infty} \mu_m(X_n).$$

Proof. Since $(\chi_{X_n})_{n=1}^{\infty}$ is an increasing sequence in $L^1[\boldsymbol{a},\boldsymbol{b}]$ converging pointwise to $\chi_{\bigcup_{k=1}^{\infty} X_k}$ on $[\boldsymbol{a},\boldsymbol{b}]$, an application of the Monotone Convergence Theorem completes the proof. □

The next theorem shows that μ_m is *countably additive.*

Theorem 3.3.22. *If $(Y_n)_{n=1}^{\infty}$ is a sequence of pairwise disjoint μ_m-measurable subsets of $[\boldsymbol{a},\boldsymbol{b}]$, then $\bigcup_{k=1}^{\infty} Y_k$ is μ_m-measurable and*

$$\mu_m\left(\bigcup_{k=1}^{\infty} Y_k\right) = \sum_{k=1}^{\infty} \mu_m(Y_k).$$

Proof. Since $(\bigcup_{k=1}^{n} Y_k)_{n=1}^{\infty}$ is an increasing sequence of μ_m-measurable subsets of $[\boldsymbol{a},\boldsymbol{b}]$, the theorem follows from Theorem 3.3.21. □

The next theorem shows that μ_m is *countably subadditive.*

Theorem 3.3.23. *If $(X_n)_{n=1}^{\infty}$ is a sequence of μ_m-measurable subsets of $[\boldsymbol{a},\boldsymbol{b}]$, then $\bigcup_{k=1}^{\infty} X_k$ is μ_m-measurable and*

$$\mu_m\left(\bigcup_{k=1}^{\infty} X_k\right) \le \sum_{k=1}^{\infty} \mu_m(X_k).$$

Proof. Set $Y_0 := \emptyset$ and $Y_n := X_n \setminus \bigcup_{k=1}^{n-1} X_k$ for every $n \in \mathbb{N}$. Then $(Y_n)_{n=1}^{\infty}$ is a sequence of pairwise disjoint μ_m-measurable subsets of $[\boldsymbol{a},\boldsymbol{b}]$ such that $\bigcup_{k=1}^{\infty} Y_k = \bigcup_{k=1}^{\infty} X_k$ and $\mu_m(Y_n) \le \mu_m(X_n)$ for $n = 1, 2, \ldots$. By Theorem 3.3.22,

$$\mu_m\left(\bigcup_{k=1}^{\infty} X_k\right) = \mu_m\left(\bigcup_{k=1}^{\infty} Y_k\right) = \sum_{k=1}^{\infty} \mu_m(Y_k) \le \sum_{k=1}^{\infty} \mu_m(X_k).$$

□

Theorem 3.3.24. *If $(X_n)_{n=1}^{\infty}$ is a decreasing sequence of μ_m-measurable subsets of $[\boldsymbol{a},\boldsymbol{b}]$, then $\bigcap_{k=1}^{\infty} X_k$ is μ_m-measurable and*

$$\mu_m\left(\bigcap_{k=1}^{\infty} X_k\right) = \lim_{n\to\infty} \mu_m(X_n).$$

Proof. Since Theorem 3.3.21 implies that $[\boldsymbol{a},\boldsymbol{b}] \setminus \bigcap_{k=1}^{\infty} X_k$ is μ_m-measurable and

$$\mu_m\left([\boldsymbol{a},\boldsymbol{b}] \setminus \bigcap_{k=1}^{\infty} X_k\right) = \lim_{n\to\infty} \mu_m([\boldsymbol{a},\boldsymbol{b}] \setminus X_n),$$

we conclude that

$$\begin{aligned}\mu_m\left(\bigcap_{k=1}^{\infty} X_k\right) &= \mu_m([\boldsymbol{a},\boldsymbol{b}]) - \mu_m\left([\boldsymbol{a},\boldsymbol{b}] \setminus \bigcap_{k=1}^{\infty} X_k\right) \\ &= \lim_{n\to\infty} \left(\mu_m([\boldsymbol{a},\boldsymbol{b}]) - \mu_m([\boldsymbol{a},\boldsymbol{b}] \setminus X_n)\right) \\ &= \lim_{n\to\infty} \mu_m(X_n).\end{aligned}$$

□

Theorem 3.3.25. *Let $(X_n)_{n=1}^{\infty}$ be a sequence of μ_m-measurable subsets of $[\boldsymbol{a}, \boldsymbol{b}]$. If $\mu_m(X_i \cap X_j) = 0$ for every $(i, j) \in \mathbb{N}^2$ satisfying $i \neq j$, then $\bigcup_{k=1}^{\infty} X_k$ is μ_m-measurable and*

$$\mu_m\left(\bigcup_{k=1}^{\infty} X_k\right) = \sum_{k=1}^{\infty} \mu_m(X_k). \tag{3.3.3}$$

Proof. According to Theorem 3.3.23, $\bigcup_{k=1}^{\infty} X_k$ is μ_m-measurable. Since our hypotheses imply that

$$\mu_m\left(\bigcup_{k=1}^{n} X_k\right) = \sum_{k=1}^{n} \mu_m(X_k)$$

for $n = 1, 2, \ldots$, and $(\bigcup_{k=1}^{n} X_k)_{n=1}^{\infty}$ is an increasing sequence of μ_m-measurable sets, an application of Theorem 3.3.21 completes the proof. □

3.4 A characterization of μ_m-measurable sets

In this section we characterize μ_m-measurable sets in terms open (or closed) sets. We begin with the following result.

Lemma 3.4.1. *If a set $Y \subseteq [\boldsymbol{a}, \boldsymbol{b}]$ is closed, then for each $\varepsilon > 0$ there exists a bounded open set G_ε such that $Y \subset G_\varepsilon$ and*

$$\mu_m(G_\varepsilon \backslash Y) < \varepsilon.$$

Proof. This is a consequence of Lemma 3.3.3 and Theorem 3.3.24. □

In view of Theorem 3.3.12, the following result is a refinement of Lemma 3.4.1.

The technique used in the proof of Theorem 3.4.2 is important.

Theorem 3.4.2. *If a set $X \subseteq [\boldsymbol{a}, \boldsymbol{b}]$ is μ_m-measurable, then for each $\varepsilon > 0$ there exists a bounded open set G_ε such that $X \subseteq G_\varepsilon$ and*

$$\mu_m(G_\varepsilon) < \mu_m(X) + \varepsilon.$$

Proof. Let $\varepsilon > 0$ and pick a gauge δ on $[\boldsymbol{a}, \boldsymbol{b}]$ such that

$$\sum_{(\boldsymbol{y}, J) \in Q} \chi_X(\boldsymbol{y}) \mu_m(J) < \mu_m(X) + \frac{\varepsilon}{4}$$

for each δ-fine Perron subpartition Q of $[\boldsymbol{a}, \boldsymbol{b}]$.

For each $n \in \mathbb{N}$, let

$$Z_n := \left\{\boldsymbol{x} \in X : \delta(\boldsymbol{x}) > \frac{1}{n}\right\},$$

and let $Y_n = \overline{Z_n} := \{\boldsymbol{x} : \operatorname{dist}(\boldsymbol{x}, Z_n) = 0\}$. We claim that Y_n is closed. If Y_n is empty, then Y_n is closed. On the other hand, suppose that Y_n is non-empty. Since Theorem 3.3.4 implies that Y_n is also closed, it follows from Lemma 3.4.1 that there exists a bounded open set G_n such that $G_n \supset Y_n$ and

$$\mu_m(G_n \backslash Y_n) < \frac{\varepsilon}{8(2^n)}.$$

Put $Y_0 = \emptyset$ and set $G_\varepsilon := \bigcup_{k=1}^{\infty}(G_k \backslash Y_{k-1})$. Then G_ε is an open set containing X:

$$X = \bigcup_{k=1}^{\infty} Z_k \subseteq \bigcup_{k=1}^{\infty} Y_k = \bigcup_{k=1}^{\infty}(Y_k \backslash Y_{k-1}) \subseteq G_\varepsilon.$$

Since $Y_0 = \emptyset$, it follows from Theorem 3.3.23 that

$$\mu_m(G_\varepsilon) \leq \sum_{k=1}^{\infty}(\mu_m(G_k \backslash Y_k) + \mu_m(Y_k \backslash Y_{k-1})) \leq \frac{\varepsilon}{8} + \lim_{n\to\infty} \mu_m(Y_n).$$

It remains to prove that

$$\lim_{n\to\infty} \mu_m(Y_n) < \mu_m(X) + \frac{\varepsilon}{2}. \tag{3.4.1}$$

If Y_n is empty for all $n \in \mathbb{N}$, then (3.4.1) holds. On the other hand, suppose that Y_N is non-empty for some positive integer N. According to our construction of the sequence $(Y_n)_{n=1}^{\infty}$ of closed sets, we may assume that N is so large that

$$\lim_{n\to\infty} \mu_m(Y_n) < \mu_m(Y_N) + \frac{\varepsilon}{8}. \tag{3.4.2}$$

Since G_N is a bounded open set containing Y_N, there exists an interval E_N in $\mathbb{R}^m$ such that $Y_N \subset G_N \subset E_N$; in particular, Y_N and $E_N \backslash G_N$ are disjoint non-empty closed sets. Hence, by Theorem 3.3.10 and Cousin's Lemma, there exists a $\min\left\{\frac{1}{N}, \operatorname{dist}(Y_N, E_N \backslash G_N)\right\}$-fine Perron subpartition $P = \{(\boldsymbol{t}_1, [\boldsymbol{u}_1, \boldsymbol{v}_1]), \ldots, (\boldsymbol{t}_p, [\boldsymbol{u}_p, \boldsymbol{v}_p])\}$ of $[\boldsymbol{a}, \boldsymbol{b}]$ such that

$$\{\boldsymbol{t} : (\boldsymbol{t}, [\boldsymbol{u}, \boldsymbol{v}]) \in P\} \subseteq Y_N \subseteq \bigcup_{(\boldsymbol{t},[\boldsymbol{u},\boldsymbol{v}])\in P} [\boldsymbol{u}, \boldsymbol{v}] \subseteq G_N.$$

By Theorem 3.3.23 again,

$$\begin{aligned}
&\mu_m(Y_N) \\
&\leq \sum_{i=1}^{p} \mu_m([\boldsymbol{u}_i, \boldsymbol{v}_i]) \\
&= \sum_{\substack{i=1 \\ (\boldsymbol{u}_i,\boldsymbol{v}_i)\cap Y_N \neq \emptyset}}^{p} \mu_m([\boldsymbol{u}_i, \boldsymbol{v}_i]) + \sum_{\substack{i=1 \\ (\boldsymbol{u}_i,\boldsymbol{v}_i)\cap Y_N = \emptyset}}^{p} \mu_m([\boldsymbol{u}_i, \boldsymbol{v}_i]). \quad (3.4.3)
\end{aligned}$$

We will next obtain an upper bound for the right-hand side of (3.4.3). For each $i \in \{1, \ldots, p\}$ satisfying $(\boldsymbol{u}_i, \boldsymbol{v}_i) \cap Y_N \neq \emptyset$, we select and fix a $\xi_i \in Z_N \cap (\boldsymbol{u}_i, \boldsymbol{v}_i)$ so that our choice δ yields

$$\sum_{\substack{i=1 \\ (\boldsymbol{u}_i,\boldsymbol{v}_i)\cap Y_N \neq \emptyset}}^{p} \mu_m([\boldsymbol{u}_i, \boldsymbol{v}_i]) < \mu_m(X) + \frac{\varepsilon}{4}. \quad (3.4.4)$$

Also, our choice of G_N yields

$$\sum_{\substack{i=1 \\ (\boldsymbol{u}_i,\boldsymbol{v}_i)\cap Y_N = \emptyset}}^{p} \mu_m([\boldsymbol{u}_i, \boldsymbol{v}_i]) < \frac{\varepsilon}{8}. \quad (3.4.5)$$

Finally, we combine (3.4.2), (3.4.3), (3.4.4) and (3.4.5) to get (3.4.1):

$$\begin{aligned}
&\lim_{n\to\infty} \mu_m(Y_n) \\
&< \mu_m(Y_N) + \frac{\varepsilon}{8} \\
&\leq \sum_{\substack{i=1 \\ (\boldsymbol{u}_i,\boldsymbol{v}_i)\cap Y_N \neq \emptyset}}^{p} \mu_m([\boldsymbol{u}_i, \boldsymbol{v}_i]) + \sum_{\substack{i=1 \\ (\boldsymbol{u}_i,\boldsymbol{v}_i)\cap Y_N = \emptyset}}^{p} \mu_m([\boldsymbol{u}_i, \boldsymbol{v}_i]) + \frac{\varepsilon}{8} \\
&< \mu_m(X) + \frac{\varepsilon}{2}.
\end{aligned}$$

□

The following theorem is the main result of this section.

Theorem 3.4.3. *A set $X \subseteq [\boldsymbol{a}, \boldsymbol{b}]$ is μ_m-measurable if and only if for each $\varepsilon > 0$ there exist a bounded open set G_ε and a closed set $Y_\varepsilon \subseteq [\boldsymbol{a}, \boldsymbol{b}]$ such that $Y_\varepsilon \subseteq X \subseteq G_\varepsilon$ and*

$$\mu_m(G_\varepsilon \backslash Y_\varepsilon) < \varepsilon. \quad (3.4.6)$$

Proof. Suppose first that X is μ_m-measurable. By Theorem 3.4.2, for each $\varepsilon > 0$ there exists a bounded open set O_1 such that $X \subseteq O_1$ and

$$\mu_m(O_1) - \mu_m(X) < \frac{\varepsilon}{2}.$$

Since the μ_m-measurability of X implies that of $[\boldsymbol{a}, \boldsymbol{b}]\backslash X$, we apply Theorem 3.4.2 to choose a bounded open set O_2 such that $[\boldsymbol{a}, \boldsymbol{b}]\backslash X \subseteq O_2$ and

$$\mu_m(O_2) - \mu_m([\boldsymbol{a}, \boldsymbol{b}]\backslash X) < \frac{\varepsilon}{2}.$$

Set $G_\varepsilon = O_1$ and $Y_\varepsilon = [\boldsymbol{a}, \boldsymbol{b}]\backslash O_2$. We have constructed a bounded open set G_ε and a closed set $Y_\varepsilon \subseteq [\boldsymbol{a}, \boldsymbol{b}]$ such that $Y_\varepsilon \subseteq X \subseteq G_\varepsilon$ and

$$\begin{aligned} \mu_m(G_\varepsilon\backslash Y_\varepsilon) &= \mu_m(G_\varepsilon\backslash X) + \mu_m(X\backslash Y_\varepsilon) \\ &= \mu_m(O_1\backslash X) + \mu_m(O_2\backslash([\boldsymbol{a}, \boldsymbol{b}]\backslash X)) \\ &< \varepsilon. \end{aligned}$$

Conversely, suppose that for each $\varepsilon > 0$ there exist a bounded open set G_ε and a closed set $Y_\varepsilon \subseteq [\boldsymbol{a}, \boldsymbol{b}]$ such that $Y_\varepsilon \subseteq X \subseteq G_\varepsilon$ and (3.4.6) holds. In particular, for each $n \in \mathbb{N}$ there exists a bounded open set G_n and a closed set $Y_n \subseteq [\boldsymbol{a}, \boldsymbol{b}]$ such that $Y_n \subseteq X \subseteq G_n$ and $\mu_m(G_n\backslash Y_n) < \frac{1}{n}$. Since it is now clear that the uniformly bounded sequence $(\chi_{G_n})_{n=1}^\infty$ converges μ_m-almost everywhere to χ_X on $[\boldsymbol{a}, \boldsymbol{b}]$, an application of Lebesgue's Dominated Convergence Theorem completes the proof. □

3.5 μ_m-measurable functions

In this section we study an important class of functions arising from μ_m-measurable sets. We begin with the following definition.

Definition 3.5.1. A real-valued function s defined on $[\boldsymbol{a}, \boldsymbol{b}]$ is a *step* function if there exists a division $\{[\boldsymbol{u}_1, \boldsymbol{v}_1], \ldots, [\boldsymbol{u}_p, \boldsymbol{v}_p]\}$ of $[\boldsymbol{a}, \boldsymbol{b}]$ such that s is constant on each $(\boldsymbol{u}_i, \boldsymbol{v}_i)$.

Lemma 3.5.2. *Let $X \subseteq [\boldsymbol{a}, \boldsymbol{b}]$ be a μ_m-measurable set. Then there exists a sequence $(s_n)_{n=1}^\infty$ of step functions on $[\boldsymbol{a}, \boldsymbol{b}]$ such that $\lim_{n\to\infty} s_n(\boldsymbol{x}) = \chi_X(\boldsymbol{x})$ for μ_m-almost all $\boldsymbol{x} \in [\boldsymbol{a}, \boldsymbol{b}]$.*

Proof. Let $n \in \mathbb{N}$ be given. Since $X \subseteq [\boldsymbol{a}, \boldsymbol{b}]$ is μ_m-measurable, we apply Theorem 3.4.3 to pick a bounded open set $G_n \supseteq X$ such that

$$\mu_m(G_n) - \mu_m(X) < \frac{1}{2^n}. \tag{3.5.1}$$

Next we apply Theorem 3.3.18 and the Monotone Convergence Theorem to choose a step function s_n on $[\boldsymbol{a}, \boldsymbol{b}]$ so that

$$0 \le \int_{[\boldsymbol{a},\boldsymbol{b}]} (\chi_{G_n} - s_n)\, d\mu_m < \frac{1}{2^n}. \tag{3.5.2}$$

Therefore (3.5.1), (3.5.2) and the triangle inequality yield

$$\int_{[\boldsymbol{a},\boldsymbol{b}]} |s_n - \chi_X| \, d\mu_m < \frac{1}{2^{n-1}}. \tag{3.5.3}$$

Since (3.5.3) holds for all $n \in \mathbb{N}$, we conclude that

$$\sum_{k=1}^{\infty} \int_{[\boldsymbol{a},\boldsymbol{b}]} |s_k - \chi_X| \, d\mu_m < \infty;$$

whence Lemma 3.2.3 implies that the series $\sum_{k=1}^{\infty}(s_k(\boldsymbol{x}) - \chi_X(\boldsymbol{x}))$ converges absolutely for μ_m-almost all $\boldsymbol{x} \in [\boldsymbol{a},\boldsymbol{b}]$; in particular, $\lim_{n\to\infty} s_n(\boldsymbol{x}) = \chi_X(\boldsymbol{x})$ for μ_m-almost all $\boldsymbol{x} \in [\boldsymbol{a},\boldsymbol{b}]$. □

Lemma 3.5.2 leads us to the following definition.

Definition 3.5.3. A function $f : [\boldsymbol{a},\boldsymbol{b}] \longrightarrow \mathbb{R} \cup \{-\infty, \infty\}$ is said to be μ_m-measurable if there exists a sequence $(s_n)_{n=1}^{\infty}$ of step functions on $[\boldsymbol{a},\boldsymbol{b}]$ such that

$$\lim_{n\to\infty} s_n(\boldsymbol{x}) = f(\boldsymbol{x})$$

for μ_m-almost all $\boldsymbol{x} \in [\boldsymbol{a},\boldsymbol{b}]$.

Lemma 3.5.4. *If a set $X \subseteq [\boldsymbol{a},\boldsymbol{b}]$ is μ_m-measurable, then χ_X is a μ_m-measurable function.*

Proof. This is a consequence of Lemma 3.5.2 and Definition 3.5.3. □

Example 3.5.5. If $\phi : [\boldsymbol{a},\boldsymbol{b}] \longrightarrow \mathbb{R}$ is a step function, then ϕ is μ_m-measurable.

Theorem 3.5.6. *If $f \in C[\boldsymbol{a},\boldsymbol{b}]$, then f is μ_m-measurable.*

Proof. Exercise. □

Theorem 3.5.7. *Let f and g be two real-valued μ_m-measurable functions defined on $[\boldsymbol{a},\boldsymbol{b}]$, and let $c \in \mathbb{R}$. Then the following functions cf, $|f|$, $f \pm g$ and fg are also μ_m-measurable.*

Proof. Since f is a real-valued μ_m-measurable function defined on $[\boldsymbol{a},\boldsymbol{b}]$, there exist a μ_m-negligible set $Z \subset [\boldsymbol{a},\boldsymbol{b}]$ and a sequence $(s_n)_{n=1}^{\infty}$ of step functions on $[\boldsymbol{a},\boldsymbol{b}]$ such that

$$\lim_{n\to\infty} s_n(\boldsymbol{x}) = f(\boldsymbol{x}) \quad \text{for all} \quad \boldsymbol{x} \in [\boldsymbol{a},\boldsymbol{b}]\backslash Z.$$

Since each $|s_n|$ is a step function on $[\boldsymbol{a}, \boldsymbol{b}]$, and

$$\lim_{n\to\infty} |s_n(\boldsymbol{x})| = |f(\boldsymbol{x})|$$

for all $\boldsymbol{x} \in [\boldsymbol{a}, \boldsymbol{b}]$ satisfying $\lim_{n\to\infty} s_n(\boldsymbol{x}) = f(\boldsymbol{x})$, we conclude that $|f|$ is μ_m-measurable.

Finally, since the sum and product of step functions are step functions, and the union of two μ_m-negligible sets is μ_m-negligible, the remaining assertions are obvious. □

Corollary 3.5.8. *Let f and g be two real-valued functions defined on $[\boldsymbol{a}, \boldsymbol{b}]$. If $f = g$ μ_m-almost everywhere on $[\boldsymbol{a}, \boldsymbol{b}]$, then f is μ_m-measurable if and only if g is μ_m-measurable.*

Proof. Since $f - g$ is μ_m-measurable, the corollary follows from Theorem 3.5.7. □

Theorem 3.5.9. *If $g : \mathbb{R} \longrightarrow \mathbb{R}$ is continuous on $\mathbb{R}$ and $f : [\boldsymbol{a}, \boldsymbol{b}] \longrightarrow \mathbb{R}$ is μ_m-measurable, then $g \circ f$ is μ_m-measurable.*

Proof. Since f is a real-valued μ_m-measurable function defined on $[\boldsymbol{a}, \boldsymbol{b}]$, there exist a μ_m-negligible set $Z \subset [\boldsymbol{a}, \boldsymbol{b}]$ and a sequence $(s_n)_{n=1}^{\infty}$ of step functions on $[\boldsymbol{a}, \boldsymbol{b}]$ such that

$$\lim_{n\to\infty} s_n(\boldsymbol{x}) = f(\boldsymbol{x}) \quad \text{for all} \quad \boldsymbol{x} \in [\boldsymbol{a}, \boldsymbol{b}] \setminus Z.$$

Since $(g \circ s_n)_{n=1}^{\infty}$ is a sequence of step functions, the continuity of g yields the μ_m-measurability of $g \circ f$:

$$\lim_{n\to\infty} g(s_n(\boldsymbol{x})) = g(f(\boldsymbol{x})) \quad \text{for all} \quad \boldsymbol{x} \in [\boldsymbol{a}, \boldsymbol{b}] \setminus Z.$$

□

Theorem 3.5.10. *Let f be a real-valued μ_m-measurable functions defined on $[\boldsymbol{a}, \boldsymbol{b}]$. Then f is μ_m-measurable if and only if $f^+ := \max\{f, 0\}$ and $f^- := -\min\{f, 0\}$ are μ_m-measurable.*

Proof. Exercise. □

The following theorem provides the link between μ_m-measurable sets and μ_m-measurable functions.

Theorem 3.5.11. *Let $f : [\boldsymbol{a}, \boldsymbol{b}] \longrightarrow \mathbb{R}$. The following conditions are equivalent.*

(i) *f is μ_m-measurable.*

(ii) *The set* $\{\boldsymbol{x} \in [\boldsymbol{a}, \boldsymbol{b}] : f(\boldsymbol{x}) < \alpha\}$ *is* μ_m*-measurable for each* $\alpha \in \mathbb{R}$.
(iii) *The set* $\{\boldsymbol{x} \in [\boldsymbol{a}, \boldsymbol{b}] : f(\boldsymbol{x}) \geq \alpha\}$ *is* μ_m*-measurable for each* $\alpha \in \mathbb{R}$.
(iv) *The set* $\{\boldsymbol{x} \in [\boldsymbol{a}, \boldsymbol{b}] : f(\boldsymbol{x}) \leq \alpha\}$ *is* μ_m*-measurable for each* $\alpha \in \mathbb{R}$.
(v) *The set* $\{\boldsymbol{x} \in [\boldsymbol{a}, \boldsymbol{b}] : f(\boldsymbol{x}) > \alpha\}$ *is* μ_m*-measurable for each* $\alpha \in \mathbb{R}$.

Proof. In view of Theorem 3.5.10, we may assume that f is non-negative.

(i) $\Longrightarrow$ (ii) Since f is a real-valued μ_m-measurable function defined on $[\boldsymbol{a}, \boldsymbol{b}]$, there exist a μ_m-negligible set $Z \subset [\boldsymbol{a}, \boldsymbol{b}]$ and a sequence $(s_n)_{n=1}^{\infty}$ of step functions on $[\boldsymbol{a}, \boldsymbol{b}]$ such that

$$\lim_{n \to \infty} s_n(\boldsymbol{x}) = f(\boldsymbol{x}) \quad \text{for all} \quad \boldsymbol{x} \in [\boldsymbol{a}, \boldsymbol{b}] \backslash Z.$$

Let $\alpha \in \mathbb{R}$ be given. We claim that

$$\{\boldsymbol{x} \in [\boldsymbol{a}, \boldsymbol{b}] \backslash Z : f(\boldsymbol{x}) < \alpha\} = \bigcup_{k=1}^{\infty} \bigcup_{r=1}^{\infty} \bigcap_{n=k}^{\infty} \left\{\boldsymbol{x} \in [\boldsymbol{a}, \boldsymbol{b}] \backslash Z : s_n(\boldsymbol{x}) < \alpha - \frac{1}{r}\right\}. \tag{3.5.4}$$

We first prove that

$$\{\boldsymbol{x} \in [\boldsymbol{a}, \boldsymbol{b}] \backslash Z : f(\boldsymbol{x}) < \alpha\} \subseteq \bigcup_{k=1}^{\infty} \bigcup_{r=1}^{\infty} \bigcap_{n=k}^{\infty} \left\{\boldsymbol{x} \in [\boldsymbol{a}, \boldsymbol{b}] \backslash Z : s_n(\boldsymbol{x}) < \alpha - \frac{1}{r}\right\}. \tag{3.5.5}$$

If $\boldsymbol{x}_0 \in [\boldsymbol{a}, \boldsymbol{b}] \backslash Z$ with $f(\boldsymbol{x}_0) < \alpha$, then there exists $N \in \mathbb{N}$ such that $f(\boldsymbol{x}_0) < \alpha - \frac{1}{N}$. Since $\lim_{n \to \infty} s_n(\boldsymbol{x}_0) = f(\boldsymbol{x}_0)$, there exists $K \in \mathbb{N}$ such that

$$n \geq K \Longrightarrow s_n(\boldsymbol{x}_0) < \alpha - \frac{1}{N}.$$

This proves (3.5.5).

Now we prove that

$$\bigcup_{k=1}^{\infty} \bigcup_{r=1}^{\infty} \bigcap_{n=k}^{\infty} \left\{\boldsymbol{x} \in [\boldsymbol{a}, \boldsymbol{b}] \backslash Z : s_n(\boldsymbol{x}) < \alpha - \frac{1}{r}\right\} \subseteq \{\boldsymbol{x} \in [\boldsymbol{a}, \boldsymbol{b}] \backslash Z : f(\boldsymbol{x}) < \alpha\}. \tag{3.5.6}$$

Let $\boldsymbol{x}_1 \in \bigcup_{k=1}^{\infty} \bigcup_{r=1}^{\infty} \bigcap_{n=k}^{\infty} \{\boldsymbol{x} \in [\boldsymbol{a}, \boldsymbol{b}] \backslash Z : s_n(\boldsymbol{x}) < \alpha - \frac{1}{r}\}$. Then there exist positive integers k_1 and r_1 such that $s_n(\boldsymbol{x}_1) < \alpha - \frac{1}{r_1}$ for all $n \geq k_1$. As n tends to infinity, we get

$$f(\boldsymbol{x}_1) \leq \alpha - \frac{1}{r_1} < \alpha$$

and so (3.5.6) holds. Since both sets Z and

$$\bigcup_{k=1}^{\infty}\bigcup_{r=1}^{\infty}\bigcap_{n=k}^{\infty}\left\{x\in[a,b]\backslash Z : s_n(x)<\alpha-\frac{1}{r}\right\}$$

are μ_m-measurable, we infer from (3.5.4), Theorems 3.3.24 and 3.3.23 that the set

$$\{x\in[a,b] : f(x)<\alpha\}$$

is μ_m-measurable. This proves that (i) implies (ii).

Statements (ii) and (iii) are equivalent because

$$\{x\in[a,b] : f(x)\geq\alpha\}=[a,b]\backslash\{x\in[a,b] : f(x)<\alpha\}$$

and

$$\{x\in[a,b] : f(x)<\alpha\}=[a,b]\backslash\{x\in[a,b] : f(x)\geq\alpha\}.$$

Statements (ii) and (iv) are equivalent because

$$\{x\in[a,b] : f(x)\leq\alpha\}=\bigcap_{k=1}^{\infty}\left\{x\in[a,b] : f(x)<\alpha+\frac{1}{k}\right\}$$

and

$$\{x\in[a,b] : f(x)<\alpha\}=\bigcup_{k=1}^{\infty}\left\{x\in[a,b] : f(x)\leq\alpha-\frac{1}{k}\right\}.$$

Statements (iv) and (v) are equivalent because

$$\{x\in[a,b] : f(x)>\alpha\}=[a,b]\backslash\{x\in[a,b] : f(x)\leq\alpha\}$$

and

$$\{x\in[a,b] : f(x)\leq\alpha\}=[a,b]\backslash\{x\in[a,b] : f(x)>\alpha\}.$$

To complete the proof, we need to prove that (ii) implies (i). For each $n\in\mathbb{N}$, we define the function $f_n : [a,b]\longrightarrow\mathbb{R}$ by setting

$$f_n(x)=\begin{cases}\frac{k-1}{2^n} \text{ if } \frac{k-1}{2^n}\leq f(x)<\frac{k}{2^n} \text{ for some } k\in\{1,\ldots,n2^n\},\\ n \quad \text{if } f(x)\geq n.\end{cases}$$

Since our hypothesis implies that the following sets

$$f^{-1}([0,2^{-n})),f^{-1}([2^{-n},2^{-n+1})),\ldots,f^{-1}([n-2^{-n},n)),f^{-1}([n,\infty))$$

are μ_m-measurable, it follows from Lemma 3.5.2 and Lebesgue's Dominated Convergence Theorem that there exists a step function s_n on $[\boldsymbol{a}, \boldsymbol{b}]$ such that

$$\int_{[\boldsymbol{a},\boldsymbol{b}]} |f_n - s_n| \, d\mu_m < \frac{1}{2^n}. \tag{3.5.7}$$

Since (3.5.7) holds for all $n \in \mathbb{N}$, and $\sum_{k=1}^{\infty} \frac{1}{2^k}$ converges, we can apply Lemma 3.2.3 to conclude that the series $\sum_{k=1}^{\infty} |f_k(\boldsymbol{x}) - s_k(\boldsymbol{x})|$ converges for μ_m-almost all $\boldsymbol{x} \in [\boldsymbol{a}, \boldsymbol{b}]$; in particular,

$$\lim_{n\to\infty} (s_n(\boldsymbol{x}) - f_n(\boldsymbol{x})) = 0 \text{ for } \mu_m\text{-almost all } \boldsymbol{x} \in [\boldsymbol{a}, \boldsymbol{b}].$$

It remains to prove that $\lim_{n\to\infty} f_n(\boldsymbol{x}) = f(\boldsymbol{x})$ for all $\boldsymbol{x} \in [\boldsymbol{a}, \boldsymbol{b}]$. For each $n \in \mathbb{N}$ we deduce from the definition of f_n that

$$0 \leq f(\boldsymbol{x}) - f_n(\boldsymbol{x}) < \frac{1}{2^n}$$

for all $\boldsymbol{x} \in [\boldsymbol{a}, \boldsymbol{b}]$ satisfying $|f(\boldsymbol{x})| < n$. Since f is real-valued, the result follows. □

Theorem 3.5.12. *Let $(f_n)_{n=1}^{\infty}$ be a sequence of real-valued μ_m-measurable functions defined on $[\boldsymbol{a}, \boldsymbol{b}]$, and let $f : [\boldsymbol{a}, \boldsymbol{b}] \longrightarrow \mathbb{R}$. If $f(\boldsymbol{x}) := \lim_{n\to\infty} f_n(\boldsymbol{x})$ for μ_m-almost all $\boldsymbol{x} \in [\boldsymbol{a}, \boldsymbol{b}]$, then f is μ_m-measurable.*

Proof. Exercise. □

A function is said to be *simple* if it takes on only a finite number of values.

Theorem 3.5.13. *If $f : [\boldsymbol{a}, \boldsymbol{b}] \longrightarrow \mathbb{R}$ is μ_m-measurable, then there exists a sequence $(f_n)_{n=1}^{\infty}$ of μ_m-measurable simple functions on $[\boldsymbol{a}, \boldsymbol{b}]$ with the following properties:*

(i) $\{|f_n|\}_{n=1}^{\infty}$ *is increasing;*
(ii) $\sup_{n\in\mathbb{N}} |f_n(\boldsymbol{x})| \leq |f(\boldsymbol{x})|$ *for all* $\boldsymbol{x} \in [\boldsymbol{a}, \boldsymbol{b}]$;
(iii) $\lim_{n\to\infty} f_n(\boldsymbol{x}) = f(\boldsymbol{x})$ *for all* $\boldsymbol{x} \in [\boldsymbol{a}, \boldsymbol{b}]$.

Proof. Following the proof of Theorem 3.5.11, there exists an increasing sequence $(f_{1,n})_{n=1}^{\infty}$ of non-negative μ_m-measurable simple functions such that $\lim_{n\to\infty} f_{1,n}(\boldsymbol{x}) = f^+(\boldsymbol{x})$ for all $\boldsymbol{x} \in [\boldsymbol{a}, \boldsymbol{b}]$. Similarly, there exists an increasing sequence $(f_{2,n})_{n=1}^{\infty}$ of non-negative μ_m-measurable simple functions such that $\lim_{n\to\infty} f_{2,n}(\boldsymbol{x}) = f^-(\boldsymbol{x})$ for all $\boldsymbol{x} \in [\boldsymbol{a}, \boldsymbol{b}]$. It is now clear that the result holds. □

3.6 Vitali Covering Theorem

The main aim of this section is to prove that if $f \in HK[\boldsymbol{a}, \boldsymbol{b}]$, then f is μ_m-measurable. We begin with some terminologies.

For each $[\boldsymbol{u}, \boldsymbol{v}] \in \mathcal{I}_m([\boldsymbol{a}, \boldsymbol{b}])$ the *regularity* of $[\boldsymbol{u}, \boldsymbol{v}]$, denoted by $\text{reg}([\boldsymbol{u}, \boldsymbol{v}])$, is the ratio of its shortest and longest sides. If $I \in \mathcal{I}_m([\boldsymbol{a}, \boldsymbol{b}])$ and $\text{reg}(I) \geq \alpha$ for some $\alpha \in (0, 1]$, we say that I is α-regular.

Let $X \subseteq [\boldsymbol{a}, \boldsymbol{b}]$ be a given set. A family $\mathcal{C} \subset \mathcal{I}_m([\boldsymbol{a}, \boldsymbol{b}])$ is a *Vitali covering* of X if given $\boldsymbol{x} \in X$ and $\varepsilon > 0$ there exists $I \in \mathcal{C}$ such that $\boldsymbol{x} \in I$ and $\text{diam}(I) < \varepsilon$, where $\text{diam}(I) := \sup\{|||\boldsymbol{s} - \boldsymbol{t}||| : \boldsymbol{s}, \boldsymbol{t} \in I\}$. We can now state and prove the following important result.

Theorem 3.6.1 (Vitali Covering Theorem). *Let $X \subseteq [\boldsymbol{a}, \boldsymbol{b}]$ be a set, and let $\alpha \in (0, 1]$. If a family $\mathcal{J}_\alpha \subset \mathcal{I}_m([\boldsymbol{a}, \boldsymbol{b}])$ of α-regular intervals covers X in the sense of Vitali, then there exists a sequence $([\boldsymbol{s}_k, \boldsymbol{t}_k])_{k=1}^{\infty}$ of pairwise disjoint intervals in $\mathcal{J}_\alpha$ such that*

$$\mu_m\left(X \setminus \bigcup_{k=1}^{\infty} [\boldsymbol{s}_k, \boldsymbol{t}_k]\right) = 0.$$

Proof. We will construct the sequence $([\boldsymbol{s}_k, \boldsymbol{t}_k])_{k=1}^{\infty}$ by induction. Suppose that $\{[\boldsymbol{s}_1, \boldsymbol{t}_1], \ldots, [\boldsymbol{s}_{n-1}, \boldsymbol{t}_{n-1}]\} \subset \mathcal{J}_\alpha$ for some integer $n \geq 2$. If $X \subseteq \bigcup_{i=1}^{n-1} [\boldsymbol{s}_i, \boldsymbol{t}_i]$, then the process stops. Henceforth, we assume that the process does not terminate at some finite stage. In this case, we can apply the Vitali condition to select $[\boldsymbol{s}_n, \boldsymbol{t}_n] \in \mathcal{J}_\alpha$ such that

$$\frac{1}{2} d_n \leq |||\boldsymbol{t}_n - \boldsymbol{s}_n||| \text{ and } [\boldsymbol{s}_n, \boldsymbol{t}_n] \cap \bigcup_{i=1}^{n-1} [\boldsymbol{s}_i, \boldsymbol{t}_i] = \emptyset,$$

where

$$d_n := \sup\left\{ |||\boldsymbol{v} - \boldsymbol{u}||| \; : [\boldsymbol{u}, \boldsymbol{v}] \in \mathcal{J}_\alpha \text{ and } [\boldsymbol{u}, \boldsymbol{v}] \cap \bigcup_{i=1}^{n-1} [\boldsymbol{s}_i, \boldsymbol{t}_i] = \emptyset \right\}.$$

It is clear that $\{[\boldsymbol{s}_n, \boldsymbol{t}_n]\}_{n=1}^{\infty}$ is a countable collection of non-overlapping subintervals of $[\boldsymbol{a}, \boldsymbol{b}]$ satisfying

$$\sum_{k=1}^{\infty} \mu_m([\boldsymbol{s}_k, \boldsymbol{t}_k]) \leq \mu_m([\boldsymbol{a}, \boldsymbol{b}]) < \infty. \tag{3.6.1}$$

Since (3.6.1) holds, it suffices to prove that

$$X \setminus \bigcup_{j=1}^{\infty} [\boldsymbol{s}_j, \boldsymbol{t}_j] \subseteq \bigcup_{k=n}^{\infty} [5\alpha^{-1}\boldsymbol{s}_k, 5\alpha^{-1}\boldsymbol{t}_k] \text{ for all } n \in \mathbb{N}\setminus\{1\}. \tag{3.6.2}$$

Since $\inf_{n\in\mathbb{N}} \mathrm{reg}([\boldsymbol{s}_n, \boldsymbol{t}_n]) \geq \alpha$, it remains to prove that

$$X \setminus \bigcup_{j=1}^{\infty} [\boldsymbol{s}_j, \boldsymbol{t}_j] \subseteq \overline{\bigcup_{k=n}^{\infty} B\left(\frac{1}{2}(\boldsymbol{s}_k + \boldsymbol{t}_k), \frac{5}{2} |||\boldsymbol{t}_k - \boldsymbol{s}_k|||\right)} \quad \text{for all } n \in \mathbb{N}\setminus\{1\}. \tag{3.6.3}$$

We will first prove that $\lim_{n\to\infty} d_n = 0$. According to our construction of the sequence $([\boldsymbol{s}_n, \boldsymbol{t}_n])_{n=1}^{\infty}$, it is enough to prove that the series $\sum_{k=1}^{\infty} |||\boldsymbol{t}_k - \boldsymbol{s}_k|||^m$ converges. But this last assertion is a direct consequence of (3.6.1), since $\inf_{n\in\mathbb{N}} \mathrm{reg}([\boldsymbol{s}_n, \boldsymbol{t}_n]) \geq \alpha$ implies

$$|||\boldsymbol{t}_n - \boldsymbol{s}_n|||^m \leq \frac{1}{\alpha^m} \mu_m([\boldsymbol{s}_n, \boldsymbol{t}_n]) \text{ for all } n \in \mathbb{N}\setminus\{1\}.$$

Finally, we prove (3.6.3). Let $\boldsymbol{x} \in X \setminus \bigcup_{j=1}^{\infty} [\boldsymbol{s}_j, \boldsymbol{t}_j]$ and let $n \in \mathbb{N}\setminus\{1\}$. By the Vitali condition there exists $[\boldsymbol{s}, \boldsymbol{t}] \in \mathscr{J}_\alpha$ such that

$$\boldsymbol{x} \in [\boldsymbol{s}, \boldsymbol{t}] \text{ and } [\boldsymbol{s}, \boldsymbol{t}] \cap \bigcup_{j=1}^{n-1} [\boldsymbol{s}_j, \boldsymbol{t}_j] = \emptyset.$$

Since $\lim_{n\to\infty} d_n = 0$, we let p be the smallest positive integer such that $d_p < \mu_m([\boldsymbol{s}, \boldsymbol{t}])$. According to our construction, $[\boldsymbol{s}, \boldsymbol{t}] \cap \bigcup_{k=1}^{p-1} [\boldsymbol{s}_k, \boldsymbol{t}_k] = \emptyset$, $[\boldsymbol{s}, \boldsymbol{t}] \cap [\boldsymbol{s}_p, \boldsymbol{t}_p] \neq \emptyset$, and $|||\boldsymbol{t} - \boldsymbol{s}||| \leq d_p \leq 2\,|||\boldsymbol{t}_p - \boldsymbol{s}_p|||$; hence (3.6.3) follows from triangle inequality:

$$\begin{aligned}
\left|\left|\left|\boldsymbol{x} - \frac{\boldsymbol{s}_p + \boldsymbol{t}_p}{2}\right|\right|\right| &\leq \left|\left|\left|\boldsymbol{x} - \frac{\boldsymbol{s} + \boldsymbol{t}}{2}\right|\right|\right| + \left|\left|\left|\frac{\boldsymbol{s} + \boldsymbol{t}}{2} - \boldsymbol{y}\right|\right|\right| + \left|\left|\left|\boldsymbol{y} - \frac{\boldsymbol{s}_p + \boldsymbol{t}_p}{2}\right|\right|\right| \\
&\qquad \text{(for some } \boldsymbol{y} \in [\boldsymbol{s}, \boldsymbol{t}] \cap [\boldsymbol{s}_p, \boldsymbol{t}_p]) \\
&\leq \frac{1}{2}(|||\boldsymbol{t} - \boldsymbol{s}|||) + \frac{1}{2}(|||\boldsymbol{t} - \boldsymbol{s}|||) + \frac{1}{2}(|||\boldsymbol{t}_p - \boldsymbol{s}_p|||) \\
&\leq \frac{5}{2}(|||\boldsymbol{t}_p - \boldsymbol{s}_p|||).
\end{aligned}$$

□

We need the following definition in order to prove Theorem 3.6.6 below.

Definition 3.6.2. Let $F : \mathcal{I}_m([\boldsymbol{a}, \boldsymbol{b}]) \longrightarrow \mathbb{R}$ and assume that $0 < \alpha \leq 1$. We define

$$\underline{F}_\alpha(\boldsymbol{x}) = \sup_{\delta>0} \inf \left\{ \frac{F(I)}{\mu_m(I)} : \boldsymbol{x} \in I \in \mathcal{I}_m([\boldsymbol{a}, \boldsymbol{b}]), \mathrm{reg}(I) \geq \alpha \text{ and } \mu_m(I) < \delta \right\}.$$

$$\overline{F}_\alpha(\boldsymbol{x}) = \inf_{\delta>0} \sup \left\{ \frac{F(I)}{\mu_m(I)} : \boldsymbol{x} \in I \in \mathcal{I}_m([\boldsymbol{a}, \boldsymbol{b}]), \mathrm{reg}(I) \geq \alpha \text{ and } \mu_m(I) < \delta \right\}.$$

Remark 3.6.3. Let F and α be given as in Definition 3.6.2. The following statements are true.

(i) $\overline{F}_\alpha$ and $\underline{F}_\alpha$ are *extended real-valued* functions defined on $[\boldsymbol{a}, \boldsymbol{b}]$:

$$-\infty \leq \underline{F}_\alpha(\boldsymbol{x}) \leq \overline{F}_\alpha(\boldsymbol{x}) \leq \infty.$$

(ii) For each $\boldsymbol{x} \in [\boldsymbol{a}, \boldsymbol{b}]$, the following limits $\overline{F}(\boldsymbol{x}) := \lim_{\alpha \to 0^+} \overline{F}_\alpha(\boldsymbol{x})$ and $\underline{F}(\boldsymbol{x}) := \lim_{\alpha \to 0^+} \underline{F}_\alpha(\boldsymbol{x})$ exist in $\mathbb{R} \cup \{-\infty, \infty\}$.

Definition 3.6.4. An interval function $F : \mathcal{I}_m([\boldsymbol{a}, \boldsymbol{b}]) \longrightarrow \mathbb{R}$ is said to be *derivable* at $\boldsymbol{x} \in [\boldsymbol{a}, \boldsymbol{b}]$ if there exists there is a real number, denoted by $F'(\boldsymbol{x})$, such that

$$F'(\boldsymbol{x}) = \underline{F}(\boldsymbol{x}) = \overline{F}(\boldsymbol{x}).$$

The following theorem is a consequence of the Vitali Covering Theorem.

Theorem 3.6.5. *Let $F : \mathcal{I}_m([\boldsymbol{a}, \boldsymbol{b}]) \longrightarrow \mathbb{R}$. If $0 < \alpha \leq 1$, then $\overline{F}_\alpha$ and $\underline{F}_\alpha$ are extended real-valued μ_m-measurable functions. In particular, if $F'(\boldsymbol{x})$ exists for μ_m-almost all $\boldsymbol{x} \in [\boldsymbol{a}, \boldsymbol{b}]$, then F' is μ_m-measurable.*

Proof. Exercise. □

The following theorem is a consequence of Theorems 3.6.1 and 3.6.5.

Theorem 3.6.6. *Let $f \in HK[\boldsymbol{a}, \boldsymbol{b}]$. If F is the indefinite Henstock-Kurzweil integral of f, then $F'(\boldsymbol{x})$ exists and $F'(\boldsymbol{x}) = f(\boldsymbol{x})$ for μ_m-almost all $\boldsymbol{x} \in [\boldsymbol{a}, \boldsymbol{b}]$. In particular, f is μ_m-measurable.*

Proof. Let X be the set of all $\boldsymbol{x} \in [\boldsymbol{a}, \boldsymbol{b}]$ for which F is not derivable at $\boldsymbol{x}$ or $F'(\boldsymbol{x}) \neq f(\boldsymbol{x})$. In view of Theorem 3.6.5, it is enough to prove that $\mu_m(X) = 0$.

For each $\boldsymbol{x} \in X$ there exists $\eta(\boldsymbol{x}) > 0$ with the following property: for each $\delta > 0$ there is an interval $I_{\boldsymbol{x}} \in \mathcal{I}_m([\boldsymbol{a}, \boldsymbol{b}])$ such that $\boldsymbol{x} \in I_{\boldsymbol{x}} \subset B(\boldsymbol{x}, \delta)$, $\operatorname{reg}(I_{\boldsymbol{x}}) \geq \eta(\boldsymbol{x})$ and

$$|f(\boldsymbol{x})\mu_m(I_{\boldsymbol{x}}) - F(I_{\boldsymbol{x}})| \geq \eta(\boldsymbol{x})\mu_m(I_{\boldsymbol{x}}).$$

For each $n \in \mathbb{N}$, let

$$X_n = \left\{\boldsymbol{x} \in X : \eta(\boldsymbol{x}) \geq \frac{1}{n}\right\}.$$

In view of the countable subadditivity of μ_m, it suffices to prove that $\mu_m(X_n) = 0$ for $n = 1, 2, \ldots$.

Fix $n \in \mathbb{N}$ and let $\varepsilon > 0$ be given. Since F is the indefinite Henstock-Kurzweil integral of f, there exists a gauge δ on $[\boldsymbol{a}, \boldsymbol{b}]$ corresponding to $\frac{\varepsilon}{n} > 0$ in the Saks-Henstock Lemma. Then

$$\{I_{\boldsymbol{x}} : \boldsymbol{x} \in X_n,\ \boldsymbol{x} \in I_{\boldsymbol{x}} \subset B(\boldsymbol{x}, \delta(\boldsymbol{x})) \text{ and } \operatorname{reg}(I_{\boldsymbol{x}}) \geq \eta(\boldsymbol{x})\}$$

covers X_n in the Vitali sense. Hence, by the Vitali Covering Theorem, there is a sequence $(I_k)_{k=1}^{\infty}$ of pairwise non-overlapping intervals in $\mathbb{R}^m$ such that

$$\{I_k\}_{k=1}^{\infty} \subset \{I_{\boldsymbol{x}} : \boldsymbol{x} \in X_n,\ \boldsymbol{x} \in I_{\boldsymbol{x}} \subset B(\boldsymbol{x}, \delta(\boldsymbol{x})) \text{ and } \operatorname{reg}(I_{\boldsymbol{x}}) \geq \eta(\boldsymbol{x})\} \tag{3.6.4}$$

and

$$\mu_m\left(X_n \setminus \bigcup_{k=1}^{\infty} I_k\right) = 0. \tag{3.6.5}$$

On the other hand, for each $k \in \mathbb{N}$ we use (3.6.4) to select $\boldsymbol{t}_k \in I_k \cap X_n$ so that $\boldsymbol{t}_k \in I_k \subset B(\boldsymbol{t}_k, \delta(\boldsymbol{t}_k))$. Thus

$$\sum_{k=1}^{\infty} \mu_m(I_k) \leq n \lim_{r\to\infty} \sum_{k=1}^{r} |f(\boldsymbol{t}_k)\mu_m(I_k) - F(I_k)| \leq \varepsilon. \tag{3.6.6}$$

Combining (3.6.5), (3.6.6) and the arbitrariness of $\varepsilon > 0$ leads to the desired result $\mu_m(X_n) = 0$. □

The proof of Theorem 3.6.6 yields the following result.

Theorem 3.6.7. *If $f \in HK[\boldsymbol{a}, \boldsymbol{b}]$ and $(HK)\int_I f = 0$ for every $I \in \mathcal{I}_m([\boldsymbol{a}, \boldsymbol{b}])$, then $f = 0$ μ_m-almost everywhere on $[\boldsymbol{a}, \boldsymbol{b}]$.*

3.7 Further properties of Lebesgue integrable functions

The aim of this section is to prove that if $f \in L^1[\boldsymbol{a}, \boldsymbol{b}]$ and g is a bounded μ_m-measurable function on $[\boldsymbol{a}, \boldsymbol{b}]$, then $fg \in L^1[\boldsymbol{a}, \boldsymbol{b}]$. Moreover, we prove that this result is, in some sense, the best possible for Lebesgue integrable functions. We begin with the following useful generalization of Theorem 3.1.4.

Theorem 3.7.1. *Let $f : [\boldsymbol{a}, \boldsymbol{b}] \longrightarrow \mathbb{R}$ be μ_m-measurable. Then $f \in L^1[\boldsymbol{a}, \boldsymbol{b}]$ if and only if there exists $g \in L^1[\boldsymbol{a}, \boldsymbol{b}]$ such that $|f(\boldsymbol{x})| \leq g(\boldsymbol{x})$ for μ_m-almost all $\boldsymbol{x} \in [\boldsymbol{a}, \boldsymbol{b}]$.*

Proof. ($\Longrightarrow$) We choose g to be $|f|$.

($\Longleftarrow$) This is a consequence of Theorem 3.5.13 and Lebesgue's Dominated Convergence Theorem. □

Theorem 3.7.2. *If $f \in L^1[\boldsymbol{a}, \boldsymbol{b}]$ and $X \subseteq [\boldsymbol{a}, \boldsymbol{b}]$ is a μ_m-measurable set, then $f\chi_X$ belongs to $L^1[\boldsymbol{a}, \boldsymbol{b}]$.*

Proof. According to Theorem 3.6.6, f is μ_m-measurable. Since Lemma 3.5.2 implies that χ_X is also μ_m-measurable, it follows from Theorem 3.5.7 that $f\chi_X$ is μ_m-measurable too. An appeal to Theorem 3.7.1 completes the proof. □

Theorem 3.7.3. *If $f \in L^1[\boldsymbol{a}, \boldsymbol{b}]$ and g is a bounded, μ_m-measurable function on $[\boldsymbol{a}, \boldsymbol{b}]$, then $fg \in L^1[\boldsymbol{a}, \boldsymbol{b}]$.*

Proof. This is a consequence of Theorems 3.5.13, 3.7.2 and Lebesgue's Dominated Convergence Theorem. □

The following theorem shows that Theorem 3.7.3 is, in some sense, sharp.

Theorem 3.7.4. *Let $g : [\boldsymbol{a}, \boldsymbol{b}] \longrightarrow \mathbb{R}$. If $fg \in L^1[\boldsymbol{a}, \boldsymbol{b}]$ for each $f \in L^1[\boldsymbol{a}, \boldsymbol{b}]$, then g is μ_m-measurable and there exists a μ_m-negligible set $Z \subset [\boldsymbol{a}, \boldsymbol{b}]$ such that $g\chi_{[\boldsymbol{a},\boldsymbol{b}]\setminus Z}$ is bounded on $[\boldsymbol{a}, \boldsymbol{b}]$.*

Proof. It is clear that $g \in L^1[\boldsymbol{a}, \boldsymbol{b}]$ and so it is μ_m-measurable (cf. Theorem 3.6.6).

Suppose that we cannot find a μ_m-negligible set Z such that $g\chi_{[\boldsymbol{a},\boldsymbol{b}]\setminus Z}$ is bounded on $[\boldsymbol{a}, \boldsymbol{b}]$. Then there exist a strictly increasing sequence $(\alpha_n)_{n=1}^{\infty}$ of positive numbers such that $\alpha_n \geq 2^n$ and $\mu_m(X_n) > 0$ for all $n \in \mathbb{N}$, where

$$X_n = \{\boldsymbol{x} \in [\boldsymbol{a}, \boldsymbol{b}] : \alpha_n < |g(\boldsymbol{x})| \leq \alpha_{n+1}\}$$

for $n = 1, 2, \ldots$.

Define the function $f_0 : [\boldsymbol{a}, \boldsymbol{b}] \longrightarrow \mathbb{R}$ by setting

$$f_0(\boldsymbol{x}) = \sum_{k=1}^{\infty} \frac{1}{\alpha_k \mu_m(X_k)} \chi_{X_k}(\boldsymbol{x}).$$

A simple application of the Monotone Convergence Theorem shows that $f_0 \in L^1[\boldsymbol{a}, \boldsymbol{b}]$. Combining this fact with our hypothesis on g, we find that $f_0 g \in L^1[\boldsymbol{a}, \boldsymbol{b}]$; hence the Monotone Convergence Theorem yields

$$\int_{[\boldsymbol{a},\boldsymbol{b}]} |f_0 g| \ d\mu_m = \sum_{k=1}^{\infty} \frac{1}{\alpha_k \mu_m(X_k)} \alpha_k \mu_m(X_k) < \infty,$$

a contradiction. This contradiction proves the theorem. □

In the proof of Theorem 3.7.4, the function $g : [\boldsymbol{a}, \boldsymbol{b}] \longrightarrow \mathbb{R}$ is said to be *essentially bounded* on $[\boldsymbol{a}, \boldsymbol{b}]$; that is, there exists a μ_m-negligible set Z such that

$$\sup_{\boldsymbol{x} \in [\boldsymbol{a}, \boldsymbol{b}] \setminus Z} |g(\boldsymbol{x})| < \infty.$$

The collection of all real-valued μ_m-measurable functions that are essentially bounded on $[\boldsymbol{a}, \boldsymbol{b}]$ will be denoted by $L^\infty[\boldsymbol{a}, \boldsymbol{b}]$. For each $g \in L^\infty[\boldsymbol{a}, \boldsymbol{b}]$, we write

$$\|g\|_{L^\infty[\boldsymbol{a},\boldsymbol{b}]} = \inf\{M : |g(\boldsymbol{x})| \leq M \text{ for } \mu_m\text{-almost all } \boldsymbol{x} \in [\boldsymbol{a}, \boldsymbol{b}]\}.$$

Also, for each $f \in L^1[\boldsymbol{a}, \boldsymbol{b}]$ we write $\|f\|_{L^1[\boldsymbol{a},\boldsymbol{b}]} = \int_{[\boldsymbol{a},\boldsymbol{b}]} |f| \, d\mu_m$. We can now state a stronger version of Theorem 3.7.4.

Theorem 3.7.5. *Let $g : [\boldsymbol{a}, \boldsymbol{b}] \longrightarrow \mathbb{R}$. Then $g \in L^\infty[\boldsymbol{a}, \boldsymbol{b}]$ if and only if $fg \in L^1[\boldsymbol{a}, \boldsymbol{b}]$ for each $f \in L^1[\boldsymbol{a}, \boldsymbol{b}]$. In this case,*

$$\|fg\|_{L^1[\boldsymbol{a},\boldsymbol{b}]} \leq \|f\|_{L^1[\boldsymbol{a},\boldsymbol{b}]} \|g\|_{L^\infty[\boldsymbol{a},\boldsymbol{b}]}$$

for each $f \in L^1[\boldsymbol{a}, \boldsymbol{b}]$.

Proof. Exercise. □

Exercise 3.7.6. Prove that if $f \in L^1[\boldsymbol{a}, \boldsymbol{b}]$, then for each $\varepsilon > 0$ there exists a step function ψ on $[\boldsymbol{a}, \boldsymbol{b}]$ such that

$$\|f - \psi\|_{L^1[\boldsymbol{a},\boldsymbol{b}]} < \varepsilon.$$

3.8 The L^p spaces

The aim of this section is to prove a useful modification of Theorem 3.7.3. We begin with the following definition.

Definition 3.8.1. For any real number $p \geq 1$, we let $L^p[\boldsymbol{a}, \boldsymbol{b}]$ be the set of all real-valued μ_m-measurable functions for which $|f|^p \in L^1[\boldsymbol{a}, \boldsymbol{b}]$.

The L^p-norm of f is given by

$$\|f\|_{L^p[\boldsymbol{a},\boldsymbol{b}]} = \left(\int_{[\boldsymbol{a},\boldsymbol{b}]} |f|^p \, d\mu_m \right)^{\frac{1}{p}}.$$

The following theorem is a useful modification of Theorem 3.7.3.

Theorem 3.8.2 (Hölder's Inequality). *Let $1 < p < \infty$ and let $q = \frac{p}{p-1}$. If $f \in L^p[\boldsymbol{a}, \boldsymbol{b}]$ and $g \in L^q[\boldsymbol{a}, \boldsymbol{b}]$, then $fg \in L^1[\boldsymbol{a}, \boldsymbol{b}]$ and*

$$\|fg\|_{L^1[a,b]} \leq \|f\|_{L^p[a,b]} \|g\|_{L^q[a,b]}.$$

Proof. If $\|f\|_{L^p[a,b]} = 0$ or $\|g\|_{L^q[a,b]} = 0$, then it follows from Theorem 3.6.7 that $fg = 0$ μ_m-a.e. on $[\boldsymbol{a}, \boldsymbol{b}]$ and so the theorem holds. Henceforth, we suppose that $\|f\|_{L^p[a,b]}\|g\|_{L^q[a,b]} > 0$.

For each $\boldsymbol{x} \in [\boldsymbol{a}, \boldsymbol{b}]$ we apply the inequality

$$\alpha\beta \leq \frac{\alpha^p}{p} + \frac{\beta^q}{q} \qquad (\alpha, \beta \geq 0)$$

with $\alpha = \dfrac{|f(\boldsymbol{x})|}{\|f\|_{L^p[a,b]}}$ and $\beta = \dfrac{|g(\boldsymbol{x})|}{\|g\|_{L^q[a,b]}}$ to obtain

$$\frac{|f(\boldsymbol{x})g(\boldsymbol{x})|}{\|f\|_{L^p[a,b]}\|g\|_{L^q[a,b]}} \leq \frac{|f(\boldsymbol{x})|^p}{p\|f\|^p_{L^p[a,b]}} + \frac{|g(\boldsymbol{x})|^q}{q\|g\|^q_{L^q[a,b]}}. \tag{3.8.1}$$

Since $f \in L^p[\boldsymbol{a}, \boldsymbol{b}]$, $g \in L^q[\boldsymbol{a}, \boldsymbol{b}]$ and (3.8.1) holds, it follows from Theorems 3.5.9 and 3.7.1 that $fg \in L^1[\boldsymbol{a}, \boldsymbol{b}]$. Finally, (3.8.1) gives

$$\frac{\|fg\|_{L^1[a,b]}}{\|f\|_{L^p[a,b]}\|g\|_{L^q[a,b]}} \leq \frac{1}{p} + \frac{1}{q} = 1$$

or $\|fg\|_{L^1[a,b]} \leq \|f\|_{L^p[a,b]}\|g\|_{L^q[a,b]}$. □

We remark that if $p = q = 2$, then Theorem 3.8.2 is known as Cauchy-Schwarz inequality.

Theorem 3.8.3 (Minkowski's Inequality). *Let $1 \leq p < \infty$. If $f, g \in L^p[\boldsymbol{a}, \boldsymbol{b}]$, then $f + g \in L^p[\boldsymbol{a}, \boldsymbol{b}]$ and*

$$\|f + g\|_{L^p[a,b]} \leq \|f\|_{L^p[a,b]} + \|g\|_{L^p[a,b]}. \tag{3.8.2}$$

Proof. By Theorems 3.5.7 and 3.5.9, $|f + g|^p$ is μ_m-measurable. Since we also have

$$|f(\boldsymbol{x}) + g(\boldsymbol{x})|^p \leq 2^p \max\{|f(\boldsymbol{x})|^p, |g(\boldsymbol{x})|^p\} \leq 2^p(|f(\boldsymbol{x})|^p + |g(\boldsymbol{x})|^p)$$

for all $\boldsymbol{x} \in [\boldsymbol{a}, \boldsymbol{b}]$, it follows from Theorem 3.7.1 that $f + g \in L^p[\boldsymbol{a}, \boldsymbol{b}]$.

It remains to prove the inequality (3.8.2). If $p = 1$ or $\|f + g\|_{L^p[a,b]} = 0$, then the inequality (3.8.2) holds. On the other hand, we suppose that $p > 1$

and $\|f+g\|_{L^p[\boldsymbol{a},\boldsymbol{b}]} \neq 0$. In this case, we infer from Hólder's inequality that $|f|\,|f+g|^{p-1}, |g|\,|f+g|^{p-1} \in L^1[\boldsymbol{a},\boldsymbol{b}]$. Consequently,

$$\int_{[\boldsymbol{a},\boldsymbol{b}]} |f+g|^p \; d\mu_m$$

$$\leq \int_{[\boldsymbol{a},\boldsymbol{b}]} |f|\,|f+g|^{p-1} \; d\mu_m + \int_{a}^{b} |g|\,|f+g|^{p-1} \; d\mu_m$$

(by triangle inequality)

$$\leq (\|f\|_{L^p[\boldsymbol{a},\boldsymbol{b}]} + \|g\|_{L^p[\boldsymbol{a},\boldsymbol{b}]}) \left(\int_{[\boldsymbol{a},\boldsymbol{b}]} |f+g|^{(p-1)q} \; d\mu_m \right)^{\frac{1}{q}}$$

(by Hölder's inequality)

$$= (\|f\|_{L^p[\boldsymbol{a},\boldsymbol{b}]} + \|g\|_{L^p[\boldsymbol{a},\boldsymbol{b}]}) \|f+g\|_{L^p[\boldsymbol{a},\boldsymbol{b}]}^{\frac{p}{q}},$$

and the required inequality follows on dividing by the non-zero number $\|f+g\|_{L^p[\boldsymbol{a},\boldsymbol{b}]}^{\frac{p}{q}}$. □

We end this section with the following result concerning the completeness of L^p spaces.

Theorem 3.8.4 (Riesz-Fischer). *Let $1 \leq p < \infty$ and let $(f_n)_{n=1}^{\infty}$ be a sequence in $L^p[\boldsymbol{a},\boldsymbol{b}]$ such that*

$$\|f_n - f_N\|_{L^p[\boldsymbol{a},\boldsymbol{b}]} \to 0 \text{ as } n, N \to \infty.$$

Then there exists $f \in L^p[\boldsymbol{a},\boldsymbol{b}]$ such that

$$\lim_{n\to\infty} \|f_n - f\|_{L^p[\boldsymbol{a},\boldsymbol{b}]} = 0.$$

The function f is unique in the sense that if $h \in L^p[\boldsymbol{a},\boldsymbol{b}]$ and $\lim_{n\to\infty} \|f_n - h\|_{L^p[\boldsymbol{a},\boldsymbol{b}]} = 0$, then $h = f$ μ_m-a.e on $[\boldsymbol{a},\boldsymbol{b}]$.

Proof. We let $f_{n_0} = 0$ and choose a subsequence $(f_{n_k})_{k=1}^{\infty}$ of $(f_n)_{n=1}^{\infty}$ such that

$$\|f_{n_{k+1}} - f_{n_k}\|_{L^p[\boldsymbol{a},\boldsymbol{b}]} < \frac{1}{2^{k+1}} \tag{3.8.3}$$

for $k = 1, 2, \ldots$. For each $n \in \mathbb{N}$ we put $g_n = \sum_{j=0}^{n} |f_{n_{j+1}} - f_{n_j}|$ so that Minkowski inequality and (3.8.3) yield $g_n \in L^p[\boldsymbol{a},\boldsymbol{b}]$ and

$$\|g_n\|_{L^p[\boldsymbol{a},\boldsymbol{b}]} < 1 + \|f_{n_1}\|_{L^p[\boldsymbol{a},\boldsymbol{b}]}.$$

Since $n \in \mathbb{N}$ is arbitrary, we infer from Hölder's inequality that $\sup_{n\in\mathbb{N}} \|g_n\|_{L^1[\boldsymbol{a},\boldsymbol{b}]}$ is finite; therefore Lemma 3.2.3 implies that the series

$$\sum_{j=0}^{\infty} (f_{n_{j+1}}(\boldsymbol{x}) - f_{n_j}(\boldsymbol{x})) \text{ converges absolutely for } \mu_m\text{- almost all } \boldsymbol{x} \in [\boldsymbol{a},\boldsymbol{b}].$$

Let

$$f(x) = \begin{cases} \sum_{j=0}^{\infty}(f_{n_{j+1}}(x) - f_{n_j}(x)) & \\ & \text{if the series } \sum_{j=0}^{\infty}\left|f_{n_{j+1}}(x) - f_{n_j}(x)\right| \text{ converges,} \\ 0 & \text{otherwise.} \end{cases}$$

Since $f_{n_0} = 0$, it is not difficult to see that

$$f(x) = \lim_{k\to\infty} f_{n_k}(x) \tag{3.8.4}$$

for μ_m-almost all $x \in [a, b]$; in particular, f is μ_m-measurable.

We will next prove that $f \in L^p[a, b]$ and $\lim_{n\to\infty} \|f_n - f\|_{L^p[a,b]} = 0$. For each $\varepsilon > 0$ we select a positive integer N so that

$$\|f_n - f_r\|_{L^p[a,b]} < \frac{\varepsilon}{2}$$

for all integers $n, r > N$. Hence, by Theorem 3.8.3 and Fatou's Lemma, $f \in L^p[a, b]$. Another application of Fatou's Lemma yields

$$\|f_n - f\|_{L^p[a,b]} \le \liminf_{k\to\infty} \|f_n - f_{n_k}\|_{L^p[a,b]} < \varepsilon$$

for all integers $n > N$. Since $\varepsilon > 0$ is arbitrary, we conclude that

$$\lim_{n\to\infty} \|f_n - f\|_{L^p[a,b]} = 0.$$

It remains to prove the last assertion of the theorem. Suppose that there exists $h \in L^p[a, b]$ such that $\|h - f_n\|_{L^p[a,b]} \to 0$ as $n \to \infty$. By Minkowski's inequality again,

$$\|f - h\|_{L^p[a,b]} \le \|f - f_n\|_{L^p[a,b]} + \|f_n - h\|_{L^p[a,b]} \to 0 \text{ as } n \to \infty;$$

that is, $\|f - h\|_{L^p[a,b]} = 0$. An appeal to Theorem 3.6.7 completes the proof. □

Exercise 3.8.5. Let $(a_n)_{n=1}^{\infty}$ and $(b_n)_{n=1}^{\infty}$ be two sequences of real numbers such that the series $\sum_{k=1}^{\infty}(a_k^2 + b_k^2)$ converges. Prove that there exists $f \in L^2([-\pi, \pi])$ such that

$$\lim_{n\to\infty} \int_{-\pi}^{\pi} \left|\sum_{k=1}^{n}(a_k \cos kt + b_k \sin kt) - f(t)\right|^2 d\mu_1(t) = 0$$

and

$$\|f\|^2_{L^2[-\pi,\pi]} = \pi \sum_{k=1}^{\infty}(a_k^2 + b_k^2).$$

3.9 Lebesgue's criterion for Riemann integrability

The aim of this section is to give a characterization of Riemann integrable functions. We first show that such kind of integrable functions must be bounded.

Theorem 3.9.1. *If $f \in R[\boldsymbol{a}, \boldsymbol{b}]$, then f is bounded on $[\boldsymbol{a}, \boldsymbol{b}]$.*

Proof. For $\varepsilon = 1$ we apply the Saks-Henstock Lemma for the Riemann integral to select a constant gauge δ on $[\boldsymbol{a}, \boldsymbol{b}]$ so that

$$\sum_{(\boldsymbol{t},I)\in P} \left| f(\boldsymbol{t})\mu_m(I) - \int_I f \right| < 1$$

for each δ-fine Perron subpartition P of $[\boldsymbol{a}, \boldsymbol{b}]$. Let N be a fixed positive integer satisfying $\max_{i=1,\ldots,m} \frac{b_i - a_i}{N} < \delta$, and let $\boldsymbol{x} \in [\boldsymbol{a}, \boldsymbol{b}]$. Since $\boldsymbol{x} \in \prod_{i=1}^m [a_i + (k_i - 1)\frac{b_i - a_i}{N}, a_i + k_i \frac{b_i - a_i}{N}] =: J_{\boldsymbol{k}}$ for some $\boldsymbol{k} \in \{1, \ldots, N\}^m$, our choice of N and δ yield

$$\left| f(\boldsymbol{x}) \prod_{i=1}^m \frac{b_i - a_i}{N} - \int_{J_{\boldsymbol{k}}} f \right| < 1,$$

which implies that

$$|f(\boldsymbol{x})| < \frac{N^m}{\mu_m([\boldsymbol{a}, \boldsymbol{b}])} \left\{ 1 + \max_{I \in \mathcal{I}_m([\boldsymbol{a},\boldsymbol{b}])} \left| (HK) \int_I f \right| \right\}.$$

Since $\boldsymbol{x} \in [\boldsymbol{a}, \boldsymbol{b}]$ is arbitrary, f is bounded on $[\boldsymbol{a}, \boldsymbol{b}]$. □

Corollary 3.9.2. *If $f \in R[\boldsymbol{a}, \boldsymbol{b}]$, then $f \in L^1[\boldsymbol{a}, \boldsymbol{b}]$ and both integrals coincide.*

Proof. This is a consequence of Theorems 3.6.6, 3.9.1 and 3.7.1. □

Definition 3.9.3. Let $f : [\boldsymbol{a}, \boldsymbol{b}] \longrightarrow \mathbb{R}$ be a bounded function. The **oscillation** of f on $[\boldsymbol{a}, \boldsymbol{b}]$ is defined by

$$\omega(f, [\boldsymbol{a}, \boldsymbol{b}]) = \sup\{f(\boldsymbol{x}) : \boldsymbol{x} \in [\boldsymbol{a}, \boldsymbol{b}]\} - \inf\{f(\boldsymbol{x}) : \boldsymbol{x} \in [\boldsymbol{a}, \boldsymbol{b}]\}.$$

Theorem 3.9.4. *Let $f : [\boldsymbol{a}, \boldsymbol{b}] \longrightarrow \mathbb{R}$ be a bounded function. Then $f \in R[\boldsymbol{a}, \boldsymbol{b}]$ if and only if for each $\varepsilon > 0$ there exists a division D_0 of $[\boldsymbol{a}, \boldsymbol{b}]$ such that*

$$\sum_{I \in D_0} \omega(f, I)\mu_m(I) < \varepsilon.$$

Proof. $(\Longrightarrow)$ For each $\varepsilon > 0$ there exists a constant gauge δ on $[\boldsymbol{a}, \boldsymbol{b}]$ such that

$$|S(f, P_0) - S(f, Q_0)| < \frac{\varepsilon}{2}$$

for each δ-fine Perron partitions P_0 and Q_0 of $[\boldsymbol{a}, \boldsymbol{b}]$. Let N be a fixed positive integer N such that $\max_{k=1,\ldots,m} \frac{b_k - a_k}{N} < \delta$ and consider the net D_0

$$:= \left\{ \prod_{i=1}^{m} \left[a_i + (k_i - 1)\frac{b_i - a_i}{N}, a_i + k_i \frac{b_i - a_i}{N}\right] : \{k_1, \ldots, k_m\} \subseteq \{1, \ldots, N\} \right\}.$$

For each $I \in D_0$ we choose $\boldsymbol{s}_I, \boldsymbol{t}_I \in I$ so that

$$\omega(f, I) < f(\boldsymbol{t}_I) - f(\boldsymbol{s}_I) + \frac{\varepsilon}{2\mathrm{card}\, D_0}.$$

Since $P_1 = \{(\boldsymbol{t}_I, I) : I \in D_0\}$ and $Q_1 = \{(\boldsymbol{s}_I, I) : I \in D_0\}$ are δ-fine Perron partitions of $[\boldsymbol{a}, \boldsymbol{b}]$, our choice of δ gives the desired result:

$$\sum_{I \in D_0} \omega(f, I)\, \mu_m(I) < S(f, P_1) - S(f, Q_1) + \frac{\varepsilon}{2} < \varepsilon.$$

$(\Longleftarrow)$ Let $\varepsilon > 0$ be given. By hypothesis, there exists a division D_1 of $[\boldsymbol{a}, \boldsymbol{b}]$ such that

$$\sum_{J \in D_1} \omega(f, J)\, \mu_m(J) < \frac{\varepsilon}{4}. \tag{3.9.1}$$

Let $\delta = \frac{1}{2} \min_{J \in D_1} \mu_m(J)$. Clearly, it suffices to prove that

$$|S(f, P) - S(f, Q)| < \varepsilon \tag{3.9.2}$$

whenever P and Q are δ-fine Perron partitions of $[\boldsymbol{a}, \boldsymbol{b}]$. For each $J \in D_1$ we choose $\boldsymbol{z}_J \in J$ to obtain

$$\begin{aligned}
&|S(f, P) - S(f, Q)| \\
&\le 2 \max_{P_0 = P, Q} \left| S(f, P_0) - \sum_{J \in D_1} f(\boldsymbol{z}_J) \mu_m(J) \right| \\
&= 2 \max_{P_0 = P, Q} \left| \sum_{(\boldsymbol{t}, I) \in P_0} \sum_{J \in D_1} f(\boldsymbol{t}) \mu_m(I \cap J) - \sum_{J \in D_1} \sum_{(\boldsymbol{t}, I) \in P_0} f(\boldsymbol{z}_J) \mu_m(I \cap J) \right| \\
&\le 2 \max_{P_0 = P, Q} \sum_{J \in D_1} \sum_{(\boldsymbol{t}, I) \in P_0} \big(|f(\boldsymbol{t}) - f(\boldsymbol{\xi}_{I,J})| + |f(\boldsymbol{\xi}_{I,J}) - f(\boldsymbol{z}_J)| \big) \mu_m(I \cap J), \\
&\qquad \text{(where } \boldsymbol{\xi}_{I,J} \in I \cap J \text{ if } I \cap J \text{ is non-empty, and } \boldsymbol{\xi}_{I,J} = \boldsymbol{a} \text{ if } I \cap J = \emptyset) \\
&\le 4 \max_{P_0 = P, Q} \sum_{J \in D_1} \omega(f, J)\, \mu_m(J) \\
&< \varepsilon.
\end{aligned}$$

□

We can now state and prove the main result of this section.

Theorem 3.9.5 (Lebesgue). *Let $f : [\boldsymbol{a}, \boldsymbol{b}] \longrightarrow \mathbb{R}$ be a bounded function. Then $f \in R[\boldsymbol{a}, \boldsymbol{b}]$ if and only if f is continuous μ_m-almost everywhere on $[\boldsymbol{a}, \boldsymbol{b}]$.*

Proof. ($\Longrightarrow$) Let $\varepsilon > 0$ be given. For each $k \in \mathbb{N}$ we let

$$X_k := \left\{\boldsymbol{x} \in [\boldsymbol{a}, \boldsymbol{b}] : \omega(f, \boldsymbol{x}) > \frac{1}{2^k}\right\},$$

where

$$\omega(f, \boldsymbol{x}) = \lim_{r \to 0+} \omega\left(f, \prod_{i=1}^{m}(x_i - r, x_i + r)\right).$$

We claim that X_k is contained in the union of a finite number of intervals whose μ_m-measure is less than $\frac{\varepsilon}{2^k}$. By Theorem 3.9.4, there is a division D_k of $[\boldsymbol{a}, \boldsymbol{b}]$ such that

$$\sum_{I \in D_k} \omega(f, I)\ \mu_m(I) < \frac{\varepsilon}{4^{k+1}}$$

and so

$$\sum_{\substack{I \in D_k \\ X_k \cap I \neq \emptyset}} \mu_m(I) \leq 2^k \sum_{I \in D_k} \omega(f, I)\ \mu_m(I) \leq \frac{\varepsilon}{2^{k+1}}.$$

Finally, since $D := \{\boldsymbol{x} \in [\boldsymbol{a}, \boldsymbol{b}] : \omega(f, \boldsymbol{x}) > 0\} = \bigcup_{k=1}^{\infty} X_k$, it follows that the set D of points of discontinuity of f is a set of μ_m-measure zero.

($\Longleftarrow$) Conversely, suppose that there exists a positive number M such that $|f(\boldsymbol{x})| \leq M$ for all $\boldsymbol{x} \in [\boldsymbol{a}, \boldsymbol{b}]$, and the set D of points of discontinuity of f is μ_m-negligible. Then for each $\varepsilon > 0$ there exists a gauge δ_1 on $[\boldsymbol{a}, \boldsymbol{b}]$ such that

$$\sum_{(\boldsymbol{t}, I) \in P} \chi_D(\boldsymbol{t})\mu_m(I) < \frac{\varepsilon}{4M + 1}$$

for each δ_1-fine Perron partition P of $[\boldsymbol{a}, \boldsymbol{b}]$. For each $\boldsymbol{t} \in [\boldsymbol{a}, \boldsymbol{b}] \backslash D$ there exists $\delta_2(\boldsymbol{t}) > 0$ such that

$$|f(\boldsymbol{x}) - f(\boldsymbol{t})| < \frac{\varepsilon}{4\mu_m([\boldsymbol{a}, \boldsymbol{b}])}$$

whenever $\boldsymbol{x} \in B(\boldsymbol{t}, \delta(\boldsymbol{t})) \cap [\boldsymbol{a}, \boldsymbol{b}]$.

Define a gauge δ on $[a,b]$ by setting

$$\delta(t) = \begin{cases} \delta_1(t) \text{ if } t \in D, \\ \delta_2(t) \text{ if } t \in [a,b]\backslash D, \end{cases}$$

and consider a δ-fine Perron partition P_0 of $[a,b]$. Then

$$\begin{aligned}
&\sum_{(t_0,I_0)\in P_0} \omega(f,I)\,\mu_m(I) \\
&= \sum_{\substack{(t_0,I_0)\in P_0 \\ t_0\in[a,b]\backslash D}} \omega(f,I)\,\mu_m(I) + \sum_{\substack{(t_0,I_0)\in P_0 \\ t_0\in D}} \omega(f,I)\,\mu_m(I) \\
&\leq \frac{\varepsilon}{4\mu_m([a,b])} \sum_{\substack{(t_0,I_0)\in P_0 \\ t_0\in[a,b]\backslash D}} \mu_m(I) + 2M \sum_{\substack{(t_0,I_0)\in P_0 \\ t_0\in D}} \mu_m(I) \\
&< \frac{\varepsilon}{4\mu_m([a,b])} \cdot \mu_m([a,b]) + 2M \cdot \frac{\varepsilon}{4M+1} \\
&< \varepsilon.
\end{aligned}$$

Since $\varepsilon > 0$ is arbitrary, an application of Theorem 3.9.4 yields the desired result. □

Exercise 3.9.6. Let $(f_n)_{n=1}^{\infty}$ be a sequence of functions in $R[a,b]$. If $f_n \to f$ uniformly on $[a,b]$, prove that $f \in R[a,b]$ and

$$\lim_{n\to\infty} \int_{[a,b]} f_n = \int_{[a,b]} f.$$

3.10 Some characterizations of Lebesgue integrable functions

If $f \in L^1[a,b]$, then it follows from Theorem 2.3.10 that the function

$$F : [u,v] \mapsto \int_{[u,v]} f \, d\mu_m : \mathcal{I}_m([a,b]) \longrightarrow \mathbb{R}$$

is additive. This interval function F is known as the indefinite Lebesgue integral of f. In this section we give a simple characterization of additive interval functions that are indefinite Lebesgue integrals. As a result, we deduce a Riemann-type definition of the Lebesgue integral.

Definition 3.10.1. An additive interval function $F : \mathcal{I}_m([\boldsymbol{a}, \boldsymbol{b}]) \longrightarrow \mathbb{R}$ is said to be *absolutely continuous* if for each $\varepsilon > 0$ there exists $\eta > 0$ such that

$$\sum_{k=1}^{q} |F(I_k)| < \varepsilon$$

whenever $\{I_1, \ldots, I_q\}$ is a collection of non-overlapping subintervals of $[\boldsymbol{a}, \boldsymbol{b}]$ with $\sum_{k=1}^{q} \mu_m(I_k) < \eta$.

Theorem 3.10.2. *Let $f \in L^1[\boldsymbol{a}, \boldsymbol{b}]$. If F is the indefinite Lebesgue integral of f, then F is absolutely continuous.*

Proof. For each $\varepsilon > 0$ we apply Theorem 3.5.13 and Lebesgue's Dominated Convergence Theorem to select a μ_m-measurable simple function ψ on $[\boldsymbol{a}, \boldsymbol{b}]$ such that

$$\int_{[\boldsymbol{a},\boldsymbol{b}]} |\psi - f| \; d\mu_m < \frac{\varepsilon}{2}.$$

If $I_1, \ldots, I_p$ are non-overlapping subintervals of $[\boldsymbol{a}, \boldsymbol{b}]$ satisfying

$$\sum_{k=1}^{p} \mu_m(I_k) < \frac{\varepsilon}{2}\left(1 + \sup_{\boldsymbol{x}\in[\boldsymbol{a},\boldsymbol{b}]} |\psi(\boldsymbol{x})|\right)^{-1},$$

then

$$\begin{aligned} \sum_{k=1}^{p} |F(I_k)| &\leq \sum_{k=1}^{p} \left| \int_{I_k} (\psi - f) \; d\mu_m \right| + \sum_{k=1}^{p} \left| \int_{I_k} \psi \; d\mu_m \right| \\ &\leq \int_{[\boldsymbol{a},\boldsymbol{b}]} |\psi - f| \; d\mu_m + \left\{ \sup_{\boldsymbol{x}\in[\boldsymbol{a},\boldsymbol{b}]} |\psi(\boldsymbol{x})| \right\} \sum_{k=1}^{p} \mu_m(I_k) \\ &< \varepsilon. \end{aligned}$$

Since F is additive, we conclude that F is absolutely continuous. □

Our next aim is to prove the converse of Theorem 3.10.2 holds; see Theorem 3.10.12. We need to prove the following assertions.

(A) If $F : \mathcal{I}_m([\boldsymbol{a}, \boldsymbol{b}]) \longrightarrow \mathbb{R}$ is absolutely continuous, then $F'(\boldsymbol{x})$ exists for μ_m-almost all $\boldsymbol{x} \in [\boldsymbol{a}, \boldsymbol{b}]$.

(B) If $F : \mathcal{I}_m([\boldsymbol{a}, \boldsymbol{b}]) \longrightarrow \mathbb{R}$ is absolutely continuous, then there exists $f \in HK[\boldsymbol{a}, \boldsymbol{b}]$ such that F is the indefinite Henstock-Kurzweil integral of f.

We need a series of lemmas.

Lemma 3.10.3. *Let $F : \mathcal{I}_m([\boldsymbol{a}, \boldsymbol{b}]) \longrightarrow \mathbb{R}$ be a non-negative additive interval function, let $\alpha \in (0, 1)$, and let $E \subseteq [\boldsymbol{a}, \boldsymbol{b}]$ be a μ_m-measurable set. If there exists $\gamma \in \mathbb{R}^+$ such that $\overline{F}_\alpha(\boldsymbol{x}) > \gamma > 0$ for all $\boldsymbol{x} \in E$, then $F([\boldsymbol{u}, \boldsymbol{v}]) > \gamma\mu_m(E)$ whenever $[\boldsymbol{u}, \boldsymbol{v}] \in \mathcal{I}_m([\boldsymbol{a}, \boldsymbol{b}])$ with $E \subseteq [\boldsymbol{u}, \boldsymbol{v}]$.*

Proof. Let $[\boldsymbol{u}, \boldsymbol{v}] \in \mathcal{I}_m([\boldsymbol{a}, \boldsymbol{b}])$ be any interval containing E. The family $\mathcal{C}$ of all α-regular intervals $C \subset [\boldsymbol{u}, \boldsymbol{v}]$ for which $F(C) > \gamma\mu_m(C)$ is a Vitali cover of $E \cap (\boldsymbol{u}, \boldsymbol{v})$. By the Vitali Covering Theorem, there exists a disjoint family $\mathcal{C}_1 \subset \mathcal{C}$ such that $(E \cap (\boldsymbol{u}, \boldsymbol{v})) \setminus \bigcup_{C \in \mathcal{C}_1} C$ is a μ_m-negligible set. Therefore

$$\gamma\mu_m(E) = \gamma\mu_m(E \cap \bigcup_{C \in \mathcal{C}_1} C) \leq \gamma \sum_{C \in \mathcal{C}_1} \mu_m(C) < \sum_{C \in \mathcal{C}_1} F(C). \quad (3.10.1)$$

Since $F(I)$ is assumed to be non-negative for all $I \in \mathcal{I}_m([\boldsymbol{a}, \boldsymbol{b}])$, we conclude that the right-hand side of (3.10.1) is less than or equal to $F([\boldsymbol{u}, \boldsymbol{v}])$. □

Lemma 3.10.4. *Let $F : \mathcal{I}_m([\boldsymbol{a}, \boldsymbol{b}]) \longrightarrow \mathbb{R}$ be a non-negative additive interval function. If $0 < \beta \leq \alpha < 1$, then $0 \leq \overline{F}_\alpha(\boldsymbol{x}) = \underline{F}_\alpha(\boldsymbol{x}) = \overline{F}_\beta(\boldsymbol{x}) = \underline{F}_\beta(\boldsymbol{x}) < \infty$ for μ_m-almost all $\boldsymbol{x} \in [\boldsymbol{a}, \boldsymbol{b}]$.*

Proof. According to our hypothesis on F, we have $0 \leq \underline{F}_\beta \leq \underline{F}_\alpha \leq \overline{F}_\alpha \leq \overline{F}_\beta$. We will next prove that the set

$$N := \{\boldsymbol{x} \in [\boldsymbol{a}, \boldsymbol{b}] : \underline{F}_\beta(\boldsymbol{x}) < \overline{F}_\beta(\boldsymbol{x})\}$$

is μ_m-negligible. Since

$$N = \bigcup_{\substack{r<s \\ r,s\in\mathbb{Q}}} \left\{\boldsymbol{x} \in [\boldsymbol{a}, \boldsymbol{b}] : \underline{F}_\beta(\boldsymbol{x}) < r < s < \overline{F}_\beta(\boldsymbol{x})\right\},$$

it suffices to prove that the set

$$E_{r,s} := \{\boldsymbol{x} \in [\boldsymbol{a}, \boldsymbol{b}] : \underline{F}_\beta(\boldsymbol{x}) < r < s < \overline{F}_\beta(\boldsymbol{x})\}$$

is μ_m-negligible whenever $(r, s) \in \mathbb{Q}^2$ with $r < s$.

Let $(r, s) \in \mathbb{Q}^2$ with $r < s$. By Theorems 3.6.5 and 3.4.3, for each $\varepsilon > 0$ there exists a bounded open set $O_{r,s} \supseteq E_{r,s}$ such that $\mu_m(O_{r,s}) < \mu_m(E_{r,s}) + \varepsilon$. The family $\mathcal{C}$ of all β-regular intervals $C \subseteq [\boldsymbol{a}, \boldsymbol{b}] \cap O_{r,s}$ for which $F(C) < r\mu_m(C)$ forms a Vitali cover of $E_{r,s}$. By the Vitali Covering Theorem, there is a countable family $\mathcal{C}_1 \subset \mathcal{C}$ such that $E_{r,s}\setminus\bigcup_{C\in\mathcal{C}_1} C$ is a μ_m-negligible set. Since our definition of $E_{r,s}$ implies that $\overline{F}_\beta(\boldsymbol{x}) > s$ for all $\boldsymbol{x} \in E_{r,s}$, it follows from Lemma 3.10.3 that $F(K) > s\mu_m(E_{r,s} \cap K)$ for all $K \in \mathcal{C}_1$. Thus

$$\begin{aligned} s\mu_m(E_{r,s}) &= s \sum_{C\in\mathcal{C}_1} \mu_m(E_{r,s} \cap C) \\ &< \sum_{C\in\mathcal{C}_1} F(C) \\ &< r \sum_{C\in\mathcal{C}_1} \mu_m(C) \\ &= r\mu_m\Bigg(\bigcup_{C\in\mathcal{C}_1} C\Bigg) \\ &\leq r\mu_m(O_{r,s}) \\ &< r\mu_m(E_{r,s}) + r\varepsilon; \end{aligned}$$

that is $\mu_m(E_{r,s}) < \dfrac{r\varepsilon}{s-r}$. Since $\varepsilon > 0$ is arbitrary, $\mu_m(E_{r,s}) = 0$.

It remains to prove that $\overline{F}_\beta(\boldsymbol{x}) \in \mathbb{R}$ for μ_m-almost all $\boldsymbol{x} \in [\boldsymbol{a}, \boldsymbol{b}]$. Clearly, it suffices to prove that the set

$$Z := \{\boldsymbol{x} \in [\boldsymbol{a}, \boldsymbol{b}] : \overline{F}_\beta(\boldsymbol{x}) = \infty\}$$

is μ_m-negligible. Proceeding towards a contradiction, suppose that Z is not μ_m-negligible. By Theorems 3.6.5 and 3.4.3, Z is a μ_m-measurable set. Since the family $\mathcal{G}_1$ of all β-regular intervals $J \subseteq [\boldsymbol{a}, \boldsymbol{b}]$ for which $F(J) > \frac{F([\boldsymbol{a},\boldsymbol{b}])}{\mu_m(Z)}\mu_m(J)$ is a Vitali cover of Z, it follows from the Vitali Covering Theorem there exists a countable family $\mathcal{G}_2 \subset \mathcal{G}_1$ such that $Z\setminus\bigcup_{J\in\mathcal{G}_2} J$ is a μ_m-negligible set. A contradiction follows:

$$\begin{aligned} F([\boldsymbol{a}, \boldsymbol{b}]) &= \frac{F([\boldsymbol{a}, \boldsymbol{b}])}{\mu_m(Z)} \sum_{J\in\mathcal{G}_2} \mu_m(Z \cap J) \\ &< \sum_{J\in\mathcal{G}_2} F(J) \\ &\leq F([\boldsymbol{a}, \boldsymbol{b}]). \end{aligned}$$

This contradiction proves the lemma. □

Definition 3.10.5. Let $F : \mathcal{I}_m([\boldsymbol{a}, \boldsymbol{b}]) \longrightarrow \mathbb{R}$ be an additive interval function, and let $I \in \mathcal{I}_m([\boldsymbol{a}, \boldsymbol{b}])$. The *total variation* of F over I is the extended number

$$V_F(I) = \sup\Bigg\{\sum_{J\in D} |F(J)| : D \text{ is a division of } I\Bigg\}.$$

Lemma 3.10.6. *Let $F : \mathcal{I}_m([\boldsymbol{a}, \boldsymbol{b}]) \longrightarrow \mathbb{R}$ be an additive interval function. If $V_F([\boldsymbol{a}, \boldsymbol{b}])$ is finite, then $V_F(I)$ is finite for all $I \in \mathcal{I}_m([\boldsymbol{a}, \boldsymbol{b}])$, and the interval function $V_F : \mathcal{I}_m([\boldsymbol{a}, \boldsymbol{b}]) \longrightarrow \mathbb{R}$ is additive.*

Proof. Exercise. □

Lemma 3.10.7. *If $F : \mathcal{I}_m([\boldsymbol{a}, \boldsymbol{b}]) \longrightarrow \mathbb{R}$ is absolutely continuous, then $V_F([\boldsymbol{a}, \boldsymbol{b}])$ is finite.*

Proof. We choose an $\eta > 0$ corresponding to $\varepsilon = 1$ in Definition 3.10.1, and fix a net $\mathcal{N}_0$ of $[\boldsymbol{a}, \boldsymbol{b}]$ so that $\max_{I \in \mathcal{N}_0} \operatorname{diam} I < \eta$. We write $F(K) = 0$ whenever K is a degenerate subinterval of $[\boldsymbol{a}, \boldsymbol{b}]$. If D is a division of $[\boldsymbol{a}, \boldsymbol{b}]$, then

$$\sum_{J \in D} |F(J)| \leq \sum_{J \in D} \sum_{I \in \mathcal{N}_0} |F(I \cap J)| = \sum_{I \in \mathcal{N}_0} \sum_{J \in D} |F(I \cap J)| \leq \operatorname{card}(\mathcal{N}_0).$$

Consequently, $V_F([\boldsymbol{a}, \boldsymbol{b}]) \leq \operatorname{card}(\mathcal{N}_0) < \infty$. □

Lemma 3.10.8. *If $F : \mathcal{I}_m([\boldsymbol{a}, \boldsymbol{b}]) \longrightarrow \mathbb{R}$ is absolutely continuous, so are $V_F + F$ and $V_F - F$.*

Proof. This is an easy consequence of Lemmas 3.10.6 and 3.10.7. □

We are now ready to prove assertion (A) stated after Theorem 3.10.2.

Theorem 3.10.9. *If $F : \mathcal{I}_m([\boldsymbol{a}, \boldsymbol{b}]) \longrightarrow \mathbb{R}$ is absolutely continuous, then $F'(\boldsymbol{x})$ exists for μ_m-almost all $\boldsymbol{x} \in [\boldsymbol{a}, \boldsymbol{b}]$.*

Proof. In view of Lemmas 3.10.8, 3.10.7 and 3.10.6, we may suppose that $F(I) \geq 0$ for every $I \in \mathcal{I}_m([\boldsymbol{a}, \boldsymbol{b}])$. An appeal to Lemma 3.10.4 completes the argument. □

To proceed further, we need the following simple lemma.

Lemma 3.10.10. *If $I \in \mathcal{I}_m([\boldsymbol{a}, \boldsymbol{b}])$, then I has a net $\mathcal{N}$ such that $\min_{J \in \mathcal{N}} reg(J) \geq \frac{1}{2}$.*

Proof. Let $[\boldsymbol{c}, \boldsymbol{d}] \in \mathcal{I}_m([\boldsymbol{a}, \boldsymbol{b}]$. For each $k \in \{1, \ldots, m\}$ we choose $\alpha_k \in \mathbb{N}$ so that $\beta_k := \frac{d_k - c_k}{\alpha_k} \in [\min_{i=1,\ldots,m}(d_i - c_i), 2\min_{i=1,\ldots,m}(d_i - c_i))$. The collection

$$\mathcal{N} := \left\{ \prod_{k=1}^{m} [c_k + (j_k - 1)\beta_k, c_k + j_k\beta_k] : j_k \in \{1, \ldots, \alpha_k\} \ (k = 1, \ldots, m) \right\}$$

of intervals is a net of $[\boldsymbol{c}, \boldsymbol{d}]$ with the desired property. □

The previous lemma leads us to the next crucial estimate.

Lemma 3.10.11. *Let* $f : [\boldsymbol{a}, \boldsymbol{b}] \longrightarrow \mathbb{R}$ *and assume that* $F : \mathcal{I}_m([\boldsymbol{a}, \boldsymbol{b}]) \longrightarrow \mathbb{R}$ *is an additive interval function. If* $\boldsymbol{t} \in [\boldsymbol{a}, \boldsymbol{b}]$, $I \in \mathcal{I}_m([\boldsymbol{a}, \boldsymbol{b}])$ *and* $X \subseteq I$, *then there exists a net* $\mathcal{N}_I$ *of* I *such that* $\min_{J \in \mathcal{N}_I} reg(J) \geq \frac{1}{2}$ *and*

$$\begin{aligned}
&|f(\boldsymbol{t})\mu_m(I) - F(I)| \\
&\leq |f(\boldsymbol{t})|\, \mu_m\Bigg(I \setminus \bigcup_{\substack{[\boldsymbol{u},\boldsymbol{v}] \in \mathcal{N}_I \\ (\boldsymbol{u},\boldsymbol{v}) \cap X \neq \emptyset}} [\boldsymbol{u}, \boldsymbol{v}]\Bigg) + \sum_{\substack{[\boldsymbol{u},\boldsymbol{v}] \in \mathcal{N}_I \\ (\boldsymbol{u},\boldsymbol{v}) \cap X \neq \emptyset}} |f(\boldsymbol{t})\mu_m([\boldsymbol{u}, \boldsymbol{v}]) - F([\boldsymbol{u}, \boldsymbol{v}])| \\
&+ \Bigg| \sum_{\substack{[\boldsymbol{u},\boldsymbol{v}] \in \mathcal{N}_I \\ (\boldsymbol{u},\boldsymbol{v}) \cap X = \emptyset}} F([\boldsymbol{u}, \boldsymbol{v}]) \Bigg| . \qquad (3.10.2)
\end{aligned}$$

Proof. This is an easy consequence of Lemma 3.10.10 and triangle inequality. □

We are now already to prove assertion (B) stated after Theorem 3.10.2. Combining this assertion with Theorem 3.1.3 leads to the converse of Theorem 3.10.2.

Theorem 3.10.12. *If* $F : \mathcal{I}_m([\boldsymbol{a}, \boldsymbol{b}]) \longrightarrow \mathbb{R}$ *is absolutely continuous, then there exists* $f \in L^1[\boldsymbol{a}, \boldsymbol{b}]$ *such that* F *is the indefinite Lebesgue integral of* f.

Proof. By Theorem 3.10.9, $F'(\boldsymbol{x})$ exists for μ_m-almost all $\boldsymbol{x} \in [\boldsymbol{a}, \boldsymbol{b}]$. Define the function $f : [\boldsymbol{a}, \boldsymbol{b}] \longrightarrow \mathbb{R}$ by setting

$$f(\boldsymbol{x}) = \begin{cases} F'(\boldsymbol{x}) & \text{if } F'(\boldsymbol{x}) \text{ exists,} \\ 0 & \text{otherwise.} \end{cases}$$

We want to prove that $f \in L^1[\boldsymbol{a}, \boldsymbol{b}]$ and F is the indefinite Lebesgue integral of f. In view of Theorem 3.1.3, it suffices to prove that $f \in HK[\boldsymbol{a}, \boldsymbol{b}]$ and F is the indefinite Henstock-Kurzweil integral of f.

According to Theorems 3.10.9 and 3.4.2, there exists a decreasing sequence $(U_n)_{n=1}^{\infty}$ of open sets such that $Z := \bigcap_{k=1}^{\infty} U_k$ is a μ_m-negligible set containing $[\boldsymbol{a},\boldsymbol{b}]\backslash(\boldsymbol{a},\boldsymbol{b})$, and $F'(\boldsymbol{x})$ exists for all $\boldsymbol{x} \in [\boldsymbol{a},\boldsymbol{b}]\backslash Z$. For each $k \in \mathbb{N}$ we set $Y_k := [\boldsymbol{a},\boldsymbol{b}]\backslash U_k$ so that $(Y_k)_{k=1}^{\infty}$ is an increasing sequence of closed subsets of $[\boldsymbol{a},\boldsymbol{b}]$.

We claim that there exists an increasing sequence $(X_n)_{n=1}^{\infty}$ of closed sets such that $\bigcup_{k=1}^{\infty} X_k = \bigcup_{k=1}^{\infty} Y_k$ and f is bounded on each X_k. For each $\varepsilon > 0$ there exists a gauge δ_1 on $[\boldsymbol{a},\boldsymbol{b}]\backslash Z$ such that

$$|f(\boldsymbol{y})\mu_m(J) - F(J)| < \frac{\varepsilon}{16(1+\mu_m([\boldsymbol{a},\boldsymbol{b}]))}\mu_m(J) \tag{3.10.3}$$

for each point-interval pair $(\boldsymbol{y}, J)$ satisfying $\boldsymbol{y} \in J \cap ([\boldsymbol{a},\boldsymbol{b}]\backslash Z)$, $J \subset B(\boldsymbol{y}, \delta_1(\boldsymbol{y}))$ and $\operatorname{reg}(J) \geq \frac{1}{2}$. Since $[\boldsymbol{a},\boldsymbol{b}]\backslash(\boldsymbol{a},\boldsymbol{b}) \subseteq Z$, we may assume that $\underline{B(\boldsymbol{x}, \delta_1(\boldsymbol{x})) \subset (\boldsymbol{a},\boldsymbol{b})}$ whenever $\boldsymbol{x} \in [\boldsymbol{a},\boldsymbol{b}]\backslash Z$. For each $k \in \mathbb{N}$ we set $X_k := \{\boldsymbol{x} \in Y_k : \delta_1(\boldsymbol{x}) \geq \frac{1}{k}\}$. Since Y_k is closed and δ_1 is a gauge on $[\boldsymbol{a},\boldsymbol{b}]\backslash Z$, the sequence $(Y_n)_{n=1}^{\infty}$ is increasing and $\bigcup_{k=1}^{\infty} X_k = \bigcup_{k=1}^{\infty} Y_k$.

Now we prove the boundedness of f on each X_k. From (3.10.3) we get

$$\begin{aligned}&\boldsymbol{x}_1\boldsymbol{x}_2 \in [\boldsymbol{a},\boldsymbol{b}]\backslash Z \text{ with } |||\boldsymbol{x}_1 - \boldsymbol{x}_2||| < \min\{\delta_1(\boldsymbol{x}_1), \delta_1(\boldsymbol{x}_2)\}\\ &\Longrightarrow |f(\boldsymbol{x}_1) - f(\boldsymbol{x}_2)| < \tfrac{\varepsilon}{8}(1+\mu_m([\boldsymbol{a},\boldsymbol{b}]))^{-1}.\end{aligned} \tag{3.10.4}$$

Let $k \in \mathbb{N}$ be fixed and let $\boldsymbol{x}_0 \in X_k$. According to the definition of X_k, there exists $\boldsymbol{x} \in Y_k$ such that $\delta_1(\boldsymbol{x}) \geq \frac{1}{k}$ and $|||\boldsymbol{x} - \boldsymbol{x}_0||| < \min\{\delta_1(\boldsymbol{x}), \frac{1}{k}\}$. Thus

$$\begin{aligned}|f(\boldsymbol{x}_0)| &< |f(\boldsymbol{x})| + \frac{\varepsilon}{8(1+\mu_m([\boldsymbol{a},\boldsymbol{b}]))}\\ &\leq \left| f(\boldsymbol{x}) - \frac{F(\prod_{i=1}^m [x_i + \frac{1}{2k}])}{\mu_m(\prod_{i=1}^m [x_i + \frac{1}{2k}])} \right| + \left| \frac{F(\prod_{i=1}^m [x_i + \frac{1}{2k}])}{\mu_m(\prod_{i=1}^m [x_i + \frac{1}{2k}])} \right| + \frac{\varepsilon}{8}\\ &\leq k^m V_F([\boldsymbol{a},\boldsymbol{b}]) + \frac{\varepsilon}{4}.\end{aligned}$$

This proves that f is bounded on X_k.

According to the hypothesis on F, there exists a sufficiently small $\eta > 0$ such that

$$\sum_{i=1}^{p} |F(I_i)| < \frac{\varepsilon}{4} \tag{3.10.5}$$

whenever $I_1, \ldots, I_p$ are non-overlapping subintervals of $[\boldsymbol{a},\boldsymbol{b}]$ satisfying $\sum_{i=1}^{p} \mu_m(I_i) < \eta$. Since the sequence $(X_n)_{n=1}^{\infty}$ of closed sets is increasing and $Z = [\boldsymbol{a},\boldsymbol{b}]\backslash \bigcup_{k=1}^{\infty} X_k$, there exists $N \in \mathbb{N}$ such that $\mu_m([\boldsymbol{a},\boldsymbol{b}]\backslash X_N) < \eta$.

For each integer $k \geq N$ we use the boundedness of f on X_k to choose a bounded open set $O_k \supseteq X_k$ so that

$$\mu_m(O_k \backslash X_k) < \frac{\varepsilon}{2^{k+3}(1 + \|f\chi_{X_k}\|_{L^\infty[\boldsymbol{a},\boldsymbol{b}]})}. \tag{3.10.6}$$

Define a gauge δ on $[\boldsymbol{a}, \boldsymbol{b}]$ by setting

$$\delta(\boldsymbol{x}) = \begin{cases} \min\{\frac{1}{2N}, \operatorname{dist}(\boldsymbol{x}, [\boldsymbol{a}, \boldsymbol{b}] \backslash O_N)\} & \text{if } \boldsymbol{x} \in X_N, \\ \min\{\operatorname{dist}(\boldsymbol{x}, X_k \cup ([\boldsymbol{a}, \boldsymbol{b}] \backslash O_{k+1})), \frac{1}{2k}\} & \\ \qquad \text{if } \boldsymbol{x} \in X_{k+1} \backslash X_k \text{ for some integer } k \geq N, & \\ \operatorname{dist}(\boldsymbol{x}, X_N) & \text{if } \boldsymbol{x} \in Z, \end{cases}$$

and consider any δ-fine Perron subpartition P of $[\boldsymbol{a}, \boldsymbol{b}]$. Since f vanishes on Z, it follows from our choice of δ that

$$\sum_{\substack{(\boldsymbol{t},I) \in P \\ \boldsymbol{t} \in Z}} |f(\boldsymbol{t})\mu_m(I) - F(I)| < \frac{\varepsilon}{4}. \tag{3.10.7}$$

Write $Z_N = X_N$ and $Z_k = X_k \backslash X_{k-1}$ for $k = N+1, N+2, \ldots$. Since $[\boldsymbol{a}, \boldsymbol{b}] = Z \cup \bigcup_{k=1}^{\infty} Z_k$, we combine Lemma 3.10.11, the boundedness of f on each X_k, and (3.10.7) to obtain

$$\begin{aligned} &\sum_{(\boldsymbol{t},I) \in P} |f(\boldsymbol{t})\mu_m(I) - F(I)| \\ &\leq \sum_{k=N}^{\infty} \Bigg\{ \sum_{\substack{(\boldsymbol{t},I) \in P \\ \boldsymbol{t} \in Z_k}} \|f\chi_{Z_k}\|_{L^\infty[\boldsymbol{a},\boldsymbol{b}]} \, \mu_m\Bigg(I \backslash \bigcup_{\substack{[\boldsymbol{u},\boldsymbol{v}] \in \mathcal{N}_I \\ (\boldsymbol{u},\boldsymbol{v}) \cap Z_k \neq \emptyset}} [\boldsymbol{u}, \boldsymbol{v}]\Bigg) \\ &\qquad + \sum_{\substack{(\boldsymbol{t},I) \in P \\ \boldsymbol{t} \in Z_k}} \sum_{\substack{[\boldsymbol{u},\boldsymbol{v}] \in \mathcal{N}_I \\ (\boldsymbol{u},\boldsymbol{v}) \cap Z_k \neq \emptyset}} |f(\boldsymbol{t})\mu_m([\boldsymbol{u}, \boldsymbol{v}]) - F([\boldsymbol{u}, \boldsymbol{v}])| \\ &\qquad + \sum_{\substack{(\boldsymbol{t},I) \in P \\ \boldsymbol{t} \in Z_k}} \Bigg| \sum_{\substack{[\boldsymbol{u},\boldsymbol{v}] \in \mathcal{N}_I \\ (\boldsymbol{u},\boldsymbol{v}) \cap Z_k = \emptyset}} F([\boldsymbol{u}, \boldsymbol{v}]) \Bigg| \Bigg\} + \frac{\varepsilon}{4}. \end{aligned} \tag{3.10.8}$$

We will next prove that (3.10.8), our choice of δ, (3.10.6) and (3.10.5) yield

$$\begin{aligned} &\sum_{(\boldsymbol{t},I) \in P} |f(\boldsymbol{t})\mu_m(I) - F(I)| \\ &< \frac{3\varepsilon}{4} + \sum_{k=N}^{\infty} \sum_{\substack{(\boldsymbol{t},I) \in P \\ \boldsymbol{t} \in Z_k}} \sum_{\substack{[\boldsymbol{u},\boldsymbol{v}] \in \mathcal{N}_I \\ (\boldsymbol{u},\boldsymbol{v}) \cap Z_k \neq \emptyset}} |f(\boldsymbol{t})\mu_m([\boldsymbol{u}, \boldsymbol{v}]) - F([\boldsymbol{u}, \boldsymbol{v}])|. \end{aligned} \tag{3.10.9}$$

Indeed,

$$\begin{aligned}
&\sum_{(t,I)\in P} |f(t)\mu_m(I) - F(I)| \\
&< \sum_{k=N}^{\infty} \|f\chi_{X_k}\|_{L^\infty[a,b]}\, \mu_m(O_k\backslash X_k) \\
&+ \sum_{k=N}^{\infty} \sum_{\substack{(t,I)\in P\\ t\in Z_k}} \sum_{\substack{[u,v]\in\mathcal{N}_I\\ (u,v)\cap Z_k\neq\emptyset}} |f(t)\mu_m([u,v]) - F([u,v])| \\
&+ \sum_{k=N}^{\infty} \sum_{\substack{(t,I)\in P\\ t\in Z_k}} \sum_{\substack{[u,v]\in\mathcal{N}_I\\ (u,v)\cap X_N=\emptyset}} |F([u,v])| + \frac{\varepsilon}{4} \\
&< \frac{3\varepsilon}{4} + \sum_{k=N}^{\infty} \sum_{\substack{(t,I)\in P\\ t\in Z_k}} \sum_{\substack{[u,v]\in\mathcal{N}_I\\ (u,v)\cap Z_k\neq\emptyset}} |f(t)\mu_m([u,v]) - F([u,v])|\,.
\end{aligned}$$

It remains to prove that the right-hand side of (3.10.8) is less than ε. For each $[u,v] \in \mathcal{N}_I$ satisfying $(u,v)\cap Z_k \neq \emptyset$ for some integer $k \geq N$, we choose $x_{[u,v]} \in Y_k \cap [u,v]$ so that (3.10.3) and (3.10.4) give

$$\left|f(x_{[u,v]}) - F([u,v])\right| < \frac{\varepsilon}{16}(1+\mu_m([a,b]))^{-1}\mu_m([u,v])$$

and

$$\left|f(x_{[u,v]}) - f(t)\right| < \frac{\varepsilon}{8}(1+\mu_m([a,b]))^{-1}$$

respectively. Therefore the right-hand side of (3.10.8) is less than ε. □

Remark 3.10.13. Let δ and P be given as in the proof of Theorem 3.10.12. We observe that the proof will still work if $t \in B(t,\delta(t)) \cap [a,b]$ for every $(t,[u,v]) \in P$.

In order to proceed further, we need some terminologies.

Definition 3.10.14.

(i) A *McShane partition* of an interval $[a,b]$ is a finite collection $\{(t_1,[u_1,v_1]),\ldots,(t_p,[u_p,v_p])\}$ of point-interval pairs, where $\{[u_k,v_k] : k = 1,\ldots,p\}$ is a division of $[a,b]$ and $t_k \in [a,b]$ for $k = 1,\ldots,p$.

(ii) Let $P = \{(t_1,[u_1,v_1]),\ldots,(t_p,[u_p,v_p])\}$ be a McShane partition of $[a,b]$ and let δ be a gauge (i.e. positive function) defined on $\{t_1,\ldots,t_p\}$. P is said to be *δ-fine* if $[u_k,v_k] \subset B(t_k,\delta(t_k))$ for $k = 1,\ldots,p$.

Definition 3.10.15. A function $f : [\boldsymbol{a}, \boldsymbol{b}] \longrightarrow \mathbb{R}$ is said to be *McShane integrable* on $[\boldsymbol{a}, \boldsymbol{b}]$ if there exists $A \in \mathbb{R}$ with the following property: given $\varepsilon > 0$ there exists a gauge δ on $[\boldsymbol{a}, \boldsymbol{b}]$ such that

$$|S(f, P) - A| < \varepsilon \tag{3.10.10}$$

for each δ-fine McShane partition P of $[\boldsymbol{a}, \boldsymbol{b}]$.

The collection of all functions that are McShane integrable on $[\boldsymbol{a}, \boldsymbol{b}]$ will be denoted by $Mc[\boldsymbol{a}, \boldsymbol{b}]$.

It is easy to see that if $f \in Mc[\boldsymbol{a}, \boldsymbol{b}]$, then $f \in HK[\boldsymbol{a}, \boldsymbol{b}]$ and their integrals coincide. In this case, we write the McShane integral of f as $(Mc) \int_{[\boldsymbol{a},\boldsymbol{b}]} f(\boldsymbol{t})\, d\mu_m(\boldsymbol{t})$ or $(Mc) \int_{[\boldsymbol{a},\boldsymbol{b}]} f\, d\mu_m$. Replacing Perron partitions by McShane partitions, Theorems 2.3.1–2.3.6, 2.3.9 and 2.3.10 are true for the McShane integral. In addition, we have the following result.

Theorem 3.10.16. *If $f \in Mc[\boldsymbol{a}, \boldsymbol{b}]$, then $|f| \in Mc[\boldsymbol{a}, \boldsymbol{b}]$.*

Proof. Let $\varepsilon > 0$ be given. By the Cauchy criterion for the McShane integral, there exists a gauge δ on $[\boldsymbol{a}, \boldsymbol{b}]$ such that

$$|S(f, P_0) - S(f, Q_0)| < \varepsilon$$

whenever P_0 and Q_0 are δ-fine McShane partitions of $[\boldsymbol{a}, \boldsymbol{b}]$.

Let $P = \{(\boldsymbol{t}_i, I_i) : i = 1, \ldots, p\}$ and $Q = \{(\boldsymbol{\xi}_k, J_k) : k = 1, \ldots, q\}$ be two δ-fine McShane partitions of $[\boldsymbol{a}, \boldsymbol{b}]$. Using the following division

$$D := \{I_i \cap J_k \in \mathcal{I}_m([\boldsymbol{a}, \boldsymbol{b}]) : i = 1, \ldots, p \text{ and } k = 1, \ldots, q\}$$

of $[\boldsymbol{a}, \boldsymbol{b}]$, we construct two δ-fine McShane partitions P_1 and Q_1 of $[\boldsymbol{a}, \boldsymbol{b}]$ as follows:

$$\begin{aligned} P_1 &= \{(\boldsymbol{t}_i, I_i \cap J_k) : f(\boldsymbol{t}_i) \geq f(\boldsymbol{\xi}_k) \text{ and } I_i \cap J_k \in D\} \\ &\quad \cup \{(\boldsymbol{\xi}_k, I_i \cap J_k) : f(\boldsymbol{t}_i) < f(\boldsymbol{\xi}_k) \text{ and } I_i \cap J_k \in D\} \end{aligned}$$

and

$$\begin{aligned} Q_1 &= \{(\boldsymbol{\xi}_k, I_i \cap J_k) : f(\boldsymbol{t}_i) \geq f(\boldsymbol{\xi}_k) \text{ and } I_i \cap J_k \in D\} \\ &\quad \cup \{(\boldsymbol{t}_i, I_i \cap J_k) : f(\boldsymbol{t}_i) < f(\boldsymbol{\xi}_k) \text{ and } I_i \cap J_k \in D\} \end{aligned}$$

Thus

$$\begin{aligned} &|S(|f|, P) - S(|f|, Q)| \\ &= \left| \sum_{i=1}^{p} \sum_{k=1}^{q} |f(\boldsymbol{t}_i)|\, \mu_m(I_i \cap J_k) - \sum_{i=1}^{p} \sum_{k=1}^{q} |f(\boldsymbol{\xi}_k)|\, \mu_m(I_i \cap J_k) \right| \\ &\leq \sum_{i=1}^{p} \sum_{k=1}^{q} |f(\boldsymbol{t}_i) - f(\boldsymbol{\xi}_k)|\, \mu_m(I_i \cap J_k) \\ &= |S(f, P_1) - S(f, Q_1)| \\ &< \varepsilon. \end{aligned}$$

By the Cauchy criterion for the McShane integral, $|f| \in Mc[\boldsymbol{a}, \boldsymbol{b}]$. □

Theorem 3.10.17. *If $f \in Mc[\boldsymbol{a}, \boldsymbol{b}]$, then $f \in L^1[\boldsymbol{a}, \boldsymbol{b}]$ and*

$$(Mc) \int_{[\boldsymbol{a},\boldsymbol{b}]} f = \int_{[\boldsymbol{a},\boldsymbol{b}]} f \, d\mu_m.$$

Proof. This follows from Theorem 3.10.16 and Definition 3.1.1. □

Combining Theorem 3.10.2, the proof of Theorem 3.10.12 and Remark 3.10.13, we get the converse of Theorem 3.10.17.

Theorem 3.10.18. *If $f \in L^1[\boldsymbol{a}, \boldsymbol{b}]$, then $f \in Mc[\boldsymbol{a}, \boldsymbol{b}]$ and*

$$\int_{[\boldsymbol{a},\boldsymbol{b}]} f \, d\mu_m = (Mc) \int_{[\boldsymbol{a},\boldsymbol{b}]} f.$$

3.11 Some results concerning one-dimensional Lebesgue integral

When $m = 1$ we can use point functions to state and prove Theorem 3.10.12. The following definition will be used.

Definition 3.11.1. Let $F : [a, b] \longrightarrow \mathbb{R}$. If

$$\sum_{k=1}^{p} |F(c_k) - F(c_{k-1})| < \infty$$

for every division $\{[c_0, c_1], \ldots, [c_{p-1}, c_p]\}$ of $[a, b]$, F is said to be *bounded variation* on $[a, b]$, and we write $F \in BV[a, b]$. In this case, the *total variation* of F over $[a, b]$ is given by

$$Var(F, [a, b]) := \sup \sum_{k=1}^{p} |F(c_k) - F(c_{k-1})|,$$

where the supremum is taken over all possible divisions $\{[c_0, c_1], \ldots, [c_{p-1}, c_p]\}$ of $[a, b]$.

Lemma 3.11.2. *If $F \in BV[a, b]$ and $c \in (a, b)$, then both $Var(F, [a, c]), Var(F, [c, b])$ are finite and*

$$Var(F, [a, b]) = Var(F, [a, c]) + Var(F, [c, b]).$$

Proof. Exercise. □

Lemma 3.11.3. *Let $F : [a, b] \longrightarrow \mathbb{R}$. If F is non-decreasing on $[a, b]$, then $F \in BV[a, b]$.*

Proof. Exercise. □

Theorem 3.11.4. *Let $F : [a, b] \longrightarrow \mathbb{R}$. Then $F \in BV[a, b]$ if and only if F can be written as the difference of two non-decreasing real-valued functions on $[a, b]$.*

Proof. ($\Longleftarrow$) This follows from Lemma 3.11.3 and triangle inequality.

($\Longrightarrow$) Conversely, suppose that $F \in BV[a, b]$. For $k = 1, 2$ we set

$$F_k(x) = \begin{cases} \frac{1}{2}(Var(g, [a, x]) + (-1)^{k-1}F(x)) & \text{if } x \in (a, b], \\ (-1)^{k-1}F(a) & \text{if } x = a. \end{cases}$$

According to Lemma 3.11.2, F_1 and F_2 are non-decreasing on $[a, b]$. Since $F(x) = F_1(x) - F_2(x)$ for all $x \in [a, b]$, the result follows. □

Corollary 3.11.5. *If $F \in BV[a, b]$, then the set of points at which F is discontinuous is countable.*

Proof. Exercise. □

When $m = 1$, the Vitali Covering Theorem is also true. In particular, Lemma 3.10.4 is applicable to prove the following result.

Theorem 3.11.6. *If $F \in BV[a, b]$, then F is differentiable μ_1-almost everywhere on $[a, b]$. Moreover, there exists $f \in L^1[a, b]$ such that $f = F'$ μ_1-almost everywhere on $[a, b]$ and*

$$\int_a^b |f(t)| \, d\mu_1(t) \leq Var(F, [a, b]). \tag{3.11.1}$$

Proof. In view of Theorem 3.11.4, we may suppose that F is non-decreasing on $[a, b]$, and $F(x) = F(b)$ for every $x > b$. Since $[u, v] \mapsto F(v) - F(u)$ is an additive interval function defined on $\mathcal{I}_1([a, b])$, and F is non-decreasing on $[a, b]$, it follows from Lemma 3.10.4 that that there exists a μ_1-negligible set $Z \subset [a, b]$ such that $F'(x)$ exists for all $x \in [a, b] \backslash Z$. Letting

$$f(x) = \begin{cases} F'(x) & \text{if } x \in [a, b] \backslash Z, \\ 0 & \text{if } x \in Z, \end{cases}$$

and observing that $f = F' \geq 0$ μ_1-almost everywhere on $[a, b]$, Fatou's Lemma gives

$$\begin{aligned}
&\int_a^b f(t)\, d\mu_1(t) \\
&\leq \liminf_{n\to\infty} \int_a^b n(F(t+\frac{1}{n}) - F(t))\, d\mu_1(t) \\
&= \liminf_{n\to\infty} \left\{ n\int_b^{b+\frac{1}{n}} F(t)\, d\mu_1(t) - n\int_a^{a+\frac{1}{n}} F(t)\, d\mu_1(t) \right\} \\
&\leq F(b) - F(a).
\end{aligned}$$

□

Remark 3.11.7. The inequality (3.11.1) can be strict; see Exercise 4.5.7.

Definition 3.11.8. A function $F : [a, b] \longrightarrow \mathbb{R}$ is said to be *absolutely continuous* on $[a, b]$ if the following condition is satisfied: for each $\varepsilon > 0$ there exists $\eta > 0$ such that

$$\sum_{k=1}^{p} |F(c_k) - F(c_{k-1})| < \varepsilon$$

whenever $\{[c_0, c_1], \ldots, [c_{p-1}, c_p]\}$ is a finite collection of pairwise non-overlapping subintervals of $[a, b]$ with $\sum_{k=1}^{p}(c_k - c_{k-1}) < \eta$.

Let $AC[a, b]$ be the space of absolutely continuous functions on $[a, b]$. The following result is a special case of Theorem 3.10.2.

Theorem 3.11.9. *If $f \in L^1[a, b]$, then the function $x \mapsto \int_a^x f\, d\mu_1$ belongs to $AC[a, b]$.*

Theorem 3.11.10. *If $F \in AC[a, b]$, then $F \in BV[a, b]$.*

Proof. Using Lemma 3.10.7 with $m = 1$, we get the result. □

The following theorem is essentially the one-dimensional version of Theorem 3.10.12.

Theorem 3.11.11. *If $F \in AC[a, b]$, then F is differentiable μ_1-almost everywhere on $[a, b]$. Moreover, there exists $f \in L^1[a, b]$ such that $f = F'$ μ_1-almost everywhere on $[a, b]$ and*

$$\int_a^x f(t)\, d\mu_1(t) = F(x) - F(a)$$

for all $x \in [a, b]$.

Proof. First, we infer from Theorems 3.11.10 and 3.11.6 that there exists a μ_1-negligible set $Z \subset [a,b]$ such that $F'(x)$ exists for all $x \in [a,b]\backslash Z$.

Define

$$f(x) = \begin{cases} F'(x) & \text{if } x \in [a,b]\backslash Z, \\ 0 & \text{if } x \in Z. \end{cases}$$

Since $F \in BV[a,b]$, it suffices to prove that $f \in HK[a,b]$ and $F(x) - F(a) = (HK)\int_a^x f(t)\,dt$ for all $x \in (a,b]$.

Let $\varepsilon > 0$ be given and choose an $\eta > 0$ corresponding to $\frac{\varepsilon}{2}$ in Definition 3.11.8. As Z is μ_1-negligible, there exists an open set $O \supset Z$ such that $\mu_1(O) < \eta$. Also, for each $[a,b]\backslash Z$ there exists $\delta_1(x) > 0$ such that

$$|f(x)(v-u) - (F(v) - F(u))| \leq \frac{\varepsilon(v-u)}{2(b-a)}$$

whenever $x \in [u,v] \subseteq [a,b] \cap (x - \delta_1(x), x + \delta_1(x))$.

Define a gauge δ on $[a,b]$ by setting

$$\delta(x) = \begin{cases} \delta_1(x) & \text{if } x \in [a,b]\backslash Z, \\ \operatorname{dist}(x, [a,b]\backslash O) & \text{if } x \in Z, \end{cases}$$

and consider any δ-fine Perron subpartition P of $[a,b]$. Then

$$\begin{aligned} &\sum_{(t,[u,v]) \in P} |f(t)(v-u) - (F(v) - F(u))| \\ &= \sum_{\substack{(t,[u,v]) \in P \\ t \in [a,b]\backslash Z}} |f(t)(v-u) - (F(v) - F(u))| + \sum_{\substack{(t,[u,v]) \in P \\ t \in Z}} |F(v) - F(u)| \\ &< \varepsilon. \end{aligned}$$

Since $\varepsilon > 0$ is arbitrary, the theorem is proved. □

3.12 Notes and Remarks

There are many excellent books on Lebesgue integration; consult, for instance, Hewitt and Stromberg [59], Royden [143], Rudin [144], Stromberg [149]. For a history of Lebesgue integration, see Hawkins [54].

A different proof of Theorem 3.6.5 can be found in [137, p.175]. Section 3.9 is based on [6, Appendix C].

Lebesgue proved Theorem 3.10.12; see [145]. The present proof of Theorem 3.10.12 is similar to that of [94, Theorem 3.5]. For other results concerning derivation of absolutely continuous interval functions, see Stokolos [148] and references therein. For other properties of The McShane integral, consult Gordon [44] or Pfeffer [137]. For the divergence theorem, the interested reader is referred to Pfeffer [136, 137, 140, 141]. Further generalizations of the McShane integral can be found in [14, 17, 140].

Change-of-variables theorems for the Lebesgue integral can be found in many books and papers; see, for example, Hewitt and Stromberg [59], Pfeffer [137], Royden [143], Rudin [144], Stromberg [149], and De Guzman [48]. On the other hand, the following example shows that we do not have a comprehensive change-of-variable theorem for multiple Henstock-Kurzweil integrals.

Example 3.12.1. For each $n \in \mathbb{N}$ we let f_n be a continuous function on $[1-\frac{1}{2^{n-1}}, 1-\frac{1}{2^n}]^2)$ such that

(i) $f_n(x_1, x_2) = 0$ for all $(x_1, x_2) \in [1-\frac{1}{2^{n-1}}, 1-\frac{1}{2^n}]^2 \backslash (1-\frac{1}{2^{n-1}}, 1-\frac{1}{2^n})^2$;
(ii) $f_n(x_1, x_2) \geq 0$ for all $(x_1, x_2) \in [1-\frac{1}{2^{n-1}}, 1-\frac{1}{2^n}]^2$ with $x_1 \geq x_2$;
(iii) $f_n(x_1, x_2) = -f_n(x_2, x_1)$ for all $(x_1, x_2) \in [1-\frac{1}{2^{n-1}}, 1-\frac{1}{2^n}]^2$;
(iv) $\int_{1-\frac{1}{2^{n-1}}}^{1-\frac{1}{2^n}} \{\int_{1-\frac{1}{2^{n-1}}}^{s} f_n(s,t)\, dt\}\, ds = \frac{1}{n}$.

Define $f : [0,1]^2 \longrightarrow \mathbb{R}$ by setting

$$f(x_1, x_2) = \begin{cases} f_n(x_1, x_2) & \text{if } (x_1, x_2) \in [1-\frac{1}{2^{n-1}}, 1-\frac{1}{2^n}]^2, \\ 0 & \text{otherwise.} \end{cases}$$

Then $f \in HK([0,1]^2)\backslash L^1([0,1]^2)$. On the other hand, if T denotes a rotation by $\frac{\pi}{4}$, $f \circ T^{-1}$ fails to be Henstock-Kurzweil integrable on any interval containing $T^{-1}([0,1]^2)$.

Although we do not have a good change of variables theorem for the multiple Henstock-Kurzweil integral, we have the following result.

Theorem 3.12.2. *Let $g : [0,1] \longrightarrow \mathbb{R}$ and let*

$$f(x,y) = \begin{cases} g(\sqrt{x^2+y^2}) & \text{if } (x,y) \in \{(s,t) \in [0,1]^2 : s^2+t^2 \leq 1\}, \\ 0 & \text{otherwise.} \end{cases}$$

If $\lim_{\delta \to 0+} \int_\delta^1 rg(r)\, d\mu_1(r)$ exists, then $f \in HK([0,1]^2)$ and

$$(HK)\int_{[0,1]^2} f = \frac{\pi}{2}\left\{\lim_{\delta \to 0+} \int_\delta^1 rg(r)\, d\mu_1(r)\right\}.$$

Proof. See [68]. □

Chapter 4

Further properties of Henstock-Kurzweil integrable functions

4.1 A necessary condition for Henstock-Kurzweil integrability

It has been proved in Chapter 3 that if $f \in L^1[\boldsymbol{a}, \boldsymbol{b}]$, then $f\chi_X \in L^1[\boldsymbol{a}, \boldsymbol{b}]$ for every μ_m-measurable set $X \subseteq [\boldsymbol{a}, \boldsymbol{b}]$. On the other hand, simple examples reveal that the above-mentioned property need not hold for Henstock-Kurzweil integrable functions; see, for example, Example 2.5.7. Thus, it is natural to ask whether the following weaker assertion holds:

if $f \in HK[\boldsymbol{a}, \boldsymbol{b}]$, then f is Lebesgue integrable on *some* subinterval of $[\boldsymbol{a}, \boldsymbol{b}]$.

In 1991, Buczolich [22] gave an affirmative answer to the above question; see Theorem 4.1.3. We begin with the following definition.

Definition 4.1.1. Let $X \subseteq [\boldsymbol{a}, \boldsymbol{b}]$. A finite collection $\mathcal{C}$ of point-interval pairs is said to be *X-tagged* if $\boldsymbol{t} \in X$ for every $(\boldsymbol{t}, I) \in \mathcal{C}$.

We need the following result concerning closed sets.

Theorem 4.1.2. *Let $X \subseteq [\boldsymbol{a}, \boldsymbol{b}]$ be a non-empty closed set. If $(X_n)_{n=1}^{\infty}$ is an increasing sequence of closed sets with $X = \bigcup_{k=1}^{\infty} X_k$, then there exist $N \in \mathbb{N}$ and $[\boldsymbol{u}, \boldsymbol{v}] \in \mathcal{I}_m([\boldsymbol{a}, \boldsymbol{b}])$ such that $X \cap [\boldsymbol{u}, \boldsymbol{v}]$ is non-empty and*

$$X \cap [\boldsymbol{u}, \boldsymbol{v}] = X_N \cap [\boldsymbol{u}, \boldsymbol{v}].$$

Proof. This follows from the Baire's category theorem ([59, (6.54)] or [143, 31. Proposition]). □

The following theorem is the main result of this section.

Theorem 4.1.3. *If $f \in HK[\boldsymbol{a}, \boldsymbol{b}]$, then there exists $[\boldsymbol{u}, \boldsymbol{v}] \in \mathcal{I}_m([\boldsymbol{a}, \boldsymbol{b}])$ such that $f \in L^1[\boldsymbol{u}, \boldsymbol{v}]$.*

Proof. For $\varepsilon = 1$ we apply the Saks-Henstock Lemma to select a gauge δ on $[\boldsymbol{a}, \boldsymbol{b}]$ so that

$$\sum_{(\boldsymbol{t},I)\in P} \left| f(\boldsymbol{t})\mu_m(I) - (HK)\int_I f \right| < 1$$

for each δ-fine Perron subpartition P of $[\boldsymbol{a}, \boldsymbol{b}]$.

For each $n \in \mathbb{N}$, let

$$Y_n = \left\{ \boldsymbol{x} \in [\boldsymbol{a}, \boldsymbol{b}] : |f(\boldsymbol{x})| < n \text{ and } \delta(\boldsymbol{x}) > \frac{1}{n} \right\}$$

and $X_n = \overline{Y_n}$. Then $(X_n)_{n=1}^{\infty}$ is an increasing sequence of closed sets with $\bigcup_{n\in\mathbb{N}} X_n = [\boldsymbol{a}, \boldsymbol{b}]$. Since $[\boldsymbol{a}, \boldsymbol{b}]$ is also closed, it follows from Theorem 4.1.2 that there exist $N \in \mathbb{N}$ and $[\boldsymbol{u}, \boldsymbol{v}] \in \mathcal{I}_m([\boldsymbol{a}, \boldsymbol{b}])$ such that $[\boldsymbol{u}, \boldsymbol{v}] \subseteq X_N$. Without loss of generality, we may assume that $|||\boldsymbol{v} - \boldsymbol{u}||| < \frac{1}{N}$.

We claim that $f \in L^1[\boldsymbol{u}, \boldsymbol{v}]$. To prove this, we let $\{[\boldsymbol{u}_1, \boldsymbol{v}_1], \ldots, [\boldsymbol{u}_q, \boldsymbol{v}_q]\}$ be an arbitrary division of $[\boldsymbol{u}, \boldsymbol{v}]$. For each $i = 1, \ldots, q$ we use the inclusion $[\boldsymbol{u}_i, \boldsymbol{v}_i] \subseteq X_N$ to pick $\boldsymbol{t}_i \in Y_N \cap (\boldsymbol{u}_i, \boldsymbol{v}_i)$. Since $|||\boldsymbol{v} - \boldsymbol{u}||| < \frac{1}{N}$, we conclude that $\{(\boldsymbol{t}_1, [\boldsymbol{u}_1, \boldsymbol{v}_1]), \ldots, (\boldsymbol{t}_q, [\boldsymbol{u}_q, \boldsymbol{v}_q])\}$ is a Y_N-tagged $\frac{1}{N}$-fine, and hence δ-fine, Perron subpartition of $[\boldsymbol{u}, \boldsymbol{v}]$. Thus

$$\sum_{i=1}^{q} \left| f(\boldsymbol{t}_i)\mu_m([\boldsymbol{u}_i, \boldsymbol{v}_i]) - (HK)\int_{[\boldsymbol{u}_i,\boldsymbol{v}_i]} f \right| < 1,$$

and so

$$\sum_{i=1}^{q} \left| (HK)\int_{[\boldsymbol{u}_i,\boldsymbol{v}_i]} f \right| < 1 + \sum_{i=1}^{q} |f(\boldsymbol{t}_i)|\, \mu_m([\boldsymbol{u}_i, \boldsymbol{v}_i]) < 1 + N\mu_m([\boldsymbol{u}, \boldsymbol{v}]).$$

Since $\{[\boldsymbol{u}_1, \boldsymbol{v}_1], \ldots, [\boldsymbol{u}_q, \boldsymbol{v}_q]\}$ is an arbitrary division of $[\boldsymbol{u}, \boldsymbol{v}]$, the result follows from Theorem 3.1.3. □

4.2 A result of Kurzweil and Jarník

Let $f \in HK[\boldsymbol{a}, \boldsymbol{b}]$ be given. According to Theorem 4.1.3, f must be Lebesgue integrable on some subinterval of $[\boldsymbol{a}, \boldsymbol{b}]$. On the other hand, it is unclear whether the sequence $(f\chi_{X_n})_{n=1}^{\infty}$ is in $L^1[\boldsymbol{a}, \boldsymbol{b}]$, where $X_1, X_2, \ldots$ are given as in the proof of Theorem 4.1.3. In this section we use the *Henstock variational measure* to prove that this is indeed the case; see Theorem 4.2.5 for details.

Definition 4.2.1. Let $F : \mathcal{I}_m([\boldsymbol{a}, \boldsymbol{b}]) \longrightarrow \mathbb{R}$. For any set $X \subseteq [\boldsymbol{a}, \boldsymbol{b}]$, the *$\delta$-variation* of F on X is given by

$$V(F, X, \delta) := \sup\left\{ \sum_{(\boldsymbol{t}, I) \in P} |F(I)| \right\},$$

where the supremum is taken over all X-tagged δ-fine Perron subpartition P of $[\boldsymbol{a}, \boldsymbol{b}]$. We let

$$V_{\mathcal{HK}}F(X) := \inf\left\{ V(F, X, \delta) : \delta \text{ is a gauge on } X \right\}.$$

The Henstock variational measure of F is the extended real-valued function

$$V_{\mathcal{HK}}F : Y \mapsto V_{\mathcal{HK}}F(Y)$$

defined for all $Y \subseteq [\boldsymbol{a}, \boldsymbol{b}]$.

We begin with the following generalization of Lemma 3.1.2.

Lemma 4.2.2. *Let $f \in HK[\boldsymbol{a}, \boldsymbol{b}]$ and let F be the indefinite Henstock-Kurzweil integral of f. If $f\chi_X \in L^1[\boldsymbol{a}, \boldsymbol{b}]$ for some μ_m-measurable set $X \subseteq [\boldsymbol{a}, \boldsymbol{b}]$, then $V_{\mathcal{HK}}F(X)$ is finite and*

$$\int_{[\boldsymbol{a}, \boldsymbol{b}]} |f\chi_X| \; d\mu_m = V_{\mathcal{HK}}F(X). \tag{4.2.1}$$

Proof. Let $\varepsilon > 0$ be given. Since $f \in HK[\boldsymbol{a}, \boldsymbol{b}]$, it follows from the Saks-Henstock Lemma that there exists a gauge δ_1 on $[\boldsymbol{a}, \boldsymbol{b}]$ such that

$$\sum_{(\boldsymbol{t}_1, I_1) \in P_1} \left| f(\boldsymbol{t}_1)\mu_m(I_1) - (HK)\int_{I_1} f \right| < \frac{\varepsilon}{3} \tag{4.2.2}$$

for each δ_1-fine Perron subpartition P_1 of $[\boldsymbol{a}, \boldsymbol{b}]$. Since $f\chi_X \in L^1[\boldsymbol{a}, \boldsymbol{b}]$, there exists a gauge δ_2 on $[\boldsymbol{a}, \boldsymbol{b}]$ such that

$$\sum_{(\boldsymbol{t}_2, I_2) \in P_2} \left| |f(\boldsymbol{t}_2)| \, \mu_m(I_2) - \int_{I_2} |f\chi_X| \; d\mu_m \right| < \frac{\varepsilon}{3} \tag{4.2.3}$$

for each X-tagged δ_2-fine Perron subpartition P_2 of $[\boldsymbol{a}, \boldsymbol{b}]$.

We first prove that

$$V(F, X, \min\{\delta_1, \delta_2\}) < \int_{[\boldsymbol{a}, \boldsymbol{b}]} |f\chi_X| \; d\mu_m + \varepsilon. \tag{4.2.4}$$

To prove (4.2.4), we let Q be an arbitrary X-tagged $\min\{\delta_1, \delta_2\}$-fine Perron subpartition of $[\boldsymbol{a}, \boldsymbol{b}]$. Using the triangle inequality, (4.2.2) and (4.2.3), we get

$$\begin{aligned}\sum_{(\boldsymbol{t}_3, I_3)\in Q} |F(I_3)| &< \sum_{(\boldsymbol{t}_3, I_3)\in Q} |f(\boldsymbol{t}_3)|\, \mu_m(I_3) + \frac{\varepsilon}{3} \\ &< \sum_{(\boldsymbol{t}_3, I_3)\in Q} \int_{I_3} |f\chi_X| \; d\mu_m + \frac{2\varepsilon}{3} \\ &\leq \int_{[\boldsymbol{a},\boldsymbol{b}]} |f\chi_X| \; d\mu_m + \frac{2\varepsilon}{3}.\end{aligned}$$

Since Q is an arbitrary X-tagged $\min\{\delta_1, \delta_2\}$-fine Perron subpartition of $[\boldsymbol{a}, \boldsymbol{b}]$, (4.2.4) follows; in particular, $V_{\mathcal{HK}}F(X)$ is finite.

We next prove that there exists a gauge δ_3 on X such that

$$V(F, X, \delta_3) \leq V_{\mathcal{HK}}F(X) + \frac{\varepsilon}{3} \tag{4.2.5}$$

and

$$\int_{[\boldsymbol{a},\boldsymbol{b}]} |f\chi_X| \; d\mu_m \leq V(F, X, \min\{\delta_1, \delta_2, \delta_3\}) + \varepsilon. \tag{4.2.6}$$

Using the finiteness of $V_{\mathcal{HK}}F(X)$ (cf. (4.2.4)), there exists a gauge δ_3 on X such that (4.2.5) holds. If P is an X-tagged $\min\{\delta_1, \delta_2, \delta_3\}$-fine Perron subpartition of $[\boldsymbol{a}, \boldsymbol{b}]$, then (4.2.6) is true:

$$\begin{aligned}\int_{[\boldsymbol{a},\boldsymbol{b}]} |f\chi_X| \; d\mu_m &< \sum_{(\boldsymbol{t}, I)\in P} |f(\boldsymbol{t})|\, \mu_m(I) + \frac{\varepsilon}{3} \\ &< \sum_{(\boldsymbol{t}, I)\in P} |F(I)| + \frac{2\varepsilon}{3} \\ &\leq V(F, X, \min\{\delta_1, \delta_2, \delta_3\}) + \frac{2\varepsilon}{3}.\end{aligned}$$

It is now clear that (4.2.1) follows from (4.2.4), (4.2.5), (4.2.6) and the arbitrariness of ε. □

The following theorem is an important generalization of Theorem 3.1.3.

Theorem 4.2.3. *Let $f \in HK[\boldsymbol{a}, \boldsymbol{b}]$, let F be the indefinite Henstock-Kurzweil integral of f, and assume that the set $X \subseteq [\boldsymbol{a}, \boldsymbol{b}]$ is μ_m-measurable. Then $f\chi_X \in L^1[\boldsymbol{a}, \boldsymbol{b}]$ if and only if $V_{\mathcal{HK}}F(X)$ is finite. In this case,*

$$\int_{[\boldsymbol{a},\boldsymbol{b}]} |f\chi_X| \; d\mu_m = V_{\mathcal{HK}}F(X).$$

Proof. If $f\chi_X \in L^1[\boldsymbol{a},\boldsymbol{b}]$, then it follows from Lemma 4.2.2 that $V_{\mathcal{HK}}F(X)$ is finite and

$$\int_{[\boldsymbol{a},\boldsymbol{b}]} |f\chi_X| \; d\mu_m = V_{\mathcal{HK}}F(X).$$

Conversely, suppose that $V_{\mathcal{HK}}F(X)$ is finite. For each $n \in \mathbb{N}$, we let

$$X_n = \{\boldsymbol{x} \in X : |f(\boldsymbol{x})| \leq n\}.$$

Since $f \in HK[\boldsymbol{a},\boldsymbol{b}]$, it follows from Theorems 3.6.6, 3.5.11, 3.7.1 and Lemma 4.2.2 that $f\chi_{X_n} \in L^1[\boldsymbol{a},\boldsymbol{b}]$ and

$$\int_{[\boldsymbol{a},\boldsymbol{b}]} |f\chi_{X_n}| \; d\mu_m = V_{\mathcal{HK}}F(X_n).$$

Finally, since f is μ_m-measurable, $\sup_{n\in\mathbb{N}} V_{\mathcal{HK}}F(X_n) \leq V_{\mathcal{HK}}F(X)$ and $V_{\mathcal{HK}}F(X)$ is finite, an application of the Monotone Convergence Theorem completes the proof. □

In order to proceed further, we need the following notation.

Notation 4.2.4. Let P be any Perron subpartition of $[\boldsymbol{a},\boldsymbol{b}]$. For each $(\boldsymbol{t},[\boldsymbol{u},\boldsymbol{v}]) \in P$, we write

$$\Lambda(\boldsymbol{t},[\boldsymbol{u},\boldsymbol{v}]) = (\lambda_1,\ldots,\lambda_m),$$

where

$$\lambda_i = \begin{cases} -1 & \text{if } u_i = t_i \in (a_i,b_i), \\ 0 & \text{if } t_i \in (u_i,v_i) \text{ or } u_i = t_i = a_i \text{ or } v_i = t_i = b_i, \\ 1 & \text{if } v_i = t_i \in (a_i,b_i) \end{cases}$$

for $i = 1,\ldots,m$.

We are now ready to state and prove the following important result of Kurzweil and Jarník [77].

Theorem 4.2.5. *If $f \in HK[\boldsymbol{a},\boldsymbol{b}]$, then there exists a sequence $(X_n)_{n=1}^{\infty}$ of closed sets such that $\bigcup_{k=1}^{\infty} X_k = [\boldsymbol{a},\boldsymbol{b}]$ and the sequence $(f\chi_{X_n})_{n=1}^{\infty}$ is in $L^1[\boldsymbol{a},\boldsymbol{b}]$.*

Proof. For $\varepsilon = 1$ we apply the Saks-Henstock Lemma to select a gauge δ on $[\boldsymbol{a},\boldsymbol{b}]$ so that

$$\sum_{(\boldsymbol{x},J)\in Q} \left| f(\boldsymbol{x})\mu_m(J) - (HK)\int_J f \right| < 1$$

for each δ-fine Perron subpartition Q of $[\boldsymbol{a}, \boldsymbol{b}]$.

For each $n \in \mathbb{N}$, we let

$$Y_n = \left\{\boldsymbol{x} \in [\boldsymbol{a}, \boldsymbol{b}] : |f(\boldsymbol{x})| < n \text{ and } \delta(\boldsymbol{x}) > \frac{1}{n}\right\},$$

and $X_n := \overline{Y_n}$. In view of Theorem 4.2.3, it suffices to prove that

$$V_{\mathcal{HK}}F(X_n) \leq 3^m(2 + n\mu_m([\boldsymbol{a}, \boldsymbol{b}])).$$

Let P be an arbitrary X_n-tagged $\frac{1}{n}$-fine Perron subpartition of $[\boldsymbol{a}, \boldsymbol{b}]$. Since

$$\sum_{(\boldsymbol{t},[\boldsymbol{u},\boldsymbol{v}])\in P} |F([\boldsymbol{u}, \boldsymbol{v}])| = \sum_{\boldsymbol{\lambda}\in\{-1,0,1\}^m} \sum_{\substack{(\boldsymbol{t},[\boldsymbol{u},\boldsymbol{v}])\in P \\ \Lambda(\boldsymbol{t},[\boldsymbol{u},\boldsymbol{v}])=\boldsymbol{\lambda}}} |F([\boldsymbol{u}, \boldsymbol{v}])|,$$

it suffices to prove that

$$\max_{\boldsymbol{\lambda}\in\{-1,0,1\}^m} \sum_{\substack{(\boldsymbol{t},[\boldsymbol{u},\boldsymbol{v}])\in P \\ \Lambda(\boldsymbol{t},[\boldsymbol{u},\boldsymbol{v}])=\boldsymbol{\lambda}}} |F([\boldsymbol{u}, \boldsymbol{v}])| \leq 2 + n\mu_m([\boldsymbol{a}, \boldsymbol{b}]). \tag{4.2.7}$$

We first prove that

$$\sum_{\substack{(\boldsymbol{t},[\boldsymbol{u},\boldsymbol{v}])\in P \\ \Lambda(\boldsymbol{t},[\boldsymbol{u},\boldsymbol{v}])=\boldsymbol{0}}} |F([\boldsymbol{u}, \boldsymbol{v}])| \leq 1 + n\mu_m([\boldsymbol{a}, \boldsymbol{b}]). \tag{4.2.8}$$

To prove (4.2.8) we may assume that

$$P_{\boldsymbol{0}} := \{(\boldsymbol{t}, [\boldsymbol{u}, \boldsymbol{v}]) \in P : \Lambda(\boldsymbol{t}, [\boldsymbol{u}, \boldsymbol{v}]) = \boldsymbol{0}\}$$

is non-empty. For each $(\boldsymbol{t}, [\boldsymbol{u}, \boldsymbol{v}]) \in P_{\boldsymbol{0}}$ we select and fix a $\boldsymbol{x}_{[\boldsymbol{u},\boldsymbol{v}]} \in Y_n \cap [\boldsymbol{u}, \boldsymbol{v}]$ so that $\{(\boldsymbol{x}_{[\boldsymbol{u},\boldsymbol{v}]}, [\boldsymbol{u}, \boldsymbol{v}]) : (\boldsymbol{t}, [\boldsymbol{u}, \boldsymbol{v}]) \in P_{\boldsymbol{0}}\}$ is a Y_n-tagged $\frac{1}{n}$-fine, and hence δ-fine, Perron subpartition of $[\boldsymbol{a}, \boldsymbol{b}]$. Consequently, (4.2.8) holds:

$$\begin{aligned}
&\sum_{\substack{(\boldsymbol{t},[\boldsymbol{u},\boldsymbol{v}])\in P \\ \Lambda(\boldsymbol{t},[\boldsymbol{u},\boldsymbol{v}])=\boldsymbol{0}}} |F([\boldsymbol{u}, \boldsymbol{v}])| \\
&\leq \sum_{\substack{(\boldsymbol{t},[\boldsymbol{u},\boldsymbol{v}])\in P \\ \Lambda(\boldsymbol{t},[\boldsymbol{u},\boldsymbol{v}])=\boldsymbol{0}}} \left\{ \left|f(\boldsymbol{x}_{[\boldsymbol{u},\boldsymbol{v}]})\mu_m([\boldsymbol{u}, \boldsymbol{v}]) - F([\boldsymbol{u}, \boldsymbol{v}])\right| + \left|f(\boldsymbol{x}_{[\boldsymbol{u},\boldsymbol{v}]})\mu_m([\boldsymbol{u}, \boldsymbol{v}])\right| \right\} \\
&< 1 + n\mu_m([\boldsymbol{a}, \boldsymbol{b}]).
\end{aligned}$$

To complete the proof of (4.2.7), we let $\boldsymbol{\lambda} \in \{-1, 0, 1\}^m \setminus \{\boldsymbol{0}\}$ and apply Theorem 2.4.9 to select a sufficiently small $\eta > 0$ so that

$$\sum_{\substack{(\boldsymbol{t},[\boldsymbol{u},\boldsymbol{v}])\in P \\ \Lambda(\boldsymbol{t},[\boldsymbol{u},\boldsymbol{v}])=\boldsymbol{\lambda}}} |F([\boldsymbol{u}, \boldsymbol{v}])| \leq \sum_{\substack{(\boldsymbol{t},[\boldsymbol{u},\boldsymbol{v}])\in P \\ \Lambda(\boldsymbol{t},[\boldsymbol{u},\boldsymbol{v}])=\boldsymbol{\lambda}}} |F([\boldsymbol{u} + \eta\boldsymbol{\lambda}, \boldsymbol{v} + \eta\boldsymbol{\lambda}])| + 1,$$

$$t \in [\boldsymbol{u}+\eta\boldsymbol{\lambda}, \boldsymbol{v}+\eta\boldsymbol{\lambda}] \subset [\boldsymbol{a}, \boldsymbol{b}] \cap B(t, \frac{1}{n}) \text{ and } \Lambda(t, [\boldsymbol{u}+\eta\boldsymbol{\lambda}, \boldsymbol{v}+\eta\boldsymbol{\lambda}]) = \mathbf{0}$$

for every $(t, [\boldsymbol{u}, \boldsymbol{v}]) \in P$ satisfying $\Lambda(t, [\boldsymbol{u}, \boldsymbol{v}]) = \boldsymbol{\lambda}$. In this case, we can follow the proof of (4.2.8) to obtain

$$\sum_{\substack{(t,[\boldsymbol{u},\boldsymbol{v}])\in P \\ \Lambda(t,[\boldsymbol{u},\boldsymbol{v}])=\boldsymbol{\lambda}}} |F([\boldsymbol{u}+\eta\boldsymbol{\lambda}, \boldsymbol{v}+\eta\boldsymbol{\lambda}])| \leq 1 + n\mu_m([\boldsymbol{a}, \boldsymbol{b}]).$$

Combining the above inequalities yields (4.2.7) to be proved. □

Our next aim is to prove some refinements of Theorem 4.2.5. For each $f \in HK[\boldsymbol{a}, \boldsymbol{b}]$, Theorem 2.4.8 tells us that the indefinite Henstock-Kurzweil integral of f is continuous in the sense that

$$\lim_{\substack{\mu_m(I)\to 0 \\ I\in\mathcal{I}_m([\boldsymbol{a},\boldsymbol{b}])}} (HK)\int_I f = 0;$$

in particular, $0 \leq \|f\|_{HK[\boldsymbol{a},\boldsymbol{b}]} < \infty$, where

$$\|f\|_{HK[\boldsymbol{a},\boldsymbol{b}]} = \sup\left\{\left|(HK)\int_I f\right| : I \in \mathcal{I}_m([\boldsymbol{a}, \boldsymbol{b}])\right\}.$$

The following result is a reformulation of Theorem 2.4.7.

Theorem 4.2.6. *Let $f \in HK[\boldsymbol{a}, \boldsymbol{b}]$. Then for each $\varepsilon > 0$ there exists a gauge δ on $[\boldsymbol{a}, \boldsymbol{b}]$ such that*

$$\sum_{(t,I)\in P} \|f(t) - f\|_{HK(I)} < \varepsilon$$

for each δ-fine Perron subpartition P of $[\boldsymbol{a}, \boldsymbol{b}]$.

Lemma 4.2.7. *Let $f \in HK[\boldsymbol{a}, \boldsymbol{b}]$ and assume that $f\chi_X \in HK[\boldsymbol{a}, \boldsymbol{b}]$ for some non-empty set $X \subseteq [\boldsymbol{a}, \boldsymbol{b}]$. Then for each $\varepsilon > 0$ there exists a gauge δ on X such that*

$$\sum_{(t,I)\in P} \|f\chi_X - f\|_{HK(I)} < \varepsilon$$

for each X-tagged δ-fine Perron subpartition P of $[\boldsymbol{a}, \boldsymbol{b}]$.

Proof. Let $\varepsilon > 0$ be given. Since $f \in HK[\boldsymbol{a}, \boldsymbol{b}]$ we use Theorem 4.2.6 to pick a gauge δ_1 on $[\boldsymbol{a}, \boldsymbol{b}]$ so that

$$\sum_{(t_1,I_1)\in P_1} \|f(t_1) - f\|_{HK(I_1)} < \frac{\varepsilon}{2}$$

for each δ_1-fine Perron subpartition P_1 of $[\boldsymbol{a}, \boldsymbol{b}]$. Similarly, there exists a gauge δ_2 on X such that

$$\sum_{(\boldsymbol{t}_2, I_2) \in P_2} \|f(\boldsymbol{t}_2) - f\chi_X\|_{HK(I_2)} < \frac{\varepsilon}{2}$$

for each X-tagged δ_2-fine Perron subpartition P_2 of $[\boldsymbol{a}, \boldsymbol{b}]$.

Define a gauge δ on X by setting $\delta(\boldsymbol{x}) = \min\{\delta_1(\boldsymbol{x}), \delta_2(\boldsymbol{x})\}$. If P is an X-tagged δ-fine Perron subpartition of $[\boldsymbol{a}, \boldsymbol{b}]$, then

$$\begin{aligned}
&\sum_{(\boldsymbol{t}, I) \in P} \|f\chi_X - f\|_{HK(I)} \\
&\leq \sum_{(\boldsymbol{t}, I) \in P} \|f(\boldsymbol{t}) - f\|_{HK(I)} + \sum_{(\boldsymbol{t}, I) \in P} \|f(\boldsymbol{t}) - f\chi_X\|_{HK(I)} \\
&< \varepsilon.
\end{aligned}$$

□

In order to state and prove a modification of Lemma 4.2.7, we need the following terminology: a real-valued function f is said to be *upper semicontinuous* on its domain $X \subseteq \mathbb{R}^m$ if it is upper semicontinuous at each $\boldsymbol{x} \in X$; that is, for each $\alpha > f(\boldsymbol{x})$ there exists $\eta > 0$ such that $\alpha > f(\boldsymbol{y})$ for each $\boldsymbol{y} \in B(\boldsymbol{x}, \eta) \cap X$.

Using Lemma 4.2.7 and following the proof of Theorem 4.2.5, we get the following result.

Theorem 4.2.8. *Let $f \in HK[\boldsymbol{a}, \boldsymbol{b}]$ and assume that $f\chi_X \in HK[\boldsymbol{a}, \boldsymbol{b}]$ for some non-empty closed set $X \subseteq [\boldsymbol{a}, \boldsymbol{b}]$. Then for each $\varepsilon > 0$ there exists an upper semicontinuous gauge δ on X such that*

$$\sum_{(\boldsymbol{t}, I) \in P} \|f\chi_X - f\|_{HK(I)} < \varepsilon$$

for each X-tagged δ-fine Perron subpartition P of $[\boldsymbol{a}, \boldsymbol{b}]$.

Proof. According to Lemma 4.2.7, for each $\varepsilon > 0$ there exists a gauge Δ on X such that

$$\sum_{(\boldsymbol{x}, J) \in Q} \|f\chi_X - f\|_{HK(J)} < \frac{\varepsilon}{3^{m+1}}$$

for each X-tagged Δ-fine Perron subpartition Q of $[\boldsymbol{a}, \boldsymbol{b}]$.

For $n = 1, 2, \ldots$, let

$$Y_n = \left\{\boldsymbol{x} \in X : \Delta(\boldsymbol{x}) \geq \frac{1}{n}\right\}$$

and let $X_n = \overline{Y_n}$. Since X is a closed set, it is clear that $(X_n)_{n=1}^{\infty}$ is an increasing sequence of closed sets and $X = \bigcup_{k=1}^{\infty} X_k$.

Define a gauge δ on X by setting

$$\delta(\boldsymbol{x}) = \begin{cases} 1 & \text{if } \boldsymbol{x} \in X_1, \\ \\ \min\{\frac{1}{1+k}, \operatorname{dist}(\boldsymbol{x}, X_k)\} & \text{if } \boldsymbol{x} \in X_{k+1} \backslash X_k \text{ for some } k \in \mathbb{N}. \end{cases}$$

Then δ is upper semicontinuous on X. To this end, we consider any X-tagged δ-fine Perron subpartition P of $[\boldsymbol{a}, \boldsymbol{b}]$. Since

$$\sum_{(\boldsymbol{t},[\boldsymbol{u},\boldsymbol{v}])\in P} \|f\chi_X - f\|_{HK[\boldsymbol{u},\boldsymbol{v}]} = \sum_{\boldsymbol{\lambda}\in\{-1,0,1\}^m} \sum_{\substack{(\boldsymbol{t},[\boldsymbol{u},\boldsymbol{v}])\in P \\ \Lambda(\boldsymbol{t},[\boldsymbol{u},\boldsymbol{v}])=\boldsymbol{\lambda}}} \|f\chi_X - f\|_{HK[\boldsymbol{u},\boldsymbol{v}]},$$

it suffices to prove that

$$\max_{\boldsymbol{\lambda}\in\{-1,0,1\}^m} \sum_{\substack{(\boldsymbol{t},[\boldsymbol{u},\boldsymbol{v}])\in P \\ \Lambda(\boldsymbol{t},[\boldsymbol{u},\boldsymbol{v}])=\boldsymbol{\lambda}}} \|f\chi_X - f\|_{HK[\boldsymbol{u},\boldsymbol{v}]} < \frac{\varepsilon}{3^m}. \tag{4.2.9}$$

We first prove that

$$\sum_{(\boldsymbol{t},[\boldsymbol{u},\boldsymbol{v}])\in P_0} \|f\chi_X - f\|_{HK[\boldsymbol{u},\boldsymbol{v}]} < \frac{\varepsilon}{3^{m+1}}, \tag{4.2.10}$$

where

$$P_0 := \{(\boldsymbol{t}, [\boldsymbol{u}, \boldsymbol{v}]) \in P : \Lambda(\boldsymbol{t}, [\boldsymbol{u}, \boldsymbol{v}]) = \boldsymbol{0}\}.$$

Clearly, we may assume that P_0 is non-empty. For each $(\boldsymbol{t}, [\boldsymbol{u}, \boldsymbol{v}]) \in P_0$, we select and fix $\boldsymbol{x}_{[\boldsymbol{u},\boldsymbol{v}]} \in Y_n \cap [\boldsymbol{u}, \boldsymbol{v}]$ so that $\{(\boldsymbol{x}_{[\boldsymbol{u},\boldsymbol{v}]}, [\boldsymbol{u}, \boldsymbol{v}]) : (\boldsymbol{t}, [\boldsymbol{u}, \boldsymbol{v}]) \in P_0\}$ is a Y_n-tagged $\frac{1}{n}$-fine, and hence Δ-fine, Perron subpartition of $[\boldsymbol{a}, \boldsymbol{b}]$. Consequently, (4.2.10) holds.

To complete the proof of (4.2.9), we let $\boldsymbol{\lambda} \in \{-1, 0, 1\}^m \backslash \{\boldsymbol{0}\}$ and use Theorem 2.4.9 to select a sufficiently small $\eta > 0$ so that

$$\begin{aligned} &\sum_{\substack{(\boldsymbol{t},[\boldsymbol{u},\boldsymbol{v}])\in P \\ \Lambda(\boldsymbol{t},[\boldsymbol{u},\boldsymbol{v}])=\boldsymbol{\lambda}}} \|f\chi_X - f\|_{HK[\boldsymbol{u},\boldsymbol{v}]} \\ &\leq \sum_{\substack{(\boldsymbol{t},[\boldsymbol{u},\boldsymbol{v}])\in P \\ \Lambda(\boldsymbol{t},[\boldsymbol{u},\boldsymbol{v}])=\boldsymbol{\lambda}}} \|f\chi_X - f\|_{HK[\boldsymbol{u}+\eta\boldsymbol{\lambda},\boldsymbol{v}+\eta\boldsymbol{\lambda}]} + \frac{\varepsilon}{3^{m+1}} \end{aligned}$$

and

$$\boldsymbol{t} \in [\boldsymbol{u} + \eta\boldsymbol{\lambda}, \boldsymbol{v} + \eta\boldsymbol{\lambda}] \subset [\boldsymbol{a}, \boldsymbol{b}] \cap B\left(\boldsymbol{t}, \frac{1}{n}\right)$$

for every $(\boldsymbol{t}, [\boldsymbol{u}, \boldsymbol{v}]) \in P$ satisfying $\Lambda(\boldsymbol{t}, [\boldsymbol{u}, \boldsymbol{v}]) = \boldsymbol{\lambda}$. Therefore we can argue as in the proof of (4.2.10) to conclude that

$$\sum_{\substack{(\boldsymbol{t},[\boldsymbol{u},\boldsymbol{v}])\in P \\ \Lambda(\boldsymbol{t},[\boldsymbol{u},\boldsymbol{v}])=\boldsymbol{\lambda}}} \|f\chi_X - f\|_{HK[\boldsymbol{u}+\eta\boldsymbol{\lambda},\boldsymbol{v}+\eta\boldsymbol{\lambda}]} < \frac{\varepsilon}{3^{m+1}}.$$

Combining the above inequalities yields (4.2.9) to be proved. □

The following theorem is another refinement of Theorem 4.2.5.

Theorem 4.2.9. *If $f \in HK[\boldsymbol{a}, \boldsymbol{b}]$, then there exists a sequence $(Y_n)_{n=1}^{\infty}$ of closed sets such that $\bigcup_{k=1}^{\infty} Y_k = [\boldsymbol{a}, \boldsymbol{b}]$, the sequence $(f\chi_{Y_n})_{n=1}^{\infty}$ is in $L^1[\boldsymbol{a}, \boldsymbol{b}]$, and*

$$\lim_{n\to\infty} \|f\chi_{Y_n} - f\|_{HK[\boldsymbol{a},\boldsymbol{b}]} = 0.$$

Proof. Since $f \in HK[\boldsymbol{a}, \boldsymbol{b}]$, it follows from Theorem 4.2.5 that there exists an increasing sequence $(X_n)_{n=1}^{\infty}$ of closed sets such that $\bigcup_{k=1}^{\infty} X_k = [\boldsymbol{a}, \boldsymbol{b}]$ and the sequence $(f\chi_{X_n})_{n=1}^{\infty}$ is in $L^1[\boldsymbol{a}, \boldsymbol{b}]$.

Let $n \in \mathbb{N}$ be given. For each $k \in \mathbb{N}$ we use Lemma 4.2.7 to choose a gauge $\delta_{k,n}$ on X_k so that

$$\sum_{(\boldsymbol{x}_1,J_1)\in Q_1} \|f\chi_{X_k} - f\|_{HK(J_1)} < \frac{1}{n(2^k)}$$

for each X_k-tagged $\delta_{k,n}$-fine Perron subpartition Q_1 of $[\boldsymbol{a}, \boldsymbol{b}]$.

We claim that there exist $k(n) \in \mathbb{N}$ and a closed set Y_n such that $X_n \subseteq Y_n \subseteq X_{k(n)}$, $f\chi_{Y_n} \in L^1[\boldsymbol{a}, \boldsymbol{b}]$ and

$$\|f\chi_{Y_n} - f\|_{HK[\boldsymbol{a},\boldsymbol{b}]} \leq \frac{1}{n}. \tag{4.2.11}$$

Define a gauge Δ_n on $[\boldsymbol{a}, \boldsymbol{b}]$ by setting

$$\Delta_n(\boldsymbol{x}) = \begin{cases} \delta_{n,n}(\boldsymbol{x}) & \text{if } \boldsymbol{x} \in X_n, \\ \\ \operatorname{dist}(\boldsymbol{x}, X_{k-1}) & \text{if } \boldsymbol{x} \in X_k \backslash X_{k-1} \text{ for some integer } k > n, \end{cases}$$

let P_n be a fixed Δ_n-fine Perron partition of $[\boldsymbol{a}, \boldsymbol{b}]$, and let

$$Y_n = X_n \cup \bigcup_{k=n+1}^{\infty} \bigcup_{\substack{(\boldsymbol{t},I)\in P_n \\ \boldsymbol{t}\in X_k\backslash X_{k-1}}} (I \cap X_k).$$

Since $(X_n)_{n=1}^{\infty}$ is a sequence of closed sets and $\text{card}(P_n)$ is finite, Y_n is closed and $X_n \subseteq Y_n \subseteq X_{k(n)}$ for some $k(n) \in \mathbb{N}$. It follows that $f\chi_{Y_n} \in L^1[\boldsymbol{a}, \boldsymbol{b}]$ and

$$\begin{aligned}
&\|f\chi_{Y_n} - f\|_{HK[a,b]} \\
&\le \sum_{\substack{(t,I)\in P_n \\ t\in X_n}} \|f\chi_{Y_n} - f\|_{HK(I)} + \sum_{k=n+1}^{\infty} \sum_{\substack{(t,I)\in P_n \\ t\in X_k\setminus X_{k-1}}} \|f\chi_{Y_n} - f\|_{HK(I)} \\
&= \sum_{\substack{(t,I)\in P_n \\ t\in X_n}} \|f\chi_{X_n} - f\|_{HK(I)} + \sum_{k=n+1}^{\infty} \sum_{\substack{(t,I)\in P_n \\ t\in X_k\setminus X_{k-1}}} \|f\chi_{X_k} - f\|_{HK(I)} \\
&\le \sum_{k=n}^{\infty} \frac{1}{n(2^k)} \\
&\le \frac{1}{n}.
\end{aligned}$$

Finally, since $X_n \subseteq Y_n \subseteq X_{k(n)}$ for every $n \in \mathbb{N}$, we can obtain a subsequence of $(Y_n)_{n=1}^{\infty}$, still denoted by $(Y_n)_{n=1}^{\infty}$, so that $Y_n \subseteq Y_{n+1}$ for all $n \in \mathbb{N}$. This completes the proof of the theorem. □

Remark 4.2.10. The converse of Theorem 4.2.9 is not true; see Example 4.5.5.

4.3 Some necessary and sufficient conditions for Henstock-Kurzweil integrability

It has been proved in Section 4.2 that if $f \in HK[\boldsymbol{a}, \boldsymbol{b}]$, then there exists a sequence $(X_n)_{n=1}^{\infty}$ of closed sets such that $\bigcup_{k=1}^{\infty} X_k = [\boldsymbol{a}, \boldsymbol{b}]$ and the sequence $(f\chi_{X_n})_{n=1}^{\infty}$ is in $L^1[\boldsymbol{a}, \boldsymbol{b}]$. However, the converse is not true; see Remark 4.2.10. In this section we obtain some necessary and sufficient conditions for a function f to be Henstock-Kurzweil integrable on $[\boldsymbol{a}, \boldsymbol{b}]$.

Theorem 4.3.1. *Let $f : [\boldsymbol{a}, \boldsymbol{b}] \longrightarrow \mathbb{R}$ and suppose that the interval function $F : \mathcal{I}_m([\boldsymbol{a}, \boldsymbol{b}]) \longrightarrow \mathbb{R}$ is additive. Then F is the indefinite Henstock-Kurzweil integral of f if and only if the following conditions are satisfied:*

(i) *there exists an increasing sequence $(X_n)_{n=1}^{\infty}$ of closed sets such that $\bigcup_{k=1}^{\infty} X_k = [\boldsymbol{a}, \boldsymbol{b}]$ and the sequence $(f\chi_{X_n})_{n=1}^{\infty}$ is in $L^1[\boldsymbol{a}, \boldsymbol{b}]$;*

(ii) *for each $n \in \mathbb{N}$ and $\varepsilon > 0$ there exists an upper semicontinuous gauge δ_n on X_n such that*

$$\sum_{(\boldsymbol{t},[\boldsymbol{u},\boldsymbol{v}])\in P} \left| \int_{[\boldsymbol{u},\boldsymbol{v}]} f\chi_{X_n}\, d\mu_m - F([\boldsymbol{u},\boldsymbol{v}]) \right| < \varepsilon$$

for each X_n-tagged δ_n-fine Perron subpartition P of $[\boldsymbol{a},\boldsymbol{b}]$.

Proof. Suppose that $f \in HK[\boldsymbol{a},\boldsymbol{b}]$ and F is the indefinite HK-integral of f. Then (i) and (ii) follow from Theorems 4.2.5 and 4.2.8 respectively.

Conversely, suppose that (i) and (ii) are satisfied. According to (ii), given $n \in \mathbb{N}$ and $\varepsilon > 0$ there exists a gauge Δ_n on X_n such that

$$V(F_n - F, X_n, \Delta_n) < \frac{\varepsilon}{2^{n+2}}, \tag{4.3.1}$$

where F_n denotes the indefinite Lebesgue integral of $f\chi_{X_n}$. In view of (i) and the Saks-Henstock Lemma, we may further assume that

$$\sum_{(\boldsymbol{t}_1,I_1)\in P_1} |f\chi_{X_n}(\boldsymbol{t}_1)\mu_m(I_1) - F_n(I_1)| < \frac{\varepsilon}{2^{n+2}} \tag{4.3.2}$$

for each X_n-tagged Δ_n-fine Perron subpartition P_1 of $[\boldsymbol{a},\boldsymbol{b}]$.

Define a gauge δ on $[\boldsymbol{a},\boldsymbol{b}]$ by setting

$$\delta(\boldsymbol{x}) = \begin{cases} \min\{\Delta_1(\boldsymbol{x})\} & \text{if } \boldsymbol{x} \in X_1, \\ \min\{\Delta_n(\boldsymbol{x}), \operatorname{dist}(\boldsymbol{x}, X_{n-1})\} & \text{if } \boldsymbol{x} \in X_n \setminus X_{n-1} \text{ for some } n \in \mathbb{N}\setminus\{1\}, \end{cases}$$

and consider an arbitrary δ-fine Perron partition P of $[\boldsymbol{a},\boldsymbol{b}]$. According to our choice of δ, (4.3.2) and (4.3.1), we have

$$\begin{aligned}
&\left| \sum_{(\boldsymbol{t},I)\in P} f(\boldsymbol{t})\mu_m(I) - F([\boldsymbol{a},\boldsymbol{b}]) \right| \\
&= \left| \sum_{(\boldsymbol{t},I)\in P} \left\{ f(\boldsymbol{t})\mu_m(I) - F(I) \right\} \right| \\
&\le \sum_{n=1}^{\infty} \sum_{\substack{(\boldsymbol{t},I)\in P \\ \boldsymbol{t}\in X_n\setminus X_{n-1}}} |f(\boldsymbol{t})\mu_m(I) - F(I)| \qquad \text{(where } X_0 := \emptyset) \\
&\le \sum_{n=1}^{\infty} \sum_{\substack{(\boldsymbol{t},I)\in P \\ \boldsymbol{t}\in X_n\setminus X_{n-1}}} |f(\boldsymbol{t})\mu_m(I) - F_n(I)| + \sum_{n=1}^{\infty} \sum_{\substack{(\boldsymbol{t},I)\in P \\ \boldsymbol{t}\in X_n\setminus X_{n-1}}} |F_n(I) - F(I)| \\
&\le \sum_{n=1}^{\infty} \frac{\varepsilon}{2^{n+1}} \\
&< \varepsilon.
\end{aligned}$$

Thus $f \in HK[a,b]$. A similar reasoning shows tha F is the indefinite Henstock-Kurzweil integral of f. □

Remark 4.3.2. Theorem 4.3.1 can be refined; Theorem 5.5.9 tells us that an additive interval function $F : \mathcal{I}_m([a,b]) \longrightarrow \mathbb{R}$ is an indefinite Henstock-Kurzweil integral if and only if $V_{\mathcal{HK}}F \ll \mu_m$; that is, the following condition is satisfied:

$$V_{\mathcal{HK}}F(X) = 0 \text{ whenever } X \subset [a,b] \text{ and } \mu_m(X) = 0.$$

4.4 Harnack extension for one-dimensional Henstock-Kurzweil integrals

The aim of this section is to sharpen the one-dimensional version of Theorem 4.2.9. We begin with the following consequence of Theorem 4.3.1.

Observation 4.4.1. If $f \in HK[a,b]$ and F is the indefinite Henstock-Kurzweil integral of f, then the following properties hold:

(i) there exists an increasing sequence $(X_n)_{n=1}^{\infty}$ of closed sets such that $\bigcup_{k=1}^{\infty} X_k = [a,b]$ and the sequence $(f\chi_{X_n})_{n=1}^{\infty}$ is in $L^1[a,b]$;
(ii) for each $n \in \mathbb{N}$ there exists a constant $\eta_n > 0$ such that

$$V(F, X_n, \eta_n) < \infty.$$

In order to proceed further, we need the following result.

Theorem 4.4.2 ([59, (6.59) Theorem]). *If $O \subset [a,b]$ is open and non-empty, then there exists a sequence $\big((u_n, v_n)\big)_{n=1}^{\infty}$ of pairwise non-overlapping intervals such that $O = \bigcup_{k=1}^{\infty}(u_k, v_k)$.*

The following lemma is an immediate consequence of Observation 4.4.1 and Theorem 4.4.2.

Lemma 4.4.3. *If $f \in HK[a,b]$, then there exists a sequence $\big((a_k, b_k)\big)_{k=1}^{\infty}$ of pairwise disjoint open subintervals of $[a,b]$ such that $f\chi_{[a,b]\setminus\bigcup_{k=1}^{\infty}(a_k,b_k)} \in L^1[a,b]$, and the series $\sum_{k=1}^{\infty} \|f\|_{HK[a_k,b_k]}$ converges.*

The following theorem shows that the converse of Lemma 4.4.3 holds.

Theorem 4.4.4 (Harnack extension). *Let $X \subset [a,b]$ be a closed set and let $\{[a_k,b_k] : k \in \mathbb{N}\}$ be the collection of pairwise disjoint intervals such that $(a,b)\backslash X = \bigcup_{k=1}^{\infty}(a_k,b_k)$. If $f\chi_X \in HK[a,b]$, $f \in HK[a_k,b_k]$ $(k = 1,2,\dots)$ and the series $\sum_{k=1}^{\infty} \|f\|_{HK[a_k,b_k]}$ converges, then $f \in HK[a,b]$ and*

$$(HK)\int_a^b f = (HK)\int_a^b f\chi_X + \sum_{k=1}^{\infty}(HK)\int_{a_k}^{b_k} f.$$

Proof. Without loss of generality, we may assume that $\{a,b\} \subseteq X$ and $f(x) = 0$ for all $x \in X$.

Let $\varepsilon > 0$. For each $k \in \mathbb{N}$ we choose a gauge δ_k on $[a_k,b_k]$ so that

$$\sum_{(t_k,[u_k,v_k])\in P_k} \left| f(t_k)(v_k-u_k) - (HK)\int_{u_k}^{v_k} f \right| < \frac{\varepsilon}{2^{k+3}} \tag{4.4.1}$$

for each δ_k-fine Perron subpartition P_k of $[a_k,b_k]$. We may further assume that $(\xi - \delta_k(\xi), \xi + \delta_k(\xi)) \subset (a_k,b_k)$ whenever $\xi \in (a_k,b_k)$. Since the series $\sum_{k=1}^{\infty} \|f\|_{HK[a_k,b_k]}$ converges, there exists $N \in \mathbb{N}$ such that

$$\sum_{k=N+1}^{\infty} \|f\|_{HK[a_k,b_k]} < \frac{\varepsilon}{8}. \tag{4.4.2}$$

Define a gauge δ on $[a,b]$ by setting

$$\delta(x) = \begin{cases} \min\{\delta_1(x), \operatorname{dist}(x, \bigcup_{k=1}^{N}[a_k,b_k]) & \text{if } x \in X\backslash \bigcup_{k=1}^{\infty}[a_k,b_k], \\ \min\{\delta_k(x), \frac{b_k-a_k}{2}\} & \text{if } x \in \{a_k,b_k\} \text{ for some } k \in \{1,\dots,N\}, \\ \min\{\delta_k(x), \frac{b_k-a_k}{2}, \operatorname{dist}(x, \bigcup_{k=1}^{N}[a_k,b_k])\} & \\ \qquad \text{if } x \in \{a_k,b_k\}\backslash \bigcup_{k=1}^{N}[a_k,b_k] & \text{for some integer } k > N, \\ \delta_k(x) & \text{if } x \in (a_k,b_k) \text{ for some } k \in \mathbb{N}, \end{cases}$$

and let P be any δ-fine Perron partition of $[a,b]$. For each $k \in \{1,\dots,N\}$ our choice of δ implies that $\{(t,[u,v]) \in P : t \in [a_k,b_k]\}$ is a δ_k-fine Perron partition of $[a_k, b_k]$. Then (4.4.1) and (4.4.2) yield

$$\begin{aligned}
&\left| \sum_{(t,[u,v])\in P} f(t)(v-u) - \sum_{k=1}^{\infty} (HK)\int_{a_k}^{b_k} f \right| \\
&\leq \left| \sum_{\substack{(t,[u,v])\in P \\ t\in \bigcup_{k=1}^{N}[a_k,b_k]}} f(t)(v-u) - \sum_{k=1}^{N} (HK)\int_{a_k}^{b_k} f \right| \\
&\quad + \left| \sum_{\substack{(t,[u,v])\in P \\ t\in \bigcup_{k=N+1}^{\infty}(a_k,b_k)}} f(t)(v-u) \right| + \sum_{k=N+1}^{\infty} \left| (HK)\int_{a_k}^{b_k} f \right| \\
&< \frac{\varepsilon}{4} + \sum_{k=N+1}^{\infty} \left| \sum_{\substack{(t,[u,v])\in P \\ t\in (a_k,b_k)}} f(t)(v-u) \right| + \frac{\varepsilon}{8}.
\end{aligned} \tag{4.4.3}$$

It remains to prove that the right-hand side of (4.4.3) is less than ε. If $\{(t,[u,v]) \in P : t \in (a_k, b_k)\}$ is non-empty for some $k \in \mathbb{N}$, then $\{(t,[u,v]) \in P : t \in (a_k, b_k)\}$ is a δ_k-fine Perron subpartition of $[a_k, b_k]$ with

$$\bigcup_{\substack{(t,[u,v])\in P \\ t\in (a_k,b_k)}} [u,v] \in \mathcal{I}_1([a_k, b_k]).$$

According to the triangle inequality, (4.4.1) and (4.4.2),

$$\begin{aligned}
&\sum_{k=N+1}^{\infty} \left| \sum_{\substack{(t,[u,v])\in P \\ t\in (a_k,b_k)}} f(t)(v-u) \right| \\
&\leq \sum_{k=N+1}^{\infty} \sum_{\substack{(t,[u,v])\in P \\ t\in (a_k,b_k)}} \left\{ \left| f(t)(v-u) - (HK)\int_u^v f \right| + \sum_{k=N+1}^{\infty} \left| (HK)\int_u^v f \right| \right\} \\
&< \sum_{k=N+1}^{\infty} \frac{\varepsilon}{2^{k+3}} + \sum_{k=N+1}^{\infty} \|f\|_{HK[a_k,b_k]} \\
&< \frac{\varepsilon}{4}
\end{aligned}$$

and so the right-hand side of (4.4.3) is less than ε. □

Theorem 4.4.5. *Let X be a closed subset of $[a,b]$ and let $\{[a_k, b_k] : k \in \mathbb{N}\}$ be the collection of pairwise disjoint intervals such that $(a,b)\setminus X = \bigcup_{k=1}^{\infty}(a_k, b_k)$. If $f \in HK[a,b]$ and the series $\sum_{k=1}^{\infty} \|f\|_{HK[a_k,b_k]}$ converges, then $f\chi_X \in HK[a,b]$. Moreover, for each $\varepsilon > 0$ there exists a constant gauge η on X such that*

$$\sum_{(t,[u,v])\in P} \left| (HK)\int_u^v f\chi_X - (HK)\int_u^v f \right| < \varepsilon$$

for each X-tagged η-fine Perron subpartition P of $[a,b]$.

Proof. The first assertion $f\chi_X \in HK[a,b]$ is a consequence of Harnack extension. To prove the second assertion we let $\varepsilon > 0$ and select $N \in \mathbb{N}$ so that

$$\sum_{k=N+1}^{\infty} \|f\|_{HK[a_k,b_k]} < \frac{\varepsilon}{4}.$$

Next we use Theorem 1.4.5 to select an $\eta > 0$ so that

$$\left| (HK)\int_u^v f \right| < \frac{\varepsilon}{4N}$$

whenever $[u,v] \subseteq [a,b]$ and $v - u < \eta$.

Let P be any X-tagged η-fine Perron subpartition of $[a,b]$. Since each interval $[a_k, b_k]$ has two endpoints, it follows from Harnack extension and triangle inequality that

$$\begin{aligned}
&\sum_{(t,[u,v])\in P} \left| (HK)\int_u^v (f - f\chi_X) \right| \\
&\leq \sum_{k=1}^{N} \sum_{(t,[u,v])\in P} \left| (HK)\int_u^v f\chi_{[a_k,b_k]} \right| + 2\sum_{k=N+1}^{\infty} \|f\|_{HK[a_k,b_k]} \\
&< \frac{2N\varepsilon}{4N} + \frac{2\varepsilon}{4} \\
&= \varepsilon.
\end{aligned}$$

□

Theorem 4.4.6. *If $f \in HK[a,b]$, then*

(i) *there exists an increasing sequence $(X_n)_{n=1}^{\infty}$ of closed sets whose union is $[a,b]$;*
(ii) *the sequence $(f\chi_{X_n})_{n=1}^{\infty}$ is in $L^1[a,b]$;*

(iii) *for each* $n \in \mathbb{N}$ *we have*

$$\sum_{i=1}^{p} \left| \int_{u_i}^{v_i} f\chi_{X_n}\, d\mu_1 - (HK)\int_{u_i}^{v_i} f \right| \leq \frac{1}{n^2}$$

whenever $[u_1, v_1], \ldots, [u_p, v_p]$ *are non-overlapping subintervals of* $[a, b]$ *with* $[u_i, v_i] \cap X_n \neq \emptyset$ *for* $i = 1, \ldots, p$.

Proof. First we infer from Observation 4.4.1 and Theorem 4.4.5 that there exists an increasing sequuence $(Y_k)_{k=1}^{\infty}$ of closed sets such that $\bigcup_{k=1}^{\infty} Y_k = [a, b]$ and the sequence $(f\chi_{Y_n})_{n=1}^{\infty}$ is in $L^1[a, b]$.

Let $n \in \mathbb{N}$ be given. For each $k \in \mathbb{N}$, we invoke Theorem 4.4.5 to select a constant gauge $\eta_{k,n}$ on Y_k such that

$$\sum_{(t_0, I_0) \in P_0} \left| \int_{I_0} f\chi_{Y_k}\, d\mu_1 - (HK)\int_{I_0} f \right| < \frac{1}{5n^2(2^k)}$$

whenever P_0 is a Y_k-tagged $\eta_{k,n}$-fine Perron subpartition of $[a, b]$. Next we define a gauge δ_n on $[a, b]$ by setting

$$\delta_n(x) = \begin{cases} \eta_{n,n} & \text{if } x \in Y_n, \\ \min\{\eta_{k,n}, \operatorname{dist}(x, Y_k)\} & \text{if } x \in Y_{k+1}\backslash Y_k \quad \text{for some integer } k \geq n, \end{cases}$$

and use Cousin's Lemma to fix a δ_n-fine Perron partition P_n of $[a, b]$. Set

$$X_n := Y_n \cup \bigcup_{k=n}^{\infty} \bigcup_{\substack{(x_n, J_n) \in P_n \\ J_n \cap Y_{k+1} \neq \emptyset \\ J_n \cap Y_k = \emptyset}} (J_n \cap Y_{k+1}).$$

Since P_n is a finite collection of point-interval pairs and each Y_k is closed, we conclude that X_n is closed. Furthermore, if $[u_1, v_1], \ldots, [u_p, v_p]$ is a finite sequence of non-overlapping subintervals of $[a, b]$ satisfying $[u_k, v_k] \cap X_n \neq \emptyset$ for $k = 1, \ldots, p$, we have

$$\begin{aligned} &\sum_{k=1}^{p} \left| \int_{u_k}^{v_k} f\chi_{X_n}\, d\mu_1 - (HK)\int_{u_k}^{v_k} f \right| \\ &= \sum_{k=1}^{p} \sum_{\substack{(t, I) \in P_n \\ I \cap I_k \neq \emptyset}} \left| \int_{I \cap [u_k, v_k]} f\chi_{X_n}\, d\mu_1 - (HK)\int_{I \cap [u_k, v_k]} f \right| \\ &< \sum_{k=1}^{p} \frac{1}{n^2 2^k} \quad \text{(since each interval } [u_k, v_k] \text{ has two endpoints)} \\ &< \frac{1}{n^2}. \end{aligned}$$

It is now easy to see that there exists a subsequence of $(X_n)_{n=1}^{\infty}$, denoted again by $(X_n)_{n=1}^{\infty}$, such that $\bigcup_{k=1}^{\infty} X_k = [a, b]$. □

Let f and $(X_n)_{n=1}^{\infty}$ be given as in Theorem 4.4.6. In order to prove a theorem concerning $(f\chi_{X_n})_{n=1}^{\infty}$, we need the following definition.

Definition 4.4.7. A sequence $(f_n)_{n=1}^{\infty}$ in $HK[a,b]$ is said to be *Henstock-Kurzweil equi-integrable* on $[a,b]$ if for each $\varepsilon > 0$ there exists a gauge δ, independent of n, on $[a,b]$ such that

$$\sup_{n\in\mathbb{N}} \left| S(f_n, P) - (HK)\int_a^b f_n \right| < \varepsilon$$

for each δ-fine Perron partition P of $[a,b]$.

Theorem 4.4.8. *Let $f : [a,b] \longrightarrow \mathbb{R}$ and let $(f_n)_{n=1}^{\infty}$ be a sequence of Henstock-Kurzweil equi- integrable functions on $[a,b]$. If $f_n \to f$ pointwise on $[a,b]$, then $f \in HK[a,b]$ and*

$$\lim_{n\to\infty} \|f_n - f\|_{HK[a,b]} = 0. \tag{4.4.4}$$

Proof. Let $\varepsilon > 0$ be given. Since $(f_n)_{n=1}^{\infty}$ is a sequence of Henstock-Kurzweil equi- integrable functions on $[a,b]$, we can follow the proof of Theorem 2.4.7 to choose a gauge δ, independent of n, on $[a,b]$ such that

$$\sup_{n\in\mathbb{N}} \sum_{(t,[u,v])\in P} \|f_n(t) - f_n\|_{HK[u,v]} < \frac{\varepsilon}{3} \tag{4.4.5}$$

for each δ-fine Perron subpartition P of $[a,b]$.

According to Cousin's Lemma we may fix a δ-fine Perron partition P_0 of $[a,b]$. Since $f_n \to f$ pointwise on $[a,b]$, we may choose and fix $N \in \mathbb{N}$ so that

$$\sum_{(t,[u,v])\in P_0} \|f_n(t) - f_p(t)\|_{HK[u,v]} < \frac{\varepsilon}{3} \tag{4.4.6}$$

for all integers $n, p \geq N$. Consequently, (4.4.5) and (4.4.6) give

$$\begin{aligned}
&\|f_n - f_p\|_{HK[a,b]} \\
&\leq \sum_{(t,[u,v])\in P_0} \|f_n(t) - f_p(t)\|_{HK[u,v]} + \sum_{(t,[u,v])\in P_0} \|f_n(t) - f\|_{HK[u,v]} \\
&+ \sum_{(t,[u,v])\in P_0} \|f_p(t) - f\|_{HK[u,v]} \\
&< \varepsilon
\end{aligned}$$

for all integers $n, p \geq N$. By the completeness of $\mathbb{R}$, there exists a function $F : [a, b] \longrightarrow \mathbb{R}$ such that

$$\lim_{n\to\infty} \sup_{x\in[a,b]} \left|(HK) \int_a^x f_n(t)\, dt - F(x)\right| = 0. \tag{4.4.7}$$

Therefore (4.4.5) yields

$$\sum_{(t,[u,v])\in P} |f(t)(v-u) - (F(v) - F(u))| < \frac{\varepsilon}{3} < \varepsilon$$

for every δ-fine Perron subpartition P of $[a, b]$; in particular, $f \in HK[a, b]$ and $F(x) = (HK) \int_a^x f(t)\, dt$ for every $x \in [a, b]$. It is now easy to see that (4.4.7) implies (4.4.4). The proof is complete. □

Corollary 4.4.9. *Let $(f_n)_{n=1}^\infty$ be a sequence in $HK[a, b]$ and suppose that $f_n \to f$ uniformly on $[a, b]$. Then $f \in HK[a, b]$ and*

$$\lim_{n\to\infty} \|f_n - f\|_{HK[a,b]} = 0.$$

Proof. Exercise. □

Theorem 4.4.10. *If $f \in HK[a, b]$, then there exists an increasing sequence $(X_n)_{n=1}^\infty$ of closed sets such that $\bigcup_{k=1}^\infty X_k = [a, b]$, the sequence $(f\chi_{X_n})_{n=1}^\infty$ is in $L^1[a, b]$, and $(f\chi_{X_n})_{n=1}^\infty$ is Henstock-Kurzweil equi-integrable on $[a, b]$.*

Proof. Let $X_1, X_2, \ldots$ be given as in Theorem 4.4.6 and let $\varepsilon > 0$ be given. For each $k \in \mathbb{N}$ we choose a gauge δ_k on $[a, b]$ such that

$$\sum_{(t_k,I_k)\in Q_k} \left|f(t_k)\chi_{X_k}(t_k)\mu_1(I_k) - \int_{I_k} f\chi_{X_k}\, d\mu_1\right| < \frac{\varepsilon}{2^{k+2}}$$

for each δ_k-fine Perron subpartition Q_k of $[a, b]$. We may further assume that for each $x \in [a, b]$, the sequence $(\delta_k(x))_{k=1}^\infty$ is non-increasing.

Let $N \geq 2$ be a fixed integer such that

$$\sum_{k=N}^\infty \frac{1}{k^2} < \frac{\varepsilon}{4},$$

let $\left([c_{N,k}, d_{N,k}]\right)_{k=1}^\infty$ be the sequence of subintervals of $[a, b]$ such that $(a, b)\backslash X_N = \bigcup_{k=1}^\infty (c_{N,k}, d_{N,k})$, and let $\eta = \dfrac{1}{4} \min_{k=1,\ldots,N} (d_{N,k} - c_{N,k})$. Next we define a gauge δ on $[a,b]$ by setting

$$\delta(x) = \begin{cases} \min\{\delta_N(x), \eta\} & \text{if } x \in X_1, \\ \min\{\delta_N(x), \operatorname{dist}(x, X_{k-1}), \eta\} & \\ \qquad \text{if } x \in X_k \backslash X_{k-1} \text{ for some } k \in \{2, \ldots, N\}, & \\ \min\{\delta_k(x), \operatorname{dist}(x, X_{k-1}), \eta\} & \\ \qquad \text{if } x \in X_k \backslash X_{k-1} \text{ for some integer } k > N. & \end{cases}$$

Let P be a δ-fine Perron partition of $[a,b]$ and let

$$N_0 = \max\{k \in \mathbb{N} : (t, [u,v]) \in P \text{ with } t \in X_k\}.$$

According to our choice of η and δ, any δ-fine cover of X_N cannot be a cover of $[c_{N,k}, d_{N,k}]$ for $k = 1, \ldots, N$. It follows that $N_0 > N$.

The following claims will enable us to prove that $(f\chi_{X_n})_{n=1}^{\infty}$ is Henstock-Kurzweil equi-integrable on $[a,b]$.

Claim 1: $\sum_{(t,[u,v])\in P} \left| f(t)(v-u) - (HK)\int_u^v f \right| < \frac{\varepsilon}{2}$.

Indeed, we have

$$\begin{aligned} &\sum_{(t,[u,v])\in P} \left| f(t)(v-u) - (HK)\int_u^v f \right| \\ &\leq \sum_{\substack{(t,[u,v])\in P \\ t\in X_N}} \left| f(t)(v-u) - \int_u^v f\chi_{X_N}\, d\mu_1 \right| \\ &\quad + \sum_{\substack{(t,[u,v])\in P \\ t\in X_N}} \left| \int_u^v f\chi_{X_N}\, d\mu_1 - (HK)\int_u^v f \right| \\ &\quad + \sum_{k=N+1}^{N_0} \sum_{\substack{(t,[u,v])\in P \\ t\in X_k\backslash X_{k-1}}} \left| f(t)(v-u) - \int_u^v f\chi_{X_k}\, d\mu_1 \right| \\ &\quad + \sum_{k=N+1}^{N_0} \left| \int_u^v f\chi_{X_k}\, d\mu_1 - (HK)\int_u^v f \right| \\ &< \frac{\varepsilon}{2^{N+2}} + \frac{1}{N^2} + \sum_{k=N+1}^{N_0} \frac{\varepsilon}{2^{k+2}} + \sum_{k=N+1}^{N_0} \frac{1}{k^2} \\ &< \frac{\varepsilon}{2}, \end{aligned}$$

and claim 1 is established.

Claim 2: If $n \in \{1, \ldots, N\}$, then

$$\sum_{(t,[u,v])\in P} \left| f(t)\chi_{X_n}(t)(v-u) - \int_u^v f\chi_{X_n}\, d\mu_1 \right| < \frac{\varepsilon}{2^{n+2}}.$$

Let $n \in \{1, \ldots, N\}$ be given. According to our definition of δ, we have $(\xi - \delta(\xi), \xi + \delta(\xi)) \subset (a,b)\setminus X_n$ whenever $\xi \notin X_n$. Thus

$$\begin{aligned}
&\sum_{(t,[u,v])\in P} \left| f(t)\chi_{X_n}(t)(v-u) - \int_u^v f\chi_{X_n}\, d\mu_1 \right| \\
&= \sum_{\substack{(t,[u,v])\in P \\ t\in X_n}} \left| f(t)\chi_{X_n}(t)(v-u) - \int_u^v f\chi_{X_n}\, d\mu_1 \right| \\
&< \frac{\varepsilon}{2^{n+2}},
\end{aligned}$$

which proves claim 2.

Claim 3: For each $n \in \mathbb{N}$ satisfying $n > N$, we have

$$\sum_{(t,[u,v])\in P} \left| f(t)\chi_{X_n}(t)(v-u) - \int_u^v f\chi_{X_n}\, d\mu_1 \right| < \varepsilon.$$

Indeed,

$$\begin{aligned}
&\sum_{(t,[u,v])\in P} \left| f(t)\chi_{X_n}(t)(v-u) - \int_u^v f\chi_{X_n}\, d\mu_1 \right| \\
&= \sum_{\substack{(t,[u,v])\in P \\ t\in X_n}} \left| f(t)\chi_{X_n}(t)(v-u) - \int_u^v f\chi_{X_n}\, d\mu_1 \right| \\
&\quad + \sum_{\substack{(t,[u,v])\in P \\ t\notin X_n}} \left| f(t)\chi_{X_n}(t)(v-u) - \int_u^v f\chi_{X_n}\, d\mu_1 \right| \\
&\le \sum_{\substack{(t,[u,v])\in P \\ t\in X_n}} \left| f(t)(v-u) - (HK)\int_u^v f \right| \\
&\quad + \sum_{\substack{(t,[u,v])\in P \\ t\in X_n}} \left| \int_u^v f\chi_{X_n}\, d\mu_1 - (HK)\int_u^v f \right| \\
&< \frac{\varepsilon}{2} + \sum_{k=N}^{\infty} \frac{1}{k^2} \qquad \text{(by claim 1)} \\
&< \varepsilon
\end{aligned}$$

by our choice of N. It follows that claim 3 holds.

Combining claims 1, 2 and 3, we conclude that $(f\chi_{X_n})_{n=1}^{\infty}$ is Henstock-Kurzweil equi-integrable on $[a,b]$. The proof is complete. □

Definition 4.4.7 and Theorem 4.4.8 translate verbatim to the higher-dimensional Henstock-Kurzweil integral. On the other hand, since the proof of Theorem 4.4.10 depends on Theorems 4.4.5 and 4.4.6 concerning one-dimensional Henstock-Kurzweil integrals, it is unclear whether Theorem 4.4.10 holds true for higher-dimensional Henstock-Kurzweil integrals.

4.5 Other results concerning one-dimensional Henstock-Kurzweil integral

The aim of this section is to present some results concerning the space $HK[a,b]$. Before we present a generalization of Theorem 1.2.5, we need the following notation.

Notation 4.5.1. For any function $F : [a,b] \longrightarrow \mathbb{R}$ we define an interval function $\Delta_F : \mathcal{I}_1([a,b]) \longrightarrow \mathbb{R}$ by setting

$$\Delta_F([u,v]) = F(v) - F(u).$$

The following theorem is a generalization of Theorem 1.2.5.

Theorem 4.5.2. *Let $F : [a,b] \longrightarrow \mathbb{R}$. The following conditions are equivalent.*

(i) *There exists $f \in HK[a,b]$ such that $F(x) - F(a) = (HK)\int_a^x f$ for all $x \in [a,b]$.*
(ii) *$V_{HK}\Delta_F \ll \mu_1$ and there exists a μ_1-negligible set $Z \subset [a,b]$ such that $F'(x) = f(x)$ for all $x \in [a,b]\backslash Z$.*

Proof. (i) $\Longrightarrow$ (ii) This follows from Theorems 4.2.3 and 3.6.6.

(ii) $\Longrightarrow$ (i) Let

$$f(x) = \begin{cases} F'(x) & \text{if } x \in [a,b]\backslash Z, \\ 0 & \text{if } x \in Z, \end{cases}$$

and let $\varepsilon > 0$. According to the definition of Z, there exists a gauge δ_1 on $[a,b]\backslash Z$ such that

$$|f(t)\mu_1(I) - \Delta_F(I)| < \frac{\varepsilon}{2(b-a)}\mu_1(I)$$

for each $I \in \mathcal{I}_1([a,b])$ satisfying $t \in I \backslash Z$ and $I \subseteq (t-\delta_1(t), t+\delta_1(t)) \cap [a,b]$. Since $V_{\mathcal{HK}}\Delta_F \ll \mu_1$, there exists a gauge δ_2 on Z such that

$$V(\Delta_F, Z, \delta_2) < \frac{\varepsilon}{2}.$$

Define a gauge $\delta : [a,b] \longrightarrow \mathbb{R}^+$ by setting

$$\delta(x) = \begin{cases} \delta_1(x) & \text{if } x \in [a,b]\backslash Z, \\ \delta_2(x) & \text{if } x \in Z, \end{cases}$$

and let P be any δ-fine Perron partition of $[a,b]$. Then

$$\begin{aligned} &\left| \sum_{(t,[u,v]) \in P} f(t)(v-u) - (F(b) - F(a)) \right| \\ &\leq \sum_{\substack{(t,[u,v]) \in P \\ t \in [a,b]\backslash Z}} |f(t)(v-u) - (F(v) - F(u))| + \sum_{\substack{(t,[u,v]) \in P \\ t \in Z}} |F(v) - F(u)| \\ &< \varepsilon. \end{aligned}$$

Since $\varepsilon > 0$ is arbitrary, we conclude that $f \in HK[a,b]$. A similar argument shows that $F(x) - F(a) = (HK)\int_a^x f$ for all $x \in [a,b]$. □

Theorem 4.5.3. *Let $F : [a,b] \longrightarrow \mathbb{R}$ be a function such that $V_{\mathcal{HK}}\Delta_F \ll \mu_1$, and let $(\delta_n)_{n=1}^{\infty}$ be a sequence of positive numbers. If there exists an increasing sequence $(Y_n)_{n=1}^{\infty}$ of closed sets such that $\bigcup_{k=1}^{\infty} Y_k = [a,b]$ and $V(\Delta_F, Y_n, \delta_n)$ is finite for $n = 1, 2, \ldots$, then there exists $f \in HK[a,b]$ such that $F(x) - F(a) = (HK)\int_a^x f$ for all $x \in (a,b]$.*

Proof. In view of Theorem 4.5.2, it suffices to prove that $F'(x)$ exists for μ_1-almost all $x \in [a,b]$.

Let $n \in \mathbb{N}$ be given. By Theorem 4.4.2, there exists a sequence $\big((c_k, d_k)\big)_{k=1}^{\infty}$ of pairwise disjoint open intervals such that $(a,b)\backslash Y_n = \bigcup_{k=1}^{\infty}(c_k, d_k)$. Next we define

$$G(x) = \begin{cases} \inf_{t \in [c_k, d_k]} F(t) - F(a) & \text{if } x \in (c_k, d_k) \text{ for some } k \in \mathbb{N}, \\ \\ F(x) - F(a) & \text{if } x \in Y_n. \end{cases}$$

$$H(x) = \begin{cases} \sup_{t \in [c_k, d_k]} F(t) - F(a) & \text{if } x \in (c_k, d_k) \text{ for some } k \in \mathbb{N}, \\ \\ F(x) - F(a) & \text{if } x \in Y_n. \end{cases}$$

Then it is easy to see that $G, H \in BV[a,b]$, $G(x) = H(x) = F(x) - F(a)$ for all $x \in Y_n$, and $G(t) \leq F(t) - F(a) \leq H(t)$ for all $t \in [a,b]$. Since

$G, H \in BV[a,b]$, an application of Theorem 3.11.6 shows that G and H are differentiable μ_1-almost everywhere on $[a,b]$ and

$$F'(x) = G'(x) = H'(x)$$

for μ_1-almost all $x \in Y_n$. Since $n \in \mathbb{N}$ is arbitrary, the theorem is proved. □

In order to proceed further, we need the following construction.

Example 4.5.4. Let $c \in [0,1)$, let $\delta_n = \frac{1-c}{3}(\frac{2}{3})^{n-1}$ for $n = 1, 2, \ldots$, and let $X_1 = [0,1]$. We will first construct a decreasing sequence $(X_n)_{n=1}^{\infty}$ of closed sets so that the following properties hold for all positive integers $n \geq 2$:

$(\text{i})_n$ the set X_n is the union of 2^{n-1} disjoint closed intervals $I_{n,1}, \ldots, I_{n,2^{n-1}}$;
$(\text{ii})_n$ $\mu_1(I_{n,k}) = \frac{1}{2^{n-1}}(1 - \sum_{k=1}^{n-1} \delta_k)$ for every $k = 1, \ldots, 2^{n-1}$.

We will first construct the set X_2 with the required properties. We delete the open interval $(u_{1,1}, v_{1,1}) := (\frac{1-\delta_1}{2}, \frac{1+\delta_1}{2})$ from $[0,1]$ and let $X_2 = [0,1]\backslash(\frac{1-\delta_1}{2}, \frac{1+\delta_1}{2})$. Then properties $(\text{i})_2$ and $(\text{ii})_2$ hold.

Now suppose that properties $(\text{i})_N$ and $(\text{ii})_N$ hold for some $N \in \mathbb{N}\backslash\{1\}$. For each $k \in \{1, \ldots, 2^{N-1}\}$, we delete an open interval $(u_{N,k}, v_{N,k})$ from $I_{N,k}$ to obtain two disjoint closed intervals $I_{N+1,2k-1}$ and $I_{N+1,2k}$ such that $I_{N,k}\backslash(u_{N,k}, v_{N,k}) = I_{N+1,2k-1} \cup I_{N+1,2k}$ and $\mu_1(I_{N+1,2k-1}) = \mu_1(I_{N+1,2k}) = \frac{1}{2^N}(1 - \sum_{k=1}^{N} \delta_k)$. Then $X_{N+1} := \bigcup_{k=1}^{2^{N+1}} I_{N+1,k}$ is a closed set satisfying properties $(\text{i})_{N+1}$ and $(\text{ii})_{N+1}$.

By induction, properties $(\text{i})_n$ and $(\text{ii})_n$ hold for all $n \in \mathbb{N}$. The set $C := \bigcap_{k=1}^{\infty} X_k$ is a closed set with $\mu_1(C) = c$. This set C is known as the *Cantor set*. A simple application of Cousin's Lemma shows that C is non-empty.

A sequence $(h_n)_{n=1}^{\infty}$ in $HK[a,b]$ is said to converge to $h \in HK[a,b]$ if $\lim_{n\to\infty} \|h_n - h\|_{HK[a,b]} = 0$. Although the convergence of $(h_n)_{n=1}^{\infty}$ in $HK[a,b]$ implies that $\|h_n - h_p\|_{HK[a,b]} \to 0$ as $n, p \to \infty$, the converse is false. In this case, we say that the space $HK[a,b]$ is not complete.

Example 4.5.5. The space $HK[0,1]$ is not complete.

Proof. Let C be constructed as in Example 4.5.4 so that $\mu_1(C) = 0$ and $(0,1)\backslash C = \bigcup_{n=1}^{\infty} \bigcup_{k=1}^{2^{n-1}} (u_{n,k}, v_{n,k})$. Define the function $f : [0,1] \longrightarrow \mathbb{R}$ by setting

$$f(x) = \begin{cases} \dfrac{1}{n(v_{n,k} - u_{n,k})} \sin\left(\dfrac{2\pi(x - u_{n,k})}{v_{n,k} - u_{n,k}}\right) & \\ \qquad \text{if } x \in [u_{n,k}, v_{n,k}] \text{ for some } n \in \mathbb{N} \text{ and } k \in \{1, \ldots, 2^{n-1}\}, & \\ 0 & \text{if } x \in C. \end{cases}$$

For each $N \in \mathbb{N}$, let

$$f_N = \sum_{n=1}^{N} \sum_{k=1}^{2^{n-1}} f\chi_{J(n,k)}, \text{ where } J(n,k) := [u_{n,k}, v_{n,k}].$$

Since $\|f\|_{HK(J(n,k))} = \frac{1}{2\pi n}$ and $\int_{J(n,k)} f = 0$ whenever $n \in \mathbb{N}$ and $k \in \{1, \ldots, 2^{n-1}\}$, we conclude that $\|f_n - f_p\|_{HK[0,1]} \to 0$ as $n, p \to \infty$.

It remains to prove that the sequence $(f_n)_{n=1}^{\infty}$ does not converge in the space $HK[0,1]$. Proceeding towards a contradiction, suppose that this sequence converges to some $g \in HK[0,1]$. Clearly, $g = f$ μ_1-almost everywhere on $[0,1]$. Since C is closed with $\mu_1(C) = 0$, it follows from Theorem 4.2.8 that there exists an upper semicontinuous gauge δ on C such that

$$V(\Delta_F, C, \delta) < 1,$$

where $F(x) := (HK) \int_a^x g$ for each $x \in [a,b]$.

For each $n \in \mathbb{N}$, let $C_n = \{x \in C : \delta(x) \geq \frac{1}{n}\}$. According to Theorem 4.1.2 and the construction of C, there exist an integer $N \geq 4$ and $[c,d] \in \{I_{N,1}, \ldots, I_{N,2^{N-1}}\}$ such that $\mu_1([c,d]) \leq \frac{1}{2^{N-1}} \leq \frac{1}{2N}$ and $C_N \cap [c,d] = C \cap [c,d]$. Also, for each $q \in \mathbb{N} \cup \{0\}$ the interval $[c,d]$ contains 2^q intervals from the set $\{J(N+q,k) : k = 1, \ldots, 2^{N+q-1}\}$, and $\bigcup_{k=1}^{2^{N+q-1}} \{u_{N+q,k}, v_{N+q,k}\} \subset C$. Since $N \geq 4$, we get

$$V(\Delta_F, C, \delta) \geq \frac{1}{2} \sum_{k=1}^{N} \frac{2^{k-1}}{N+k-1} \geq \frac{2^N - 1}{2(2N-1)} \geq 1,$$

a contradiction. □

Corollary 4.5.6. *Let C be given as in Example 4.5.4. Then C is uncountable.*

Proof. If $\mu_1(C) > 0$, then C is uncountable. Henceforth, we suppose that $\mu_1(C) = 0$. Proceeding towards a contradiction, suppose that C is countable. Let F be given as in Example 4.5.5. Since C is assumed to be countable, the continuity of F implies $V_{\mathcal{HK}}\Delta_F(C) = 0$. On the other hand, since $V_{\mathcal{HK}}\Delta_F(C) = 0$ and $F'(t)$ exists for all $t \in (0,1)\backslash C$, we infer from Theorem 4.5.2 that F is the indefinite HK-integral of some function $f \in HK[0,1]$, contrary to Example 4.5.5. This contradiction shows that C must be uncountable. □

Exercise 4.5.7. Let C, $(X_n)_{n=1}^{\infty}$ and $(\delta_n)_{n=1}^{\infty}$ be given as in Example 4.5.5. Show that there exists a sequence $(F_n)_{n=1}^{\infty}$ in $AC[0,1]$ such that $\max_{x\in[0,1]} |F_n(x) - F_N(x)| \to 0$ as $n, N \to \infty$. However, the limiting function F is not $AC[0,1]$. This function is known as the Cantor singular function.

Let $\Delta[a,b]$ be the space of differentiable functions on $[a,b]$, and let $\Delta'[a,b]$ be the space of derivatives on $[a,b]$.

Example 4.5.8. Let $C \subset [0,1]$ be the Cantor set and let $\big((c_k, d_k)\big)_{k=1}^{\infty}$ be the sequence of pairwise disjoint open intervals such that $(0,1)\backslash C = \bigcup_{k=1}^{\infty}(c_k, d_k)$. For each $\alpha > 1$ the function $F_\alpha : [0,1] \longrightarrow \mathbb{R}$ is given by

$$F_\alpha(x) = \begin{cases} (x-c_k)^2(d_k-x)^2 \sin \dfrac{1}{(x-c_k)^\alpha(d_k-x)^\alpha} & \\ \qquad\qquad\qquad\qquad \text{if } x \in (c_k, d_k) \text{ for some } k \in \mathbb{N}, \\ 0 \qquad\qquad\qquad\qquad \text{if } x \in C. \end{cases}$$

Then $F_\alpha \in \Delta[0,1]$. Furthermore, the following statements are true.

(i) $F_1' \in L^\infty[0,1]\backslash C[0,1]$.
(ii) $F_{\frac{3}{2}}' \in L^1[0,1]\backslash L^\infty[0,1]$.
(iii) $F_2' \in HK[0,1]\backslash L^1[0,1]$.
(iv) If $\mu_1(C) > 0$, then $F_2' \in \Delta'[0,1]\backslash R[0,1]$.

Proof. Exercise. □

4.6 Notes and Remarks

The proof of Theorem 4.1.3 is due to Lee [97]. Several related results have been established in [23, 24, 98].

Lemma 4.2.2 and Theorem 4.2.3 are due to Lee [91, 102]. For a more general version of Theorem 4.2.5, see [77, Theroem 2.10]. The present proof of Theorem 4.2.9 is taken from the paper [111].

Sections 4.3 and 4.4 are based on the papers [91] and [90] respectively. For the equivalence of Theorem 4.4.8 with the Controlled Convergence Theorem ([85, Theorem 7.6]), see [77].

Example 4.5.8 is essentially due to Bullen [27]. Further examples of Henstock-Kurzweil integrable functions can be found in [17, 20, 69].

Chapter 5

The Henstock variational measure

The main aim of this chapter is to obtain a simple characterization of additive interval functions that are indefinite Henstock-Kurzweil integrals; see Theorem 5.5.9 for details. In order to prove this result, it is necessary to have a better understanding of the Henstock variational measure. We begin with a special case of the Henstock variational measure.

5.1 Lebesgue outer measure

For any set $X \subseteq [\boldsymbol{a}, \boldsymbol{b}]$, we let

$$\mu_m^*(X) = \inf_{\delta} \sup_{P} \sum_{(\boldsymbol{t}, I) \in P} \mu_m(I),$$

where δ is a gauge on X, and P is an X-tagged δ-fine Perron subpartition of $[\boldsymbol{a}, \boldsymbol{b}]$. The following lemma is a special case of Lemma 4.2.2.

Lemma 5.1.1. *If $X \subseteq [\boldsymbol{a}, \boldsymbol{b}]$ is μ_m-measurable, then $\mu_m^*(X) = \mu_m(X)$.*

The proof of the following result is left to the reader.

Theorem 5.1.2. *If $Y \subseteq Z \subseteq [\boldsymbol{a}, \boldsymbol{b}]$, then $\mu_m^*(Y) \leq \mu_m^*(Z)$.*

The following theorem is the main result of this section.

Theorem 5.1.3. *If $X \subseteq [\boldsymbol{a}, \boldsymbol{b}]$, then*

$$\mu_m^*(X) = \inf \left\{ \sum_{k=1}^{\infty} \mu_m((\boldsymbol{u}_k, \boldsymbol{v}_k)) : X \subseteq \bigcup_{k=1}^{\infty} (\boldsymbol{u}_k, \boldsymbol{v}_k) \right\}.$$

Proof. First we observe that if $((\boldsymbol{u}_k, \boldsymbol{v}_k))_{k=1}^{\infty}$ is any sequence of open intervals such that $X \subseteq \bigcup_{k=1}^{\infty}(\boldsymbol{u}_k, \boldsymbol{v}_k)$, then Theorem 5.1.2, Lemma 5.1.1, and the countable subadditivity of μ_m yield

$$\mu_m^*(X) \leq \mu_m^*\left(\bigcup_{k=1}^{\infty}(\boldsymbol{u}_k, \boldsymbol{v}_k)\right) = \mu_m\left(\bigcup_{k=1}^{\infty}(\boldsymbol{u}_k, \boldsymbol{v}_k)\right) \leq \sum_{k=1}^{\infty}\mu_m((\boldsymbol{u}_k, \boldsymbol{v}_k)).$$

Thus

$$\mu_m^*(X) \leq \inf\left\{\sum_{k=1}^{\infty}\mu_m((\boldsymbol{u}_k, \boldsymbol{v}_k)) : X \subseteq \bigcup_{k=1}^{\infty}(\boldsymbol{u}_k, \boldsymbol{v}_k)\right\}.$$

It remains to prove that the reverse inequality holds. To do this, we let $\varepsilon > 0$ and follow the proof of Theorem 3.4.2 to construct a bounded open set G_ε such that $G_\varepsilon \supseteq X$ and

$$\mu_m(G_\varepsilon) < \mu_m^*(X) + \frac{\varepsilon}{2}.$$

Next we apply Theorem 3.3.18 to select a sequence $((\boldsymbol{s}_k, \boldsymbol{t}_k))_{k=1}^{\infty}$ of open intervals such that

$$\sum_{k=1}^{\infty}\mu_m((\boldsymbol{s}_k, \boldsymbol{t}_k)) < \mu_m(G_\varepsilon) + \frac{\varepsilon}{2}.$$

Consequently,

$$\inf\left\{\sum_{k=1}^{\infty}\mu_m((\boldsymbol{u}_k, \boldsymbol{v}_k)) : X \subseteq \bigcup_{k=1}^{\infty}(\boldsymbol{u}_k, \boldsymbol{v}_k)\right\} < \mu_m^*(X) + \varepsilon,$$

and the desired inequality follows from the arbitrariness of ε. □

Remark 5.1.4. For each set $X \subseteq [\boldsymbol{a}, \boldsymbol{b}]$, the Lebesgue outer measure of X is defined by

$$\inf\left\{\sum_{k=1}^{\infty}\mu_m((\boldsymbol{u}_k, \boldsymbol{v}_k)) : X \subseteq \bigcup_{k=1}^{\infty}(\boldsymbol{u}_k, \boldsymbol{v}_k)\right\}.$$

Theorem 5.1.3 tells us that $\mu_m^*(X)$ is precisely the Lebesgue outer measure of X.

Theorem 5.1.5. *If X_1 and X_2 are subsets of $[\boldsymbol{a}, \boldsymbol{b}]$, then*

$$\mu_m^*(X_1 \cup X_2) \leq \mu_m^*(X_1) + \mu_m^*(X_2).$$

Proof. Exercise. □

Theorem 5.1.6. *If $X \subseteq [\boldsymbol{a}, \boldsymbol{b}]$, then there exists a μ_m-measurable set X_0 such that $X \subseteq X_0 \subseteq [\boldsymbol{a}, \boldsymbol{b}]$ and $\mu_m(X_0 \cap Y) = \mu_m^*(X \cap Y)$ for every μ_m-measurable set $Y \subseteq [\boldsymbol{a}, \boldsymbol{b}]$.*

Proof. For each $n \in \mathbb{N}$ we apply Theorem 5.1.3 to select a bounded open set $G_n \supseteq X$ such that

$$\mu_m(G_n) < \mu_m^*(X) + \frac{1}{n}.$$

Set $X_0 = \bigcap_{n=1}^{\infty} G_n$. Then X_0 is a μ_m-measurable set with $X \subseteq X_0 \subseteq [\boldsymbol{a}, \boldsymbol{b}]$ and $\mu_m(X_0) = \mu_m^*(X)$:

$$\mu_m^*(X) \le \mu_m(X_0) \le \liminf_{n\to\infty} \mu_m(G_n) \le \lim_{n\to\infty} (\mu_m^*(X) + \frac{1}{n}) = \mu_m^*(X).$$

It remains to prove the last assertion of the theorem. Let $Y \subseteq [\boldsymbol{a}, \boldsymbol{b}]$ be any μ_m-measurable set. Since $X \subseteq X_0 \subseteq [\boldsymbol{a}, \boldsymbol{b}]$ with $\mu_m(X_0) = \mu_m^*(X)$, it follows from Theorem 5.1.2, Lemma 5.1.1 and Theorem 5.1.5 that $\mu_m(X_0 \cap Y) = \mu_m^*(X \cap Y)$:

$$\begin{aligned} \mu_m^*(X \cap Y) &\le \mu_m(X_0 \cap Y) \\ &= \mu_m(X_0) - \mu_m(X_0 \backslash Y) \\ &\le \mu_m^*(X) - \mu_m^*(X \backslash Y) \\ &\le \mu_m^*(X \cap Y). \end{aligned}$$ □

Theorem 5.1.7. *Let $X \subseteq [\boldsymbol{a}, \boldsymbol{b}]$. The following statements are equivalent.*

(i) *X is μ_m-measurable.*
(ii) *$\mu_m^*(A) = \mu_m^*(A \cap X) + \mu_m^*(A \backslash X)$ for every set $A \subseteq [\boldsymbol{a}, \boldsymbol{b}]$.*

Proof. (i) $\Longrightarrow$ (ii) For any set $A \subseteq [\boldsymbol{a}, \boldsymbol{b}]$, we use Theorem 5.1.6 to choose a μ_m-measurable set A_0 such that $A \subseteq A_0 \subseteq [\boldsymbol{a}, \boldsymbol{b}]$ and $\mu_m(A_0 \cap Y) = \mu_m^*(A \cap Y)$ for every μ_m-measurable set $Y \subseteq [\boldsymbol{a}, \boldsymbol{b}]$. Thus $\mu_m^*(A) = \mu_m(A_0) = \mu_m(A_0 \cap X) + \mu_m(A_0 \backslash X) = \mu_m^*(A \cap X) + \mu_m^*(A \backslash X)$.

Conversely, suppose that (ii) holds. By Theorem 5.1.6, there exists a μ_m-measurable set X_0 such that $X \subseteq X_0 \subseteq [\boldsymbol{a}, \boldsymbol{b}]$ and $\mu_m(X_0) = \mu_m^*(X)$. Hence, using the equality $\mu_m(X_0) = \mu_m^*(X)$ and (ii) with $A = X_0$, we get

$$\mu_m^*(X_0 \backslash X) = \mu_m(X_0) - \mu_m^*(X) = 0$$

and so $X_0 \backslash X$ is μ_m-measurable. Since X_0 is also μ_m-measurable, we conclude that $X = X_0 \backslash (X_0 \backslash X)$ is μ_m-measurable too. □

Remark 5.1.8. According to Theorem 3.3.23, μ_m is countably additive. On the other hand, μ_m^* is not countable additive. See, for example, [143, p.66].

5.2 Basic properties of the Henstock variational measure

In this section we will prove some basic measure-theoretic properties of the Henstock variational measure. We begin with the following result.

Theorem 5.2.1. *Let $F : \mathcal{I}_m([\boldsymbol{a}, \boldsymbol{b}]) \longrightarrow \mathbb{R}$. Then the following statements are true.*

(i) $V_{\mathcal{HK}}F(\emptyset) = 0$.
(ii) *If* $X \subseteq Y \subseteq [\boldsymbol{a}, \boldsymbol{b}]$, *then* $V_{\mathcal{HK}}F(X) \leq V_{\mathcal{HK}}F(Y)$.
(iii) *If* $(X_n)_{n=1}^{\infty}$ *is a sequence of subsets of* $[\boldsymbol{a}, \boldsymbol{b}]$, *then* $V_{\mathcal{HK}}F(\bigcup_{k=1}^{\infty} X_k) \leq \sum_{k=1}^{\infty} V_{\mathcal{HK}}F(X_k)$.
(iv) *If* $X, Y \subseteq [\boldsymbol{a}, \boldsymbol{b}]$ *are non-empty sets with* $\text{dist}(X, Y) > 0$, *then*

$$V_{\mathcal{HK}}F(X \cup Y) = V_{\mathcal{HK}}F(X) + V_{\mathcal{HK}}F(Y).$$

Proof. Statements (i) and (ii) are clearly true.

To prove (iii), we may further assume that each X_k is non-empty, the series $\sum_{k=1}^{\infty} V_{\mathcal{HK}}F(X_k)$ converges, and $X_i \cap X_j = \emptyset$ whenever i and j are distinct positive integers. For each $k \in \mathbb{N}$ there exists a gauge δ_k on X_k such that

$$V(F, X_k, \delta_k) < V_{\mathcal{HK}}F(X_k) + \frac{\varepsilon}{2^k}.$$

Define a gauge δ on $\bigcup_{k=1}^{\infty} X_k$ by setting

$$\delta(\boldsymbol{x}) = \delta_k(\boldsymbol{x}) \quad (\boldsymbol{x} \in X_k),$$

and select an arbitrary $(\bigcup_{k=1}^{\infty} X_k)$-tagged δ-fine Perron subpartition of $[\boldsymbol{a}, \boldsymbol{b}]$. Then

$$\begin{aligned} V(F, \bigcup_{k=1}^{\infty} X_k, \delta) &\leq \sum_{k=1}^{\infty} V(F, X_k, \delta_k) \\ &< \sum_{k=1}^{\infty} \left(V_{\mathcal{HK}}F(X_k) + \frac{\varepsilon}{2^k}\right) \\ &\leq \sum_{k=1}^{\infty} V_{\mathcal{HK}}F(X_k) + \varepsilon. \end{aligned}$$

Since $\varepsilon > 0$ is arbitrary, we get (iii).

To prove (iv), we let $X, Y \subseteq [\boldsymbol{a}, \boldsymbol{b}]$ be non-empty sets with $\text{dist}(X, Y) > 0$, and let $\varepsilon > 0$. Then there exists a gauge δ_1 on X such that

$$V(F, X, \delta_1) < \sum_{(\boldsymbol{t}_1, I_1) \in P_1} |F(I_1)| + \frac{\varepsilon}{3}$$

for some X-tagged δ_1-fine Perron subpartition P_1 of $[\boldsymbol{a}, \boldsymbol{b}]$. Similarly, there exists a gauge δ_2 on Y such that

$$V(F, Y, \delta_2) < \sum_{(\boldsymbol{t}_2, I_2) \in P_2} |F(I_2)| + \frac{\varepsilon}{3}$$

for some Y-tagged δ_2-fine Perron subpartition P_2 of $[\boldsymbol{a}, \boldsymbol{b}]$. We also choose a gauge δ_3 on $X \cup Y$ so that

$$V(F, X \cup Y, \delta_3) < V_{\mathcal{HK}}F(X \cup Y) + \frac{\varepsilon}{3}.$$

Since $\mathrm{dist}(X, Y) > 0$, we can define a gauge δ on X by setting

$$\delta(\boldsymbol{x}) = \begin{cases} \min\{\delta_1(\boldsymbol{x}), \delta_3(\boldsymbol{x}), \mathrm{dist}(X, Y)\} \text{ if } \boldsymbol{x} \in X, \\ \min\{\delta_2(\boldsymbol{x}), \delta_3(\boldsymbol{x}), \mathrm{dist}(X, Y)\} \text{ if } \boldsymbol{x} \in Y. \end{cases}$$

Then

$$\begin{aligned} V_{\mathcal{HK}}F(X) + V_{\mathcal{HK}}F(Y) &\leq \sum_{(\boldsymbol{t}_1, I_1) \in P_1} |F(I_1)| + \frac{\varepsilon}{3} + \sum_{(\boldsymbol{t}_2, I_2) \in P_2} |F(I_2)| + \frac{\varepsilon}{3} \\ &\leq V(F, X \cup Y, \delta) + \frac{2\varepsilon}{3} \\ &< V_{\mathcal{HK}}F(X \cup Y) + \varepsilon. \end{aligned}$$

Since $\varepsilon > 0$ is arbitrary, we conclude that

$$V_{\mathcal{HK}}F(X) + V_{\mathcal{HK}}F(Y) \leq V_{\mathcal{HK}}F(X \cup Y).$$

Combining the last inequality with (iii), we get (iv). □

According to Remark 5.1.8, μ_m^* is not countably additive. Thus, $V_{\mathcal{HK}}F$ cannot be countably additive. Our aim is to describe a sufficiently large class $\mathcal{B}$ of subsets of $[\boldsymbol{a}, \boldsymbol{b}]$ such that

$$V_{\mathcal{HK}}F\left(\bigcup_{k=1}^{\infty} Y_k\right) = \sum_{k=1}^{\infty} V_{\mathcal{HK}}F(Y_k)$$

for every pairwise disjoint sequence $(Y_n)_{n=1}^{\infty}$ of sets belonging to $\mathcal{B}$. We begin with the following lemma.

Lemma 5.2.2. *Let* $F : \mathcal{I}_m([\boldsymbol{a}, \boldsymbol{b}]) \longrightarrow \mathbb{R}$. *If* $X \subseteq [\boldsymbol{a}, \boldsymbol{b}]$ *is closed, then*

$$V_{\mathcal{HK}}F(A) = V_{\mathcal{HK}}F(A \cap X) + V_{\mathcal{HK}}F(A \backslash X)$$

for every set $A \subseteq [\boldsymbol{a}, \boldsymbol{b}]$.

Proof. Let $A \subseteq [\boldsymbol{a}, \boldsymbol{b}]$. In view of Theorem 5.2.1(iii), we may suppose that $V_{\mathcal{HK}}F(A)$ is finite. For each $n \in \mathbb{N}$, let

$$A_n = \left\{ \boldsymbol{x} \in A : \mathrm{dist}(\boldsymbol{x}, X) \geq \frac{1}{n} \right\}.$$

Since X is closed and the sequence $(\frac{1}{n})_{n=1}^{\infty}$ is decreasing, we conclude that $A_n \subseteq A_{n+1} \subseteq A \backslash X$ for all $n \in \mathbb{N}$.

We next establish the inequality

$$\lim_{n \to \infty} V_{\mathcal{HK}}F(A_n) = V_{\mathcal{HK}}F(A \backslash X). \tag{5.2.1}$$

To prove (5.2.1), we need to prove that

$$V_{\mathcal{HK}}F(A_{n+1}) = \sum_{k=0}^{n} V_{\mathcal{HK}}F(A_{k+1} \backslash A_k) \text{ for all } n \in \mathbb{N}, \tag{5.2.2}$$

where $A_0 := \emptyset$. To do this, we let $n \in \mathbb{N}$ and observe that if j and k are integers satisfying $0 \leq j < k \leq n$ and $A_{j+1} \backslash A_j \neq \emptyset \neq A_{k+1} \backslash A_k$, then

$$\begin{aligned}
\mathrm{dist}(A_{k+1} \backslash A_k, A_{j+1} \backslash A_j) &\geq \mathrm{dist}(A_{k+1} \backslash A_k, A_{j+1}) \\
&\geq \mathrm{dist}(A_{j+1}, X) - \mathrm{dist}(A_{k+1} \backslash A_k, X) \\
&> \frac{1}{j+1} - \frac{1}{k} \\
&\geq 0
\end{aligned}$$

and so (5.2.2) follows from Theorem 5.2.1(iv). Therefore, the obvious equality $A \backslash X = \bigcup_{k=0}^{\infty} (A_{k+1} \backslash A_k)$, Theorem 5.2.1(iii), (5.2.2), and Theorem 5.2.1(ii) give (5.2.1):

$$V_{\mathcal{HK}}F(A \backslash X) \leq \sum_{k=0}^{\infty} V_{\mathcal{HK}}F(A_{k+1} \backslash A_k) = \lim_{n \to \infty} V_{\mathcal{HK}}F(A_{n+1}) \leq V_{\mathcal{HK}}F(A \backslash X).$$

Finally, since $\mathrm{dist}(A_n, A \cap X) \geq \mathrm{dist}(A_n, X) \geq \frac{1}{n}$ for all $n \in \mathbb{N}$, the conclusion follows from Theorem 5.2.1(iii), (5.2.1), Theorems 5.2.1(iv) and 5.2.1(ii):

$$\begin{aligned}
V_{\mathcal{HK}}F(A) &\leq V_{\mathcal{HK}}F(A \cap X) + V_{\mathcal{HK}}F(A \backslash X) \\
&= V_{\mathcal{HK}}F(A \cap X) + \lim_{n \to \infty} V_{\mathcal{HK}}F(A_n) \\
&= \lim_{n \to \infty} V_{\mathcal{HK}}F(A_n \cup (A \cap X)) \\
&\leq V_{\mathcal{HK}}F(A).
\end{aligned}$$

□

Recall that Theorem 5.1.7 tells us that a set $X \subseteq [\boldsymbol{a}, \boldsymbol{b}]$ is μ_m-measurable if and only if $\mu_m^*(A) = \mu_m^*(A \cap X) + \mu_m^*(A \backslash X)$ for every set $A \subseteq [\boldsymbol{a}, \boldsymbol{b}]$. Thus, it is natural to have the following definition.

Definition 5.2.3. Let $F : \mathcal{I}_m([\boldsymbol{a}, \boldsymbol{b}]) \longrightarrow \mathbb{R}$. A set $Z \subseteq [\boldsymbol{a}, \boldsymbol{b}]$ is said to be $V_{\mathcal{HK}}F$-measurable if $V_{\mathcal{HK}}F(A) = V_{\mathcal{HK}}F(A \cap Z) + V_{\mathcal{HK}}F(A \backslash Z)$ for every set $A \subseteq [\boldsymbol{a}, \boldsymbol{b}]$.

We will next establish two basic properties of $V_{\mathcal{HK}}F$-measurable sets.

Lemma 5.2.4. *Let $F : \mathcal{I}_m([\boldsymbol{a}, \boldsymbol{b}]) \longrightarrow \mathbb{R}$. If X and Y are $V_{\mathcal{HK}}F$-measurable subsets of $[\boldsymbol{a}, \boldsymbol{b}]$, so is $X \cup Y$.*

Proof. Let $A \subseteq [\boldsymbol{a}, \boldsymbol{b}]$ be any set. Then

$$\begin{aligned}
&V_{\mathcal{HK}}F(A) \\
&\leq V_{\mathcal{HK}}F(A \cap (X \cup Y)) + V_{\mathcal{HK}}F(A \backslash (X \cup Y)) \qquad \text{(by Theorem 5.2.1(iii))} \\
&\leq V_{\mathcal{HK}}F(A \cap X) + V_{\mathcal{HK}}F((A \backslash X) \cap Y) + V_{\mathcal{HK}}F((A \backslash X) \backslash Y) \\
&\qquad\qquad\qquad\qquad\qquad\qquad\qquad\qquad \text{(by Theorem 5.2.1(iii) again)} \\
&= V_{\mathcal{HK}}F(A \cap X) + V_{\mathcal{HK}}F(A \backslash X) \qquad \text{(since } Y \text{ is } V_{\mathcal{HK}}F\text{-measurable)} \\
&= V_{\mathcal{HK}}F(A),
\end{aligned}$$

where the last equality holds because X is $V_{\mathcal{HK}}F$-measurable. Thus, $X \cup Y$ is $V_{\mathcal{HK}}F$-measurable. □

Lemma 5.2.5. *Let $F : \mathcal{I}_m([\boldsymbol{a}, \boldsymbol{b}]) \longrightarrow \mathbb{R}$, and let $A \subseteq [\boldsymbol{a}, \boldsymbol{b}]$ be any set. If $X_1, \ldots, X_n$ are pariwise disjoint $V_{\mathcal{HK}}F$-measurable subsets of $[\boldsymbol{a}, \boldsymbol{b}]$, then $\bigcup_{k=1}^n X_k$ is $V_{\mathcal{HK}}F$-measurable and*

$$V_{\mathcal{HK}}F\left(A \cap \bigcup_{k=1}^n X_k\right) = \sum_{k=1}^n V_{\mathcal{HK}}F(A \cap X_k).$$

Proof. Applying Lemma 5.2.4 repeatedly, we see that $\bigcup_{k=1}^n X_k$ is $V_{\mathcal{HK}}F$-measurable. To prove the last assertion, we write $X_0 := \emptyset$ and observe that our hypotheses yield

$$\begin{aligned}
&V_{\mathcal{HK}}F\left(A \cap \bigcup_{j=1}^k X_j\right) \\
&= V_{\mathcal{HK}}F\left(\left(A \cap \bigcup_{j=0}^k X_j\right) \cap X_k\right) + V_{\mathcal{HK}}F\left(\left(A \cap \bigcup_{j=0}^k X_j\right) \backslash X_k\right) \\
&= V_{\mathcal{HK}}F(A \cap X_k) + V_{\mathcal{HK}}F\left(A \cap \bigcup_{j=0}^{k-1} X_j\right)
\end{aligned}$$

for $k = 1, \dots, n$. Consequently,

$$\begin{aligned}\sum_{k=1}^{n} V_{\mathcal{HK}}F(A \cap X_k) &= \sum_{k=1}^{n} \left\{ V_{\mathcal{HK}}F\left(A \cap \bigcup_{j=0}^{k} X_j\right) - V_{\mathcal{HK}}F\left(A \cap \bigcup_{j=0}^{k-1} X_j\right) \right\} \\ &= V_{\mathcal{HK}}F\left(A \cap \bigcup_{j=1}^{n} X_j\right) - V_{\mathcal{HK}}F(\emptyset) \\ &= V_{\mathcal{HK}}F\left(A \cap \bigcup_{j=1}^{n} X_j\right).\end{aligned}$$

□

Definition 5.2.6. A *σ-algebra* of subsets of $[\boldsymbol{a}, \boldsymbol{b}]$ is a collection $\mathcal{M}$ of sets satisfying the following conditions:

(i) $\emptyset \in \mathcal{M}$,
(ii) $X \in \mathcal{M}$ implies $[\boldsymbol{a}, \boldsymbol{b}] \setminus X \in \mathcal{M}$,
(iii) if $X_n \in \mathcal{M}$ for every $n \in \mathbb{N}$, then $\bigcup_{k=1}^{\infty} X_k \in \mathcal{M}$.

Lemma 5.2.7. *Let $F : \mathcal{I}_m([\boldsymbol{a}, \boldsymbol{b}]) \longrightarrow \mathbb{R}$. Then the collection of $V_{\mathcal{HK}}F$-measurable sets forms a σ-algebra.*

Proof. First it is easy to see that the empty set is $V_{\mathcal{HK}}F$-measurable.

Next it is obvious that a set $X \subseteq [\boldsymbol{a}, \boldsymbol{b}]$ is $V_{\mathcal{HK}}F$-measurable if and only if $[\boldsymbol{a}, \boldsymbol{b}] \setminus X$ is $V_{\mathcal{HK}}F$-measurable.

Finally, we let $(X_n)_{n=1}^{\infty}$ be any sequence of $V_{\mathcal{HK}}F$-measurable subsets of $[\boldsymbol{a}, \boldsymbol{b}]$. Then $\bigcup_{k=1}^{\infty} X_k$ is $V_{\mathcal{HK}}F$-measurable. Indeed, for any $A \subseteq [\boldsymbol{a}, \boldsymbol{b}]$, Theorem 5.2.1(iii) and Lemma 5.2.5 imply $V_{\mathcal{HK}}F(A) = V_{\mathcal{HK}}F(A \cap \bigcup_{k=1}^{\infty} X_k) + V_{\mathcal{HK}}F(A \setminus \bigcup_{k=1}^{\infty} X_k)$:

$$\begin{aligned}&V_{\mathcal{HK}}F(A) \\ &\leq V_{\mathcal{HK}}F\left(A \cap \bigcup_{k=1}^{\infty} X_k\right) + V_{\mathcal{HK}}F\left(A \setminus \bigcup_{k=1}^{\infty} X_k\right) \\ &\leq \sum_{k=1}^{\infty} V_{\mathcal{HK}}F(A \cap X_k) + V_{\mathcal{HK}}F\left(A \setminus \bigcup_{k=1}^{\infty} X_k\right) \\ &\leq \lim_{n\to\infty} \left\{ \sum_{k=1}^{n} V_{\mathcal{HK}}F(A \cap X_k) + V_{\mathcal{HK}}F\left(A \setminus \bigcup_{k=1}^{n} X_k\right) \right\} \\ &= \lim_{n\to\infty} \left\{ V_{\mathcal{HK}}F\left(A \cap \bigcup_{k=1}^{n} X_k\right) + V_{\mathcal{HK}}F\left(A \setminus \bigcup_{k=1}^{n} X_k\right) \right\} \\ &= V_{\mathcal{HK}}F(A).\end{aligned}$$

□

Definition 5.2.8. The class $\mathcal{B}$ of Borel subsets of $[\boldsymbol{a}, \boldsymbol{b}]$ is the smallest σ-algebra that contains all closed sets in $[\boldsymbol{a}, \boldsymbol{b}]$.

Lemma 5.2.9. *Let $F : \mathcal{I}_m([\boldsymbol{a}, \boldsymbol{b}]) \longrightarrow \mathbb{R}$. Then every Borel subset of $[\boldsymbol{a}, \boldsymbol{b}]$ is $V_{\mathcal{HK}}F$-measurable.*

Proof. This is a consequence of Lemmas 5.2.2, 5.2.7 and Definition 5.2.8. □

Theorem 5.2.10. *Let $F : \mathcal{I}_m([\boldsymbol{a}, \boldsymbol{b}]) \longrightarrow \mathbb{R}$. If $(X_k)_{k=1}^{\infty}$ is a sequence of pairwise disjoint Borel subsets of $[\boldsymbol{a}, \boldsymbol{b}]$, then*

$$V_{\mathcal{HK}}F\Big(\bigcup_{k=1}^{\infty} X_k\Big) = \sum_{k=1}^{\infty} V_{\mathcal{HK}}F(X_k).$$

Proof. First we observe that $\bigcup_{k=1}^{\infty} X_k$ is a Borel set and so it is $V_{\mathcal{HK}}F$-measurable by Lemma 5.2.9. Since Theorem 5.2.1(iii) implies

$$V_{\mathcal{HK}}F\Big(\bigcup_{k=1}^{\infty} X_k\Big) \leq \sum_{k=1}^{\infty} V_{\mathcal{HK}}F(X_k), \tag{5.2.3}$$

it remains to prove that

$$\sum_{k=1}^{\infty} V_{\mathcal{HK}}F(X_k) \leq V_{\mathcal{HK}}F\Big(\bigcup_{k=1}^{\infty} X_k\Big). \tag{5.2.4}$$

Using Lemma 5.2.5 with $A = [\boldsymbol{a}, \boldsymbol{b}]$, we get

$$V_{\mathcal{HK}}F\Big(\bigcup_{k=1}^{n} X_k\Big) = \sum_{k=1}^{n} V_{\mathcal{HK}}F(X_k) \tag{5.2.5}$$

for every positive integers n. It is now clear that (5.2.4) follows from (5.2.5) and the monotonicity of $V_{\mathcal{HK}}F$:

$$\sum_{k=1}^{\infty} V_{\mathcal{HK}}F(X_k) = \lim_{n\to\infty} V_{\mathcal{HK}}F\Big(\bigcup_{k=1}^{n} X_k\Big) \leq V_{\mathcal{HK}}F\Big(\bigcup_{k=1}^{\infty} X_k\Big).$$

□

Theorem 5.2.11. *Let $F : \mathcal{I}_m([\boldsymbol{a}, \boldsymbol{b}]) \longrightarrow \mathbb{R}$. If $(X_n)_{n=1}^{\infty}$ is an increasing sequence of Borel subsets of $[\boldsymbol{a}, \boldsymbol{b}]$, then*

$$V_{\mathcal{HK}}F\Big(\bigcup_{k=1}^{\infty} X_k\Big) = \lim_{n\to\infty} V_{\mathcal{HK}}F(X_n).$$

Proof. By Theorem 5.2.10,

$$\begin{aligned} V_{\mathcal{HK}}F\Big(\bigcup_{k=1}^{\infty} X_k\Big) &= V_{\mathcal{HK}}F\Big(\bigcup_{k=1}^{\infty}(X_k \backslash X_{k-1})\Big), \text{ where } X_0 = \emptyset, \\ &= \sum_{k=1}^{\infty} V_{\mathcal{HK}}F(X_k \backslash X_{k-1}) \\ &= \sum_{k=1}^{\infty}(V_{\mathcal{HK}}F(X_k) - V_{\mathcal{HK}}F(X_{k-1})) \\ &= \lim_{n\to\infty} V_{\mathcal{HK}}F(X_n). \end{aligned}$$

□

Theorem 5.2.12. *Let* $F : \mathcal{I}_m([\boldsymbol{a}, \boldsymbol{b}]) \longrightarrow \mathbb{R}$. *If* $(X_k)_{k=1}^{\infty}$ *is a decreasing sequence of Borel subsets of* $[\boldsymbol{a}, \boldsymbol{b}]$ *and* $V_{\mathcal{HK}}F(X_1)$ *is finite, then*

$$V_{\mathcal{HK}}F\Big(\bigcap_{k=1}^{\infty} X_k\Big) = \lim_{n\to\infty} V_{\mathcal{HK}}F(X_n). \tag{5.2.6}$$

Proof. First, we observe that Theorem 5.2.1(ii) and our hypotheses imply

$$V_{\mathcal{HK}}F\Big(\bigcap_{k=1}^{\infty} X_k\Big) \le \lim_{n\to\infty} V_{\mathcal{HK}}F(X_n) \le V_{\mathcal{HK}}F(X_1) < \infty. \tag{5.2.7}$$

It remains to prove that

$$\lim_{n\to\infty} V_{\mathcal{HK}}F(X_n) \le V_{\mathcal{HK}}F\Big(\bigcap_{k=1}^{\infty} X_k\Big) \tag{5.2.8}$$

Using the finiteness of $V_{\mathcal{HK}}F(X_1)$ and (5.2.7) again, we see that (5.2.8) holds if and only if

$$V_{\mathcal{HK}}F(X_1) - \lim_{n\to\infty} V_{\mathcal{HK}}F(X_n) \ge V_{\mathcal{HK}}F(X_1) - V_{\mathcal{HK}}F\Big(\bigcap_{k=1}^{\infty} X_k\Big). \tag{5.2.9}$$

In view of Theorem 5.2.10, (5.2.9) holds if and only if

$$\lim_{n\to\infty} V_{\mathcal{HK}}F(X_1 \backslash X_n) \ge V_{\mathcal{HK}}F\Big(\bigcup_{k=1}^{\infty}(X_1 \backslash X_k)\Big). \tag{5.2.10}$$

The proof is now complete because (5.2.10) follows from Theorem 5.2.11.

□

The following example shows that the conclusion (5.2.6) need not hold if $(X_n)_{n=1}^{\infty}$ is a decreasing sequence of Borel subsets of $[\boldsymbol{a}, \boldsymbol{b}]$ satisfying $V_{\mathcal{HK}}F(X_n) = \infty$ for every $n \in \mathbb{N}$.

Example 5.2.13. Let f be given as in Example 2.5.6 and let F denote the indefinite Henstock-Kurzweil integral of f. For each $n \in \mathbb{N}$, we have $f \in HK([0, \frac{1}{n}]^2)\backslash L^1([0, \frac{1}{n}]^2)$ and so $V_{\mathcal{HK}}F([0, \frac{1}{n}]^2) = \infty$ by Theorem 4.2.3. Consequently,

$$V_{\mathcal{HK}}F\Big(\bigcap_{k=1}^{\infty}[0, \frac{1}{k}]^2\Big) = V_{\mathcal{HK}}F(\{(0,0)\}) = 0 \neq \lim_{n\to\infty} V_{\mathcal{HK}}F\Big([0, \frac{1}{n}]^2\Big).$$

5.3 Another characterization of Lebesgue integrable functions

In this section we will state and prove a refinement of Theorems 3.10.2 and 3.10.12 via the Henstock variational measure. We begin with two useful lemmas concerning the Henstock variational measure.

Lemma 5.3.1. *Let $F : \mathcal{I}_m([\boldsymbol{a}, \boldsymbol{b}]) \longrightarrow \mathbb{R}$ be an additive interval function. If $X \subseteq [\boldsymbol{a}, \boldsymbol{b}]$ and $V_{\mathcal{HK}}F(X)$ is finite, then for each $\varepsilon > 0$ there exists a gauge δ on X such that*

$$\sum_{i=1}^{p} |F(I_i)| < V_{\mathcal{HK}}F\Big(X \cap \bigcup_{i=1}^{p} I_i\Big) + \varepsilon$$

for each X-tagged δ-fine Perron subpartition $\{(\boldsymbol{t}_1, I_1), \ldots, (\boldsymbol{t}_p, I_p)\}$ of $[\boldsymbol{a}, \boldsymbol{b}]$.

Proof. For each $\varepsilon > 0$ we choose a gauge δ on X such that

$$V(F, X, \delta) < V_{\mathcal{HK}}F(X) + \frac{\varepsilon}{2}.$$

Consider any X-tagged δ-fine Perron subpartition $\{(\boldsymbol{t}_1, I_1), \ldots, (\boldsymbol{t}_p, I_p)\}$ of $[\boldsymbol{a}, \boldsymbol{b}]$. If $X\backslash\bigcup_{i=1}^{p} I_i$ is empty, then there is nothing to prove. On the other hand, suppose that $X\backslash\bigcup_{i=1}^{p} I_i$ is non-empty. Then the map $\boldsymbol{x} \mapsto \Delta(\boldsymbol{x}) := \min\{\delta(\boldsymbol{x}), \mathrm{dist}(\boldsymbol{x}, \bigcup_{i=1}^{p} I_i)\}$ is a gauge on $X\backslash\bigcup_{i=1}^{p} I_i$; by Cousin's Lemma we fix a $(X\backslash\bigcup_{i=1}^{p} I_i)$-tagged Δ-fine Perron subpartition $\{(\boldsymbol{y}_1, J_1), \ldots, (\boldsymbol{y}_q, J_q)\}$ of $[\boldsymbol{a}, \boldsymbol{b}]$ such that

$$V_{\mathcal{HK}}F\Big(X\backslash\bigcup_{i=1}^{p} I_i\Big) < \sum_{i=1}^{q} |F(J_i)| + \frac{\varepsilon}{2}.$$

Since it is clear that $\{(\boldsymbol{t}_1, I_1), \ldots, (\boldsymbol{t}_p, I_p)\} \cup \{(\boldsymbol{y}_1, J_1), \ldots, (\boldsymbol{y}_q, J_q)\}$ is an X-tagged δ-fine Perron subpartition of $[\boldsymbol{a}, \boldsymbol{b}]$, the previous inequalities and the countable subadditivity of $V_{\mathcal{HK}}F$ yield

$$\begin{aligned}
&\sum_{i=1}^{p} |F(I_i)| \\
&< V_{\mathcal{HK}}F(X) - \sum_{i=1}^{q} |F(J_i)| + \frac{\varepsilon}{2} \\
&< V_{\mathcal{HK}}F(X) - V_{\mathcal{HK}}F\left(X \setminus \bigcup_{i=1}^{p} I_i\right) + \varepsilon \\
&\leq V_{\mathcal{HK}}F\left(X \cap \bigcup_{i=1}^{p} I_i\right) + \varepsilon.
\end{aligned}$$

□

Lemma 5.3.2. *Let $F : \mathcal{I}_m([\boldsymbol{a}, \boldsymbol{b}]) \longrightarrow \mathbb{R}$ be an interval function such that $V_{\mathcal{HK}}F \ll \mu_m$. If $V_{\mathcal{HK}}F([\boldsymbol{a}, \boldsymbol{b}])$ is finite, then for each $\varepsilon > 0$ there exists $\delta > 0$ such that $V_{\mathcal{HK}}F(Y) < \varepsilon$ whenever $Y \subseteq [\boldsymbol{a}, \boldsymbol{b}]$ with $\mu_m^*(Y) < \delta$.*

Proof. Suppose not. Then there exists $\varepsilon_0 > 0$ with the following property: for each $n \in \mathbb{N}$ there exists $Y_n \subseteq [\boldsymbol{a}, \boldsymbol{b}]$ such that $\mu_m^*(Y_n) < \frac{1}{2^n}$ but $V_{\mathcal{HK}}F(Y_n) \geq \varepsilon_0$. According to the proof of Theorem 5.1.6, each Y_n may be assumed to be a Borel set. Set $Y = \bigcap_{k=1}^{\infty} \bigcup_{k=n}^{\infty} Y_k$. Then Y is a μ_m-negligible Borel set:

$$0 \leq \mu_m(Y) = \lim_{n\to\infty} \mu_m\left(\bigcup_{k=n}^{\infty} Y_k\right) \leq \lim_{n\to\infty} \sum_{k=n}^{\infty} \mu_m(Y_k) = 0.$$

Since $V_{\mathcal{HK}}F \ll \mu_m$, we conclude that $V_{\mathcal{HK}}F(Y) = 0$. On the other hand, Theorems 5.2.12 and 5.2.1(ii) yield

$$V_{\mathcal{HK}}F(Y) = \lim_{n\to\infty} V_{\mathcal{HK}}F\left(\bigcup_{k=n}^{\infty} Y_k\right) \geq \limsup_{n\to\infty} V_{\mathcal{HK}}F(Y_n) \geq \varepsilon_0,$$

which contradicts the equality $V_{\mathcal{HK}}F(Y) = 0$. This contradiction completes the proof. □

The following result is a consequence of Lemmas 5.3.1 and 5.3.2.

Theorem 5.3.3. *Let $F : \mathcal{I}_m([\boldsymbol{a}, \boldsymbol{b}]) \longrightarrow \mathbb{R}$ be an additive interval function. If $V_{\mathcal{HK}}F([\boldsymbol{a}, \boldsymbol{b}])$ is finite, then $V_{\mathcal{HK}}F \ll \mu_m$ if and only if F is absolutely continuous.*

Proof. ($\Longrightarrow$) Suppose that $V_{\mathcal{HK}}F \ll \mu_m$. Since $V_{\mathcal{HK}}F([\boldsymbol{a}, \boldsymbol{b}])$ is finite, for each $\varepsilon > 0$ we apply Lemma 5.3.2 to pick $\eta > 0$ so that

$$Z \subseteq [\boldsymbol{a}, \boldsymbol{b}] \text{ with } \mu_m^*(Z) < \eta \implies V_{\mathcal{HK}}F(Z) < \frac{\varepsilon}{2}.$$

Let $J_1, \ldots, J_q$ be a finite sequence of non-overlapping subintervals of $[\boldsymbol{a}, \boldsymbol{b}]$ such that $\sum_{k=1}^{q} \mu_m(J_k) < \eta$. We want to prove that $\sum_{k=1}^{q} |F(J_k)| < \varepsilon$. To do this, we apply Lemma 5.3.1 to choose a gauge δ on $[\boldsymbol{a}, \boldsymbol{b}]$ so that

$$\sum_{i=1}^{p} |F(I_i)| < V_{\mathcal{HK}}F\Big(\bigcup_{i=1}^{p} I_i\Big) + \frac{\varepsilon}{2}$$

for each δ-fine Perron subpartition $\{(\boldsymbol{t}_1, I_1), \ldots, (\boldsymbol{t}_p, I_p)\}$ of $[\boldsymbol{a}, \boldsymbol{b}]$. Finally, for each $k \in \{1, \ldots, q\}$ we apply Cousin's Lemma to fix a δ-fine Perron partition P_k of J_k. As F is additive, a direct computation yields the desired inequality:

$$\sum_{k=1}^{q} |F(J_k)| \leq \sum_{k=1}^{q} \sum_{(\boldsymbol{t}_k, I_k) \in P_k} |F(I_k)| \leq V_{\mathcal{HK}}F\Big(\bigcup_{k=1}^{q} \bigcup_{(\boldsymbol{t}_k, I_k) \in P_k} I_k\Big) + \frac{\varepsilon}{2} < \varepsilon.$$

($\Longleftarrow$) Conversely, suppose that F is absolutely continuous. Let $Z \subset [\boldsymbol{a}, \boldsymbol{b}]$ be any μ_m-negligible set. For each $n \in \mathbb{N}$ there exists $\eta_n > 0$ such that if $\{[\boldsymbol{u}_k, \boldsymbol{v}_k]\}_{k=1}^{r_n}$ is a finite collection of non-overlapping subintervals of $[\boldsymbol{a}, \boldsymbol{b}]$ with $\sum_{k=1}^{r_n} \mu_m([\boldsymbol{u}_k, \boldsymbol{v}_k]) < \eta_n$, then $\sum_{k=1}^{r_n} |F([\boldsymbol{u}_k, \boldsymbol{v}_k])| < \frac{1}{n}$. Next, we choose a gauge δ_n on Z such that

$$\sum_{(\boldsymbol{t}_n, I_n) \in P_n} \mu_m(I_n) < \eta_n$$

whenever P_n is a Z-tagged δ_n-fine Peron subpartition of $[\boldsymbol{a}, \boldsymbol{b}]$. Thus

$$V_{\mathcal{HK}}F(Z) \leq V(F, Z, \delta_n) \leq \frac{1}{n}$$

for all $n \in \mathbb{N}$. As $n \in \mathbb{N}$ is arbitrary, $V_{\mathcal{HK}}F(Z) = 0$. The proof is complete. □

We can now state and prove the main result of this section.

Theorem 5.3.4. *Let $F : \mathcal{I}_m([\boldsymbol{a}, \boldsymbol{b}]) \longrightarrow \mathbb{R}$ be an additive interval function. If $V_{\mathcal{HK}}F([\boldsymbol{a}, \boldsymbol{b}])$ is finite, then the following conditions are equivalent.*

(i) *F is the indefinite Lebesgue integral of some $f \in L^1[\boldsymbol{a}, \boldsymbol{b}]$.*
(ii) *$V_{\mathcal{HK}}F \ll \mu_m$.*

Proof. This follows from Theorems 3.10.2, 3.10.12 and 5.3.3. □

5.4 A result of Kurzweil and Jarník revisited

In Section 4.2 we prove that if $f \in HK[\boldsymbol{a}, \boldsymbol{b}]$ and F is the indefinite Henstock-Kurzweil integral of f, then there exists a sequence $(X_n)_{n=1}^{\infty}$ of closed sets such that $\bigcup_{k=1}^{\infty} X_k = [\boldsymbol{a}, \boldsymbol{b}]$ and $V_{\mathcal{HK}}F(X_n)$ is finite for every $n \in \mathbb{N}$. The aim of this section is to prove that if $F : \mathcal{I}_m([\boldsymbol{a}, \boldsymbol{b}]) \longrightarrow \mathbb{R}$ is an additive interval function such that $V_{\mathcal{HK}}F \ll \mu_m$, then there exists a sequence $(X_n)_{n=1}^{\infty}$ of closed sets such that $\bigcup_{k=1}^{\infty} X_k = [\boldsymbol{a}, \boldsymbol{b}]$ and $V_{\mathcal{HK}}F(X_n)$ is finite for every $n \in \mathbb{N}$. Our first goal is to establish an analogous version of Theorem 2.4.9.

Definition 5.4.1. Let $I \in \mathcal{I}_m([\boldsymbol{a}, \boldsymbol{b}])$ and let $F : \mathcal{I}_m(I) \longrightarrow \mathbb{R}$. The *oscillation* of F on I is the extended number

$$\omega(F, I) = \sup\left\{ |F(J)| : J \in \mathcal{I}_m(I) \right\}.$$

Lemma 5.4.2. *Let $F : \mathcal{I}_m([\boldsymbol{a}, \boldsymbol{b}]) \longrightarrow \mathbb{R}$ be an additive interval function such that $V_{\mathcal{HK}}F \ll \mu_m$. If $Z \subset [\boldsymbol{a}, \boldsymbol{b}]$ is μ_m-negligible, then for each $\varepsilon > 0$ there exists a gauge δ on Z such that*

$$\sum_{(\boldsymbol{t}, I) \in P} \omega(F, I) < \varepsilon$$

for each Z-tagged δ-fine Perron subpartition P of $[\boldsymbol{a}, \boldsymbol{b}]$.

Proof. This is an immediate consequence of Lemma 2.4.6. □

In order to formulate the next lemma, we need the following useful notation.

Notation 5.4.3. For any set $\prod_{i=1}^{m} X_i \subseteq [\boldsymbol{a}, \boldsymbol{b}]$, $k \in \{1, \ldots, m\}$, and $T \subseteq [a_k, b_k]$, we write $\Phi_{\prod_{i=1}^{m} X_i, k}(T) := \prod_{i=1}^{m} Y_i$, where $Y_k = T$ and $Y_i = X_i$ for all $i \in \{1, \ldots, m\} \backslash \{k\}$.

Lemma 5.4.4. *Let $F : \mathcal{I}_m([\boldsymbol{a}, \boldsymbol{b}]) \longrightarrow \mathbb{R}$ be an additive interval function such that $V_{\mathcal{HK}}F \ll \mu_m$. If $x \in [a_k, b_k]$ for some $k \in \{1, \ldots, m\}$, then for each $\varepsilon > 0$ there exists a gauge δ on $\Phi_{[\boldsymbol{a}, \boldsymbol{b}], k}(\{x\})$ such that*

$$\sum_{(\boldsymbol{t}, I) \in P} \omega(F, I) < \varepsilon$$

for each $\Phi_{[\boldsymbol{a}, \boldsymbol{b}], k}(\{x\})$-tagged δ-fine Perron subpartition P of $[\boldsymbol{a}, \boldsymbol{b}]$.

Proof. Since $\mu_m\big(\Phi_{[\boldsymbol{a},\boldsymbol{b}],k}(\{x\})\big) = 0$, the lemma follows from Lemma 5.4.2. □

We are ready to state and prove an analogous version of Theorem 2.4.9 for the Henstock variational measure.

Theorem 5.4.5. *Let $F : \mathcal{I}_m([\boldsymbol{a},\boldsymbol{b}]) \longrightarrow \mathbb{R}$ be an additive interval function. If $V_{\mathcal{HK}}F \ll \mu_m$, then for each $\varepsilon > 0$ there exists $\eta_0 > 0$ such that*

$$|F(J) - F(K)| < \varepsilon$$

for each $J, K \in \mathcal{I}_m([\boldsymbol{a},\boldsymbol{b}])$ with $\mu_m(J\Delta K) < \eta_0$.

Proof. Following the proof of Theorem 2.4.8, it suffices to prove that if $k \in \{1,\ldots,m\}$ and $\varepsilon > 0$, then there exists $\eta_k > 0$ such that $\omega(F, \Phi_{[\boldsymbol{a},\boldsymbol{b}],k}([u,v])) < \varepsilon$ whenever $[u,v] \in \mathcal{I}_1([a_k,b_k])$ with $v - u < \eta_k$.

For each $x \in [a_k, b_k]$ we apply Lemma 5.4.4 and Cousin's Lemma (Theorem 2.2.2) to choose $\delta_k(x) > 0$ such that

$$\omega(F, \Phi_{[\boldsymbol{a},\boldsymbol{b}],k}([u',v'])) < \frac{\varepsilon}{2}$$

whenever $[u',v'] \in \mathcal{I}_1([a_k,b_k])$ with $x \in [u',v'] \subset (x-\delta_k(x), x+\delta_k(x))$. Next we use Cousin's Lemma (Theorem 1.1.5) to fix a δ_k-fine Perron partition P_k of $[a_k,b_k]$. Finally, we set $\eta_k = \min\{v - u : (t,[u,v]) \in P_k\}$ to complete the argument. □

Our next goal is to establish a modification of Lemma 5.3.1; see Theorem 5.4.9. We begin with the following lemma whose proof is similar to that of Lemma 5.3.2.

Lemma 5.4.6. *Let $F : \mathcal{I}_m([\boldsymbol{a},\boldsymbol{b}]) \longrightarrow \mathbb{R}$ be an interval function such that $V_{\mathcal{HK}}F \ll \mu_m$. If $V_{\mathcal{HK}}F(X)$ is finite for some $X \subseteq [\boldsymbol{a},\boldsymbol{b}]$, then for each $\varepsilon > 0$ there exists $\delta_0 > 0$ such that $V_{\mathcal{HK}}F(W) < \varepsilon$ for each set $W \subseteq X$ with $\mu_m^*(W) < \delta_0$.*

Proof. Exercise. □

The proof of the following lemma is similar to that of Theorem 5.1.6.

Lemma 5.4.7. *Let $F : \mathcal{I}_m([\boldsymbol{a},\boldsymbol{b}]) \longrightarrow \mathbb{R}$ be an interval function such that $V_{\mathcal{HK}}F \ll \mu_m$. If $X \subseteq [\boldsymbol{a},\boldsymbol{b}]$ and $V_{\mathcal{HK}}F(X)$ is finite, then there exists a Borel set $Z \subseteq [\boldsymbol{a},\boldsymbol{b}]$ such that $X \subseteq Z$ and $V_{\mathcal{HK}}F(Z \cap A) = V_{\mathcal{HK}}F(X \cap A)$ for every $A \subseteq [\boldsymbol{a},\boldsymbol{b}]$.*

Proof. Exercise. □

A set $Z \subseteq [\boldsymbol{a}, \boldsymbol{b}]$ is said to be an F_σ-set if it can be written as a countable union of closed sets.

Theorem 5.4.8. *Let $F : \mathcal{I}_m([\boldsymbol{a}, \boldsymbol{b}]) \longrightarrow \mathbb{R}$ be an additive interval function such that $V_{\mathcal{HK}}F \ll \mu_m$. If $X \subseteq [\boldsymbol{a}, \boldsymbol{b}]$ and $\varepsilon > 0$, then there exist an F_σ-set Y and an upper semicontinuous gauge Δ on Y such that $X \subseteq Y \subseteq [\boldsymbol{a}, \boldsymbol{b}]$ and*

$$V(F, Y, \Delta) \leq V_{\mathcal{HK}}F(X) + \varepsilon.$$

Proof. If $X \subseteq [\boldsymbol{a}, \boldsymbol{b}] \backslash (\boldsymbol{a}, \boldsymbol{b})$ or $V_{\mathcal{HK}}F(X) = \infty$, then the theorem is obviously true. Henceforth, we assume that X is a Borel set with $X \cap (\boldsymbol{a}, \boldsymbol{b}) \neq \emptyset$ and $V_{\mathcal{HK}}F(X)$ is finite.

Let $\varepsilon > 0$ be given. By Lemma 5.3.1 there exists a gauge δ_0 on X such that

$$\sum_{i=1}^{p} |F(I_i)| < V_{\mathcal{HK}}F\left(X \cap \bigcup_{i=1}^{p} I_i\right) + \frac{\varepsilon}{4(3^m)}$$

for each X-tagged δ-fine Perron subpartition $\{(\boldsymbol{t}_1, I_1), \ldots, (\boldsymbol{t}_p, I_p)\}$ of $[\boldsymbol{a}, \boldsymbol{b}]$. Moreover, we may assume that $B(\boldsymbol{x}, \delta_0(\boldsymbol{x})) \subset (\boldsymbol{a}, \boldsymbol{b})$ whenever $\boldsymbol{x} \in (\boldsymbol{a}, \boldsymbol{b}) \cap X$.

For each $\boldsymbol{\lambda} \in \{-1, 0, 1\}^m$, let $P_{\boldsymbol{\lambda}} := \{(\boldsymbol{t}, [\boldsymbol{u}, \boldsymbol{v}]) \in P : \Lambda(\boldsymbol{t}, [\boldsymbol{u}, \boldsymbol{v}]) = \boldsymbol{\lambda}\}$. Since X is a Borel set and $V_{\mathcal{HK}}F \ll \mu_m$, the following inequality

$$\sum_{\boldsymbol{\lambda} \in \{-1,0,1\}^m} V_{\mathcal{HK}}F\left(X \cap \bigcup_{(\boldsymbol{t}, [\boldsymbol{u}, \boldsymbol{v}]) \in P_{\boldsymbol{\lambda}}} [\boldsymbol{u}, \boldsymbol{v}]\right) \leq V_{\mathcal{HK}}F(X) \tag{5.4.1}$$

follows from Theorems 5.2.1 and 5.2.10:

$$\begin{aligned} V_{\mathcal{HK}}F(X) &\geq V_{\mathcal{HK}}F\left(X \cap \bigcup_{\boldsymbol{\lambda} \in \{-1,0,1\}^m} \bigcup_{(\boldsymbol{t}, [\boldsymbol{u}, \boldsymbol{v}]) \in P_{\boldsymbol{\lambda}}} (\boldsymbol{u}, \boldsymbol{v})\right) \\ &= \sum_{\boldsymbol{\lambda} \in \{-1,0,1\}^m} V_{\mathcal{HK}}F\left(X \cap \bigcup_{(\boldsymbol{t}, [\boldsymbol{u}, \boldsymbol{v}]) \in P_{\boldsymbol{\lambda}}} (\boldsymbol{u}, \boldsymbol{v})\right) \\ &= \sum_{\boldsymbol{\lambda} \in \{-1,0,1\}^m} V_{\mathcal{HK}}F\left(X \cap \bigcup_{(\boldsymbol{t}, [\boldsymbol{u}, \boldsymbol{v}]) \in P_{\boldsymbol{\lambda}}} [\boldsymbol{u}, \boldsymbol{v}]\right). \end{aligned}$$

For each $k \in \mathbb{N}$, set

$$X_k = \left\{\boldsymbol{x} \in [\boldsymbol{a}, \boldsymbol{b}] : \delta_0(\boldsymbol{x}) \geq \frac{1}{k}\right\} \text{ and } Y_k = \overline{X_k}.$$

We claim that $Y := \bigcup_{k=1}^{\infty} Y_k$ has the desired properties.

Define a gauge Δ on Y by setting

$$\Delta(\boldsymbol{x}) = \begin{cases} 1 & \text{if } \boldsymbol{x} \in Y_1, \\ \min\{\frac{1}{k}, \operatorname{dist}(\boldsymbol{x}, Y_{k-1})\} & \text{if } \boldsymbol{x} \in Y_k \backslash Y_{k-1} \text{ for some integer } k \geq 2 \end{cases}$$

Then Δ is upper semicontinuous on Y.

Let $Y_0 = \emptyset$ and let P be any Y-tagged Δ-fine Perron subpartition of $[\boldsymbol{a}, \boldsymbol{b}]$. If $P_{\boldsymbol{\lambda}}$ is non-empty for some $\boldsymbol{\lambda} \in \{-1, 0, 1\}^m$, we choose $\eta_1 > 0$ (independent of $\boldsymbol{\lambda} \in \{-1, 0, 1\}^m$) so that the following conditions are met for every $\eta \in (0, \eta_1)$:

(A) $\{[\boldsymbol{u} + \eta\boldsymbol{\lambda}, \boldsymbol{v} + \eta\boldsymbol{\lambda}] : (\boldsymbol{t}, [\boldsymbol{u}, \boldsymbol{v}]) \in P_{\boldsymbol{\lambda}}\}$ is a finite collection of non-overlapping intervals in $\mathcal{I}_m([\boldsymbol{a}, \boldsymbol{b}])$.

(B) If $(\boldsymbol{t}, [\boldsymbol{u}, \boldsymbol{v}]) \in P_{\boldsymbol{\lambda}}$ and $\boldsymbol{t} \in Y_k \backslash Y_{k-1}$ for some $k \in \mathbb{N}$, then $X_k \cap [\boldsymbol{u} + \eta\boldsymbol{\lambda}, \boldsymbol{v} + \eta\boldsymbol{\lambda}] \neq \emptyset$ and $Y_{k-1} \cap [\boldsymbol{u} + \eta\boldsymbol{\lambda}, \boldsymbol{v} + \eta\boldsymbol{\lambda}] = \emptyset$.

Since $F : \mathcal{I}_m([\boldsymbol{a}, \boldsymbol{b}]) \longrightarrow \mathbb{R}$ is an additive interval function and $V_{\mathcal{HK}}F \ll \mu_m$, we use (A) and Theorem 5.4.5 to choose an $\eta_2 \in (0, \eta_1)$ so that the following condition holds for every $\boldsymbol{\lambda} \in \{-1, 0, 1\}^m$:

(C) If $\eta \in (0, \eta_2)$, then

$$|F([\boldsymbol{u} + \eta\boldsymbol{\lambda}, \boldsymbol{v} + \eta\boldsymbol{\lambda}]) - F([\boldsymbol{u}, \boldsymbol{v}])| < \frac{\varepsilon}{4(3^m)\operatorname{card}(P)}$$

whenever $(\boldsymbol{t}, [\boldsymbol{u}, \boldsymbol{v}]) \in P_{\boldsymbol{\lambda}}$.

Since $F : \mathcal{I}_m([\boldsymbol{a}, \boldsymbol{b}]) \longrightarrow \mathbb{R}$ is an interval function such that $V_{\mathcal{HK}}F \ll \mu_m$ and $V_{\mathcal{HK}}F(X)$ is finite, we use (A) and Lemma 5.4.6 to choose $\eta_3 \in (0, \eta_1)$ so that the following condition holds for every $\boldsymbol{\lambda} \in \{-1, 0, 1\}^m$:

(D) If $\eta \in (0, \eta_3)$, then

$$V_{\mathcal{HK}}F\Big(X \cap \bigcup_{(\boldsymbol{t},[\boldsymbol{u},\boldsymbol{v}]) \in P_{\boldsymbol{\lambda}}} [\boldsymbol{u} + \eta\boldsymbol{\lambda}, \boldsymbol{v} + \eta\boldsymbol{\lambda}]\Big) < V_{\mathcal{HK}}F\Big(X \cap \bigcup_{(\boldsymbol{t},[\boldsymbol{u},\boldsymbol{v}]) \in P_{\boldsymbol{\lambda}}} [\boldsymbol{u}, \boldsymbol{v}]\Big) + \frac{\varepsilon}{4}.$$

Set $\eta_0 := \frac{1}{2}\min\{\eta_1, \eta_2, \eta_3\}$. Then (A), (C) and the triangle inequality give

$$\sum_{(\boldsymbol{t},[\boldsymbol{u},\boldsymbol{v}]) \in P} |F([\boldsymbol{u}, \boldsymbol{v}])| < \frac{\varepsilon}{4} + \sum_{\boldsymbol{\lambda} \in \{-1,0,1\}^m} \sum_{(\boldsymbol{t},[\boldsymbol{u},\boldsymbol{v}]) \in P_{\boldsymbol{\lambda}}} |F([\boldsymbol{u} + \eta_0\boldsymbol{\lambda}, \boldsymbol{v} + \eta_0\boldsymbol{\lambda}])|. \tag{5.4.2}$$

It remains to prove that the right-hand side of (5.4.2) is less than $V_{\mathcal{HK}}F(X)+\varepsilon$. Suppose that $P_{\boldsymbol{\lambda}}$ is non-empty for some $\boldsymbol{\lambda} \in \{-1,0,1\}^m$. Then for each $(\boldsymbol{t},[\boldsymbol{u},\boldsymbol{v}]) \in P_{\boldsymbol{\lambda}}$ with $\boldsymbol{t} \in Y_k \backslash Y_{k-1}$ and $k \in \mathbb{N}$, we use (B) to select $\boldsymbol{z} \in (X_k \backslash Y_{k-1}) \cap [\boldsymbol{u}+\eta_0\boldsymbol{\lambda}, \boldsymbol{v}+\eta_0\boldsymbol{\lambda}]$ so that $\Delta(\boldsymbol{t}) \leq \frac{1}{k} \leq \delta_0(\boldsymbol{z})$. As $P_{\boldsymbol{\lambda}}$ is Δ-fine, we conclude that $|||(\boldsymbol{v}+\eta_0\boldsymbol{\lambda})-(\boldsymbol{u}+\eta_0\boldsymbol{\lambda})||| < \Delta(\boldsymbol{t}) \leq \frac{1}{k} \leq \delta_0(\boldsymbol{z})$. Consequently, our choice of δ_0 yields

$$\sum_{(\boldsymbol{t},[\boldsymbol{u},\boldsymbol{v}])\in P_{\boldsymbol{\lambda}}} |F([\boldsymbol{u}+\eta_0\boldsymbol{\lambda}, \boldsymbol{v}+\eta_0\boldsymbol{\lambda}])|$$

$$< V_{\mathcal{HK}}F\Bigg(X \cap \bigcup_{(\boldsymbol{t},[\boldsymbol{u},\boldsymbol{v}])\in P_{\boldsymbol{\lambda}}} [\boldsymbol{u}+\eta_0\boldsymbol{\lambda}, \boldsymbol{v}+\eta_0\boldsymbol{\lambda}]\Bigg) + \frac{\varepsilon}{4(3^m)}. \tag{5.4.3}$$

Finally, we combine (5.4.2), (5.4.3), (D) and (5.4.1) to obtain the desired result:

$$\sum_{(\boldsymbol{t},[\boldsymbol{u},\boldsymbol{v}])\in P} |F([\boldsymbol{u},\boldsymbol{v}])| < V_{\mathcal{HK}}F(X)+\varepsilon.$$

□

The following theorem is a useful modification of Lemma 5.3.1.

Theorem 5.4.9. *Let $F : \mathcal{I}_m([\boldsymbol{a},\boldsymbol{b}]) \longrightarrow \mathbb{R}$ be an additive interval function such that $V_{\mathcal{HK}}F \ll \mu_m$. If $X \subseteq [\boldsymbol{a},\boldsymbol{b}]$ is closed and $V_{\mathcal{HK}}F(X)$ is finite, then for each $\varepsilon > 0$ there exists an upper semicontinuous gauge δ on X such that*

$$\sum_{i=1}^{p} |F(I_i)| < V_{\mathcal{HK}}F\left(X \cap \bigcup_{i=1}^{p} I_i\right) + \varepsilon$$

for each X-tagged δ-fine Perron subpartition $\{(\boldsymbol{t}_1, I_1), \ldots, (\boldsymbol{t}_p, I_p)\}$ of $[\boldsymbol{a},\boldsymbol{b}]$.

Proof. In view of Theorem 5.4.8, the proof is similar to that of Lemma 5.3.1. □

A set $X \subseteq [\boldsymbol{a},\boldsymbol{b}]$ is said to be perfect if for each $\boldsymbol{x} \in X$ there exists a sequence $(\boldsymbol{x}_n)_{n=1}^{\infty}$ in $X\backslash\{\boldsymbol{x}\}$ such that $\lim_{n\to\infty} |||\boldsymbol{x}_n - \boldsymbol{x}||| = 0$. The following result is useful for proving Theorem 5.4.11 below.

Theorem 5.4.10 ([59, Theorem 6.66]). *If $X \subseteq [\boldsymbol{a},\boldsymbol{b}]$ is closed, then X contains a perfect set X_0 and a countable set Y_0 such that $X = X_0 \cup Y_0$.*

The following theorem is the main result of this section.

Theorem 5.4.11. *Let $F : \mathcal{I}_m([\boldsymbol{a},\boldsymbol{b}]) \longrightarrow \mathbb{R}$ be an additive interval function such that $V_{\mathcal{HK}}F \ll \mu_m$. Then there exists a sequence $(X_n)_{n=1}^{\infty}$ of closed sets such that $[\boldsymbol{a},\boldsymbol{b}] = \bigcup_{k=1}^{\infty} X_k$ and $V_{\mathcal{HK}}F(X_n)$ is finite for all $n \in \mathbb{N}$.*

Proof. A point $\boldsymbol{x} \in [\boldsymbol{a}, \boldsymbol{b}]$ is said to be regular if there exists an open interval $I_{\boldsymbol{x}} \subseteq [\boldsymbol{a}, \boldsymbol{b}]$ with the following properties:

(i) $\boldsymbol{x} \in I_{\boldsymbol{x}}$;
(ii) there exists a sequence $(Z_{\boldsymbol{x},n})_{n=1}^{\infty}$ of closed sets such that $I_{\boldsymbol{x}} = \bigcup_{k=1}^{\infty} Z_{\boldsymbol{x},k}$;
(iii) $V_{\mathcal{HK}}F(Z_{\boldsymbol{x},k})$ is finite for all $k \in \mathbb{N}$.

Then the set X of all irregular points of $[\boldsymbol{a}, \boldsymbol{b}]$ is closed.

We claim that X is empty. Proceeding towards a contradiction, suppose that X is non-empty. Since $\mu_m([\boldsymbol{a}, \boldsymbol{b}] \backslash (\boldsymbol{a}, \boldsymbol{b})) = 0$ and $V_{\mathcal{HK}}F \ll \mu_m$, we may suppose that $X \subset (\boldsymbol{a}, \boldsymbol{b})$. In view of Theorem 5.4.10, we may further assume that X is perfect. We first construct an X-tagged δ_1-fine Perron subpartition P_1 of $[\boldsymbol{a}, \boldsymbol{b}]$ so that the following conditions are satisfied:

$(\text{i})_1$ $\text{card}(P_1) \geq 2$;
$(\text{ii})_1$ $[\boldsymbol{u}_1, \boldsymbol{v}_1] \subset (\boldsymbol{a}, \boldsymbol{b})$ whenever $(\boldsymbol{t}_1, [\boldsymbol{u}_1, \boldsymbol{v}_1]) \in P_1$;
$(\text{iii})_1$ $\boldsymbol{t}_1 \in (\boldsymbol{u}_1, \boldsymbol{v}_1)$ whenever $(\boldsymbol{t}_1, [\boldsymbol{u}_1, \boldsymbol{v}_1]) \in P_1$;
$(\text{iv})_1$

$$\sum_{(\boldsymbol{t}_1, [\boldsymbol{u}_1, \boldsymbol{v}_1]) \in P_1} \mu_m([\boldsymbol{u}_1, \boldsymbol{v}_1]) < \frac{1}{2};$$

$(\text{v})_1$ $\sum_{(\boldsymbol{t}_1, [\boldsymbol{u}_1, \boldsymbol{v}_1]) \in P_1} |F([\boldsymbol{u}_1, \boldsymbol{v}_1])| > 1.$

Since our construction of X implies that $V_{\mathcal{HK}}F(X \cap I) = \infty$ for every $I \in \mathcal{I}_m([\boldsymbol{a}, \boldsymbol{b}])$ satisfying $\mu_m(X \cap I) > 0$, there exist a gauge δ_1 on X, $\boldsymbol{\lambda} \in \{-1, 0, 1\}^m$ and an X-tagged δ_1-fine Perron subpartition Q_1 such that the following conditions hold:

(A) $\text{card}(Q_1) \geq 2$;
(B) $[\boldsymbol{u}, \boldsymbol{v}] \subset (\boldsymbol{a}, \boldsymbol{b})$ whenever $(\boldsymbol{t}, [\boldsymbol{u}, \boldsymbol{v}]) \in Q_1$;
(C) $\Lambda(\boldsymbol{t}, [\boldsymbol{u}, \boldsymbol{v}]) = \boldsymbol{\lambda}$ whenever $(\boldsymbol{t}, [\boldsymbol{u}, \boldsymbol{v}]) \in Q_1$;
(D)

$$\sum_{(\boldsymbol{t}, [\boldsymbol{u}, \boldsymbol{v}]) \in Q_1} \mu_m([\boldsymbol{u}, \boldsymbol{v}]) < \frac{1}{2};$$

(E) $\sum_{(\boldsymbol{t}, [\boldsymbol{u}, \boldsymbol{v}]) \in Q_1} |F([\boldsymbol{u}, \boldsymbol{v}])| > 1.$

Next we use (B), (C), (E) and Theorem 5.4.5 to select a sufficient small $\eta > 0$ so that

$$\left| \sum_{(\boldsymbol{t},[\boldsymbol{u},\boldsymbol{v}])\in Q_1} |F([\boldsymbol{u},\boldsymbol{v}])| - \sum_{(\boldsymbol{t},[\boldsymbol{u},\boldsymbol{v}])\in Q_1} |F([\boldsymbol{u}+\eta\boldsymbol{\lambda},\boldsymbol{v}+\eta\boldsymbol{\lambda}])| \right| < \sum_{(\boldsymbol{t},[\boldsymbol{u},\boldsymbol{v}])\in Q_1} |F([\boldsymbol{u},\boldsymbol{v}])| - 1,$$

and the following conditions hold for every $(\boldsymbol{t},[\boldsymbol{u},\boldsymbol{v}]) \in Q_1$:

(F) $\boldsymbol{t} \in (\boldsymbol{u}+\eta\boldsymbol{\lambda}, \boldsymbol{v}+\eta\boldsymbol{\lambda})$; in particular, $(\boldsymbol{u}+\eta\ \boldsymbol{\lambda}, \boldsymbol{v}+\eta\boldsymbol{\lambda})\cap X$ is uncountable;
(G) $[\boldsymbol{u}+\eta\boldsymbol{\lambda}, \boldsymbol{v}+\eta\boldsymbol{\lambda}] \subset (\boldsymbol{a},\boldsymbol{b}) \cap B(\boldsymbol{t},\delta_1(\boldsymbol{t}))$.

Set $P_1 := \{(\boldsymbol{t},[\boldsymbol{u}+\eta\boldsymbol{\lambda}, \boldsymbol{v}+\eta\boldsymbol{\lambda}]) : (\boldsymbol{t},[\boldsymbol{u},\boldsymbol{v}]) \in Q_1\}$. It is now clear that conditions $(\text{i})_1$–$(\text{v})_1$ follow from conditions (A) – (G).

We proceed by induction. Suppose that $n \geq 2$ is a positive integer such that P_{n-1} is an X-tagged δ_1-fine Perron subpartition of $[\boldsymbol{a},\boldsymbol{b}]$ and

$$V_{\mathcal{HK}}F([\boldsymbol{u}_{n-1},\boldsymbol{v}_{n-1}]\cap X) = \infty$$

for every $(\boldsymbol{t}_{n-1},[\boldsymbol{u}_{n-1},\boldsymbol{v}_{n-1}]) \in P_{n-1}$. By modifying the above construction of P_1, we obtain an X-tagged δ_1-fine Perron subpartition P_n of $[\boldsymbol{a},\boldsymbol{b}]$ so that the following conditions hold:

$(\text{i})_n$ if $(\boldsymbol{t}_{n-1},[\boldsymbol{u}_{n-1},\boldsymbol{v}_{n-1}]) \in P_{n-1}$, then

$$\text{card}\{(\boldsymbol{t}_n,[\boldsymbol{u}_n,\boldsymbol{v}_n]) \in P_n : [\boldsymbol{u}_n,\boldsymbol{v}_n] \subseteq [\boldsymbol{u}_{n-1},\boldsymbol{v}_{n-1}]\} \geq 2;$$

$(\text{ii})_n$ if $(\boldsymbol{t}_n,[\boldsymbol{u}_n,\boldsymbol{v}_n]) \in P_n$, then $[\boldsymbol{u}_n,\boldsymbol{v}_n] \subset (\boldsymbol{u}'_{n-1},\boldsymbol{v}'_{n-1})$ for some $(\boldsymbol{t}'_{n-1},[\boldsymbol{u}'_{n-1},\boldsymbol{v}'_{n-1}]) \in P_{n-1}$;
$(\text{iii})_n$ $\boldsymbol{t}_n \in (\boldsymbol{u}_n,\boldsymbol{v}_n)$ whenever $(\boldsymbol{t}_n,[\boldsymbol{u}_n,\boldsymbol{v}_n]) \in P_n$;
$(\text{iv})_n$

$$\sum_{(\boldsymbol{t}_n,[\boldsymbol{u}_n,\boldsymbol{v}_n])\in P_n} \mu_m([\boldsymbol{u}_n,\boldsymbol{v}_n]) < \frac{1}{2^n};$$

$(\text{v})_n$ if $(\boldsymbol{t}_{n-1},[\boldsymbol{u}_{n-1},\boldsymbol{v}_{n-1}]) \in P_{n-1}$, then

$$\sum_{\substack{(\boldsymbol{t}_n,[\boldsymbol{u}_n,\boldsymbol{v}_n])\in P_n \\ [\boldsymbol{u}_n,\boldsymbol{v}_n]\subseteq[\boldsymbol{u}_{n-1},\boldsymbol{v}_{n-1}]}} |F([\boldsymbol{u}_n,\boldsymbol{v}_n])| > 1.$$

Set $Z = \bigcap_{k=1}^{\infty} \bigcup_{(t,I) \in P_k} I$. Since property $(\text{iv})_n$ is true for every $n \in \mathbb{N}$, the set Z is μ_m-negligible and closed, and a simple application of Cousin's Lemma shows that Z is non-empty. By Theorem 5.4.9, there exists an upper semicontinuous gauge δ on Z such that

$$V(F, Z, \delta) < 1.$$

On the other hand, we apply Theorem 4.1.2 to select a positive integer N and $[\boldsymbol{\alpha}, \boldsymbol{\beta}] \in \mathcal{I}_m([\boldsymbol{a}, \boldsymbol{b}])$ such that

$$\left\{\boldsymbol{x} \in Z : \delta(\boldsymbol{x}) \geq \frac{1}{N}\right\} \cap [\boldsymbol{\alpha}, \boldsymbol{\beta}] = Z \cap [\boldsymbol{\alpha}, \boldsymbol{\beta}].$$

Since property $(\text{i})_n$ holds for every $n \in \mathbb{N}$, we can assume that $|||\boldsymbol{\beta} - \boldsymbol{\alpha}||| < \frac{1}{2N}$ and $(\boldsymbol{z}, [\boldsymbol{\alpha}, \boldsymbol{\beta}]) \in P_N$ for some $\boldsymbol{z} \in Z$. Therefore property $(\text{v})_{N+1}$ gives the desired contradiction:

$$V(F, Z, \delta) \geq \sum_{\substack{(\boldsymbol{t}_{N+1}, [\boldsymbol{u}_{N+1}, \boldsymbol{v}_{N+1}]) \in P_{N+1} \\ [\boldsymbol{u}_{N+1}, \boldsymbol{v}_{N+1}] \subseteq [\boldsymbol{\alpha}, \boldsymbol{\beta}]}} |F([\boldsymbol{u}_{N+1}, \boldsymbol{v}_{N+1}])| > 1.$$

The proof is complete. □

We end this section with the following improvement of Theorem 4.5.3.

Theorem 5.4.12. *Let $F : \mathcal{I}_1([a, b]) \longrightarrow \mathbb{R}$ be an additive interval function such that $V_{\mathcal{HK}} F \ll \mu_1$. Then there exists $f \in HK[a, b]$ such that F is the indefinite HK-integral of f.*

Proof. In view of Theorem 4.5.3, it suffices to construct an increasing sequence $(X_n)_{n=1}^{\infty}$ of closed sets with the following properties:

(i) $\bigcup_{k=1}^{\infty} X_k = [a, b]$;
(ii) for each $n \in \mathbb{N}$ there exists a positive constant η_n such that $V(F, X_n, \eta_n)$ is finite.

First we apply Theorem 5.4.11 and the countable subadditivity of $V_{\mathcal{HK}} F$ (Theorem 5.2.1(iii)) to choose an increasing sequence $(Y_n)_{n=1}^{\infty}$ of closed sets such that $\bigcup_{k=1}^{\infty} Y_k = [a, b]$ and $V_{\mathcal{HK}} F(Y_n)$ is finite for all $n \in \mathbb{N}$. Next for each $n \in \mathbb{N}$ we apply Theorem 5.4.9 to select an upper semicontinuous gauge δ_n on Y_n such that

$$V(F, Y_n, \delta_n) < V_{\mathcal{HK}} F(Y_n) + 1.$$

Define an upper semicontinuous gauge δ on $[a, b]$ by setting

$$\delta(x) = \begin{cases} \delta_1(x) & \text{if } x \in Y_1, \\ \min\{\frac{1}{k}, \delta_{k-1}(x), \operatorname{dist}(x, Y_{k-1})\} & \text{if } x \in Y_k \setminus Y_{k-1} \text{ for some } k \in \mathbb{N} \setminus \{1\}. \end{cases}$$

It is now clear that $\left(\delta^{-1}([\frac{1}{n}, \infty))\right)_{n=1}^{\infty}$ is a sequence of closed sets with the desired properties. □

The proof of Theorem 5.4.12 depends on Theorem 4.4.2. In the next section we will use a different method to obtain a higher-dimensional analogue of Theorem 5.4.12.

5.5 A measure-theoretic characterization of the Henstock-Kurzweil integral

The aim of this section is to establish an m-dimensional analogue of Theorem 5.4.12, where $m \geq 2$. We begin with the following result.

Theorem 5.5.1 (Ward). *Let $F : \mathcal{I}_m([\boldsymbol{a}, \boldsymbol{b}]) \longrightarrow \mathbb{R}$ be an additive interval function, and let $\alpha \in (0,1)$. Then F is α-derivable at μ_m-almost all the points $\boldsymbol{x}$ at which either $\overline{F}_\alpha(\boldsymbol{x}) < \infty$ or $\underline{F}_\alpha(\boldsymbol{x}) > -\infty$.*

Proof. See [145, Theorem 11.15]. □

Before we state and prove the main result of this chapter (Theorem 5.5.9), we need seven lemmas.

Lemma 5.5.2. *Let $F : \mathcal{I}_m([\boldsymbol{a}, \boldsymbol{b}]) \longrightarrow \mathbb{R}$ be an additive interval function, and let $X \subset [\boldsymbol{a}, \boldsymbol{b}]$ be a μ_m-measurable set. If $\alpha \in (0,1)$, $\mu_m(X) > 0$ and $V_{\mathcal{HK}}F(X)$ is finite, then $F'_\alpha(\boldsymbol{x})$ exists for μ_m-almost all $\boldsymbol{x} \in X$.*

Proof. According to Theorems 3.6.5, 3.5.11 and 3.3.24, the set

$$X_\alpha := \{\boldsymbol{x} \in X : \overline{F}_\alpha(\boldsymbol{x}) = \infty\}$$

is μ_m-measurable. In view of Ward's theorem, it remains to prove that $\mu_m(X_\alpha) = 0$.

Suppose that $\mu_m(X_\alpha) > 0$. Then the family $\mathcal{G}_1$ of all α-regular intervals $I \subseteq [\boldsymbol{a}, \boldsymbol{b}]$ for which $F(I) > \frac{V_{\mathcal{HK}}F(X_\alpha)+2}{\mu_m(X_\alpha)}\mu_m(I)$ is a Vitali cover of X_α; hence we can apply the Vitali covering theorem to choose a countable family $\mathcal{G}_2 \subset \mathcal{G}_1$ such that $X_\alpha \setminus \bigcup_{I \in \mathcal{G}_2} I$ is μ_m-negligible and

$$\sum_{I \in \mathcal{G}_2} F(I) \leq V_{\mathcal{HK}}F(X_\alpha) + 1.$$

A contradiction follows:

$$\begin{aligned} V_{\mathcal{HK}}F(X_\alpha) + 2 &\leq \frac{V_{\mathcal{HK}}F(X_\alpha) + 2}{\mu_m(X_\alpha)} \sum_{I \in \mathcal{G}_2} \mu_m(I \cap X_\alpha) \\ &\leq \sum_{I \in \mathcal{G}_2} F(I) \\ &\leq V_{\mathcal{HK}}F(X_\alpha) + 1. \end{aligned}$$

□

The following three lemmas will enable us to generalize Lemma 3.10.4; see Theorem 5.5.6.

Lemma 5.5.3. *Let $0 < \alpha < 1$, let $r > 0$, and let $u, v \in \mathbb{R}$ such that $0 < v - u \leq (1-\alpha)r$. If $\xi \in [u,v]$, then $\alpha r \leq \xi - v + r \leq r$ and $\alpha r \leq u + r - \xi \leq r$.*

Proof. Exercise. □

Lemma 5.5.4. *If $t \in [a,b]$ and $r \geq 0$, then the following implications hold.*

(i) $\xi \in [u,v] \subseteq [t,b] \cap [t-r, t+r]$ *implies* $v - r \leq t \leq \xi$.
(ii) $\xi \in [u,v] \subseteq [a,t] \cap [t-r, t+r]$ *implies* $\xi \leq t \leq u + r$.

Proof. Exercise. □

The following lemma is due to Kurzweil and Jarník [78].

Lemma 5.5.5. *Let $G : \mathcal{I}_m([\boldsymbol{a},\boldsymbol{b}]) \longrightarrow \mathbb{R}$ be an additive interval function. If $\boldsymbol{t} \in (\boldsymbol{a},\boldsymbol{b})$, $0 < \alpha < 1$, and there exists $r > 0$ such that $\prod_{i=1}^m [t_i - r, t_i + r] \subseteq [\boldsymbol{a},\boldsymbol{b}]$, then there exists a positive integer N_α such that*

$$\begin{aligned}&\Omega_r\\ &\leq \sup\left\{ |G([\boldsymbol{u},\boldsymbol{v}])| : [\boldsymbol{u},\boldsymbol{v}] \in \mathcal{I}_m([\boldsymbol{a},\boldsymbol{b}]), [\boldsymbol{u},\boldsymbol{v}] \subseteq \prod_{i=1}^m [t_i - r, t_i + r] \right\}\\ &\leq N_\alpha^{m-1}\Omega_r,\end{aligned}$$

where Ω_r denotes

$$\sup\left\{ |G([\boldsymbol{u},\boldsymbol{v}])| : \boldsymbol{t} \in [\boldsymbol{u},\boldsymbol{v}] \in \mathcal{I}_m([\boldsymbol{a},\boldsymbol{b}]), \alpha r \leq v_i - u_i \leq r \ (i = 1, \ldots, m) \right\}.$$

Proof. Let $[\boldsymbol{u},\boldsymbol{v}] \in \mathcal{I}_m([\boldsymbol{a},\boldsymbol{b}])$ be given. We consider two cases.

Case 1: $[\boldsymbol{u},\boldsymbol{v}] \subseteq [\boldsymbol{t},\boldsymbol{b}] \cap \prod_{i=1}^m [t_i - r, t_i + r]$ with $|||\boldsymbol{v} - \boldsymbol{u}||| \leq (1-\alpha)r$.

If $\boldsymbol{\xi} \in [\boldsymbol{u},\boldsymbol{v}]$, then it follows from Lemmas 5.5.3 and 5.5.4 that $\boldsymbol{t} \in \prod_{i=1}^m [v_i - r, \xi_i]$ and $\alpha r \leq \xi_i - (v_i - r) \leq r$ for $i = 1, \ldots, m$. Thus

$$|G([\boldsymbol{u},\boldsymbol{v}])| \leq \sum_{\substack{\boldsymbol{\xi} \in [\boldsymbol{u},\boldsymbol{v}] \\ \xi_i \in \{u_i, v_i\}\, \forall\, i \in \{1,\ldots,m\}}} \left| G\Big(\prod_{i=1}^m [v_i - r, \xi_i]\Big)\right| \leq 2^m \Omega_r.$$

Case 2: $[\boldsymbol{u},\boldsymbol{v}] \in \mathcal{I}_m([\boldsymbol{a},\boldsymbol{b}])$ with $[\boldsymbol{u},\boldsymbol{v}] \subseteq [\boldsymbol{t},\boldsymbol{b}] \cap \prod_{i=1}^m [t_i - r, t_i + r]$.

In this case, we have $\max_{i=1,\ldots,m}(v_i - u_i) \leq r$ and so $\max_{i=1,\ldots,m} \frac{v_i-u_i}{K_\alpha} \leq r(1-\alpha)$, where $K_\alpha = \lceil \frac{1}{1-\alpha} \rceil$. As G is additive, the proof of case 1 gives

$$|G([\boldsymbol{u}, \boldsymbol{v}])| \leq (2K_\alpha)^m \Omega_r.$$

Finally, the proofs of cases 1 and 2 hold for any one of the orthants in $[\boldsymbol{a}, \boldsymbol{b}]$ (with $\boldsymbol{t}$ as the origin); therefore

$$|G([\boldsymbol{u}, \boldsymbol{v}])| \leq (4K_\alpha)^m \Omega_r.$$

□

The following result of Kurzweil and Jarník [78] generalizes Lemma 3.10.4.

Theorem 5.5.6. *Let $0 < \beta < \alpha < 1$ and let $F : \mathcal{I}_m([\boldsymbol{a}, \boldsymbol{b}]) \longrightarrow \mathbb{R}$ be an additive interval function. If $F'_\alpha(\boldsymbol{x})$ exists for some $\boldsymbol{x} \in (\boldsymbol{a}, \boldsymbol{b})$, then $F'_\beta(\boldsymbol{x})$ exists and*

$$F'_\beta(\boldsymbol{x}) = F'_\alpha(\boldsymbol{x}).$$

Proof. This is an immediate consequence of Lemma 5.5.5. □

The following result is a consequence of Lemma 5.5.5 and Theorem 5.5.6.

Lemma 5.5.7. *Let $F : \mathcal{I}_m([\boldsymbol{a}, \boldsymbol{b}]) \longrightarrow \mathbb{R}$ be an additive interval function and suppose that F is derivable at each point of a non-empty closed set $X \subset (\boldsymbol{a}, \boldsymbol{b})$. Then given $\varepsilon > 0$ there exists an upper semicontinuous gauge Δ on X such that*

$$\sup_{I \in \mathcal{I}_m([\boldsymbol{u}, \boldsymbol{v}])} |F'(\boldsymbol{t})\mu_m(I) - F(I)| < \varepsilon\mu_m([\boldsymbol{u}, \boldsymbol{v}]) \tag{5.5.1}$$

for each point-interval pair $(\boldsymbol{t}, [\boldsymbol{u}, \boldsymbol{v}])$ satisfying $\boldsymbol{t} \in X$, $\boldsymbol{t} \in [\boldsymbol{u}, \boldsymbol{v}] \subset B(\boldsymbol{t}, \Delta(\boldsymbol{t}))$ and $reg([\boldsymbol{u}, \boldsymbol{v}]) \geq \frac{1}{2}$.

Proof. According to our assumptions, Theorem 5.5.6 and Lemma 5.5.5, for each $\varepsilon > 0$ there exists a gauge δ_0 on X such that $\delta_0(\boldsymbol{z}) < \text{dist}(\boldsymbol{z}, [\boldsymbol{a}, \boldsymbol{b}] \backslash (\boldsymbol{a}, \boldsymbol{b}))$ and

$$\sup_{0 < r(\boldsymbol{z}) < \delta_0(\boldsymbol{z})} \sup_{[\boldsymbol{c}, \boldsymbol{d}] \in \mathcal{I}_m(\prod_{i=1}^m [z_i - r(\boldsymbol{z}), z_i + r(\boldsymbol{z})])} \frac{|F'(\boldsymbol{z})\mu_m([\boldsymbol{c}, \boldsymbol{d}]) - F([\boldsymbol{c}, \boldsymbol{d}])|}{(r(\boldsymbol{z}))^m} < \frac{\varepsilon}{6} \tag{5.5.2}$$

for each $\boldsymbol{z} \in X$.

For each $k \in \mathbb{N}$ we let $X_k = \overline{\{\boldsymbol{x} \in X : \delta_0(\boldsymbol{x}) \geq \frac{1}{k}\}}$. Since $X \subset (\boldsymbol{a}, \boldsymbol{b})$ is closed, a simple calculation reveals that the sequence $(X_n)_{n=1}^\infty$ is increasing with $\bigcup_{k=1}^\infty X_k = X$.

Define an upper semicontinuous gauge Δ on X by setting

$$\Delta(\boldsymbol{x}) = \begin{cases} \min\{\frac{1}{2}, \operatorname{dist}(\boldsymbol{x}, [\boldsymbol{a},\boldsymbol{b}]\backslash(\boldsymbol{a},\boldsymbol{b}))\} \text{ if } \boldsymbol{x} \in X_1, \\ \min\{\frac{1}{2k}, \frac{1}{2}\operatorname{dist}(\boldsymbol{x}, X_{k-1}), \operatorname{dist}(\boldsymbol{x}, [\boldsymbol{a},\boldsymbol{b}]\backslash(\boldsymbol{a},\boldsymbol{b}))\} \\ \qquad\qquad \text{if } \boldsymbol{x} \in X_k\backslash X_{k-1} \text{ for some integer } k \geq 2, \end{cases}$$

and consider any point-interval pair $(\boldsymbol{t}, [\boldsymbol{u},\boldsymbol{v}])$ such that $\boldsymbol{t} \in X$, $\boldsymbol{t} \in [\boldsymbol{u},\boldsymbol{v}] \subset B(\boldsymbol{t}, \Delta(\boldsymbol{t}))$ and $\operatorname{reg}([\boldsymbol{u},\boldsymbol{v}]) \geq \frac{1}{2}$. Clearly, it suffices to prove that

$$|F'(\boldsymbol{t})\mu_m(I) - F(I)| < \frac{2\varepsilon}{3}\mu_m([\boldsymbol{u},\boldsymbol{v}]) \tag{5.5.3}$$

for every $I \in \mathcal{I}_m([\boldsymbol{u},\boldsymbol{v}])$.

Let q be the minimum positive integer such that $\boldsymbol{t} \in X_q$, and let $\eta \in (0, \frac{1}{2}\min\{\frac{1}{q}, \delta_0(\boldsymbol{t})\})$ be sufficiently small so that the following conditions are satisfied:

(A) $\prod_{i=1}^m [u_i - \eta, v_i + \eta] \subset B(\boldsymbol{t}, \Delta(\boldsymbol{t}))$;
(B) $\boldsymbol{t} \in X_q \cap \prod_{i=1}^m (u_i - \eta, v_i + \eta)$;
(C) $\mu_m(\prod_{i=1}^m [u_i - \eta, v_i + \eta]) \leq 2\mu_m([\boldsymbol{u},\boldsymbol{v}])$.

Now we use (B) we select a point $\boldsymbol{y} \in X$ so that the following conditions are satisfied:

(D) $\delta_0(\boldsymbol{y}) \geq \frac{1}{q}$;
(E) $\boldsymbol{y} \in X \cap \prod_{i=1}^m (u_i - \eta, v_i + \eta)$.

Using (A), (B), (D), (E) and (5.5.2), we get

$$\begin{aligned} &|F'(\boldsymbol{t}) - F'(\boldsymbol{y})| \\ &\leq \left| F'(\boldsymbol{t}) - \frac{F\left(\prod_{i=1}^m [u_i - \eta, v_i + \eta]\right)}{\mu_m\left(\prod_{i=1}^m [u_i - \eta, v_i + \eta]\right)} \right| + \left| F'(\boldsymbol{y}) - \frac{F\left(\prod_{i=1}^m [u_i - \eta, v_i + \eta]\right)}{\mu_m\left(\prod_{i=1}^m [u_i - \eta, v_i + \eta]\right)} \right| \\ &< \frac{\varepsilon}{3}. \end{aligned} \tag{5.5.4}$$

Finally, we use the triangle inequality, (5.5.4), (D), (E), (5.5.2) and (C) to obtain

$$\begin{aligned}
&|F'(\boldsymbol{t})\mu_m(I) - F(I)| \\
&\leq |F'(\boldsymbol{t})\mu_m(I) - F'(\boldsymbol{y})\mu_m(I)| + |F'(\boldsymbol{y})\mu_m(I) - F(I)| \\
&< \frac{\varepsilon}{3}\mu_m(I) + \frac{\varepsilon}{6}\mu_m\left(\prod_{i=1}^{m}[u_i - \eta, v_i + \eta]\right) \\
&\leq \frac{2\varepsilon}{3}\mu_m([\boldsymbol{u}, \boldsymbol{v}]).
\end{aligned}$$

□

Lemma 5.5.8. *Let $F : \mathcal{I}_m([\boldsymbol{a}, \boldsymbol{b}]) \longrightarrow \mathbb{R}$ be an additive interval function such that $V_{\mathcal{HK}}F \ll \mu_m$. If F is derivable at each point of a non-empty closed set $X \subseteq (\boldsymbol{a}, \boldsymbol{b})$, $F'|_X$ is bounded and $V_{\mathcal{HK}}F(X)$ is finite, then for each $\varepsilon > 0$ there exists an upper semicontinuous gauge δ on X such that*

$$\sum_{(\boldsymbol{t},I)\in P} |F'(\boldsymbol{t})\mu_m(I) - F(I)| < \varepsilon$$

for each X-tagged δ-fine Perron subpartition P of $[\boldsymbol{a}, \boldsymbol{b}]$.

Proof. Let $\varepsilon > 0$ and write $\varepsilon_0 := \frac{\varepsilon}{5}(m + \mu_m([\boldsymbol{a}, \boldsymbol{b}]) + \sup_{\boldsymbol{x}\in X} |F'(\boldsymbol{x})|)^{-1}$. According to Theorem 5.4.9, there exists an upper semicontinuous gauge δ_1 on X such that

$$\sum_{i=1}^{p} |F(I_i)| < V_{\mathcal{HK}}F\left(X \cap \bigcup_{i=1}^{p} I_i\right) + \varepsilon_0$$

for each X-tagged δ-fine Perron subpartition $\{(\boldsymbol{t}_1, I_1), \ldots, (\boldsymbol{t}_p, I_p)\}$ of $[\boldsymbol{a}, \boldsymbol{b}]$.

By Lemma 5.5.7 there exists an upper semicontinuous gauge δ_2 on X such that

$$|F'(\boldsymbol{x})\mu_m([\boldsymbol{c}, \boldsymbol{d}]) - F([\boldsymbol{c}, \boldsymbol{d}])| < \varepsilon_0\mu_m([\boldsymbol{c}, \boldsymbol{d}])$$

for each point-interval pair $(\boldsymbol{x}, [\boldsymbol{c}, \boldsymbol{d}])$ satisfying $\boldsymbol{x} \in X$, $\boldsymbol{x} \in [\boldsymbol{c}, \boldsymbol{d}] \cap B(\boldsymbol{x}, \delta_2(\boldsymbol{x}))$, and $\operatorname{reg}([\boldsymbol{c}, \boldsymbol{d}]) \geq \frac{1}{2}$.

According to our assumptions, we can choose $N \in \mathbb{N}$ so that the following properties hold:

(a) the set $X_N := \{\boldsymbol{x} \in X : \min\{\delta_1(\boldsymbol{x}), \delta_2(\boldsymbol{x})\} \geq \frac{1}{N}\}$ is closed and non-empty;
(b) $\min\{\mu_m(X \backslash X_N), V_{\mathcal{HK}}F(X \backslash X_N)\} < \varepsilon_0$;
(c) there exists an open set $G \supset X$ such that $\mu_m(G \backslash X_N) < \varepsilon_0$.

Now define an upper semicontinuous gauge δ on X as follows:

$$\delta(\boldsymbol{x}) = \begin{cases} \frac{1}{N} & \text{if } \boldsymbol{x} \in X_N, \\ \min\{\delta_1(\boldsymbol{x}), \operatorname{dist}(\boldsymbol{x}, X_N \cup ([\boldsymbol{a}, \boldsymbol{b}] \backslash G))\} & \text{if } \boldsymbol{x} \in X \backslash X_N. \end{cases}$$

If P is an X-tagged δ-fine Perron subpartition of $[\boldsymbol{a}, \boldsymbol{b}]$, the triangle inequality yields

$$\begin{aligned} &\sum_{(\boldsymbol{t},I)\in P} |F'(\boldsymbol{t})\mu_m(I) - F(I)| \\ &\le \sum_{\substack{(\boldsymbol{t},I)\in P \\ \boldsymbol{t}\in X\backslash X_N}} |F'(\boldsymbol{t})\mu_m(I) - F(I)| + \sum_{\substack{(\boldsymbol{t},I)\in P \\ \boldsymbol{t}\in X_N}} |F'(\boldsymbol{t})\mu_m(I) - F(I)| \\ &= S_1 + S_2, \end{aligned}$$

say.

Using (a), (b), (c) and our choice of δ, we conclude that $S_1 < \sup_{\boldsymbol{x}\in X} |F'(\boldsymbol{x})| + 2\varepsilon_0$:

$$\begin{aligned} S_1 &\le \sum_{\substack{(\boldsymbol{t},I)\in P \\ \boldsymbol{t}\in X\backslash X_N}} |F'(\boldsymbol{t})\mu_m(I)| + \sum_{\substack{(\boldsymbol{t},I)\in P \\ \boldsymbol{t}\in X\backslash X_N}} |F(I)| \\ &< \varepsilon_0 \sup_{\boldsymbol{x}\in X} |F'(\boldsymbol{x})| + V_{\mathcal{HK}}F(X\backslash X_N) + \varepsilon_0 \\ &< \varepsilon_0 \sup_{\boldsymbol{x}\in X} |F'(\boldsymbol{x})| + 2\varepsilon_0. \end{aligned}$$

We will next obtain an upper bound for S_2. Following the proof Theorem 3.10.12, we fix a $\frac{1}{2}$-regular net $\mathcal{N}(I)$ of I. For each $J \in \mathcal{N}(I)$ satisfying $J \cap X_N \neq \emptyset$, we fix a point $\boldsymbol{x}_J \in J \cap X_N$. Then

$$\begin{aligned} &S_2 \\ &= \sum_{\substack{(\boldsymbol{t},I)\in P \\ \boldsymbol{t}\in X_N}} |F'(\boldsymbol{t})\mu_m(I) - F(I)| \\ &\le \Bigg\{ \sum_{\substack{(\boldsymbol{t},I)\in P \\ \boldsymbol{t}\in X_N}} \sum_{\substack{J\in\mathcal{N}(I) \\ J\cap X_N\neq\emptyset}} |F'(\boldsymbol{x}_J)\mu_m(J) - F(J)| \\ &+ \sum_{\substack{(\boldsymbol{t},I)\in P \\ \boldsymbol{t}\in X_N}} \sum_{\substack{J\in\mathcal{N}(I) \\ J\cap X_N\neq\emptyset}} |F'(\boldsymbol{x}_J) - F'(\boldsymbol{t})|\, \mu_m(J) \Bigg\} \\ &+ \sum_{\substack{(\boldsymbol{t},I)\in P \\ \boldsymbol{t}\in X_N}} \sum_{\substack{J\in\mathcal{N}(I) \\ J\cap X_N=\emptyset}} |F'(\boldsymbol{t})|\, \mu_m(J) + \sum_{\substack{(\boldsymbol{t},I)\in P \\ \boldsymbol{t}\in X_N}} \Bigg| \sum_{\substack{J\in\mathcal{N}(I) \\ J\cap X_N=\emptyset}} F(J) \Bigg| \\ &= S_3 + S_4 + S_5, \end{aligned}$$

say.

Since $\delta(\boldsymbol{x}) \leq \frac{1}{N} \leq \delta_2(\boldsymbol{x})$ for all $\boldsymbol{x} \in X_N$, and $|||\boldsymbol{x}_J - \boldsymbol{t}||| \leq \operatorname{diam}(I) < \frac{1}{N}$, our choice of δ_2 yields

$$\begin{aligned} &S_3 \\ &= \sum_{\substack{(\boldsymbol{t},I)\in P\\ \boldsymbol{t}\in X_N}} \sum_{\substack{J\in\mathcal{N}(I)\\ J\cap X_N\neq\emptyset}} |F'(\boldsymbol{x}_J)\mu_m(J) - F(J)| \\ &+ \sum_{\substack{(\boldsymbol{t},I)\in P\\ \boldsymbol{t}\in X_N}} \sum_{\substack{J\in\mathcal{N}(I)\\ J\cap X_N\neq\emptyset}} |F'(\boldsymbol{x}_J) - F'(\boldsymbol{t})|\,\mu_m(J) \\ &< 3\varepsilon_0\mu_m([\boldsymbol{a},\boldsymbol{b}]). \end{aligned}$$

Also, it follows from (c) that

$$S_4 = \sum_{\substack{(\boldsymbol{t},I)\in P\\ \boldsymbol{t}\in X_N}} \sum_{\substack{J\in\mathcal{N}(I)\\ J\cap X_N=\emptyset}} |F'(\boldsymbol{t})|\,\mu_m(J) \leq \varepsilon_0 \sup_{\boldsymbol{x}\in X} |F'(\boldsymbol{x})|\,.$$

It remains to prove that $S_5 < 2m\varepsilon_0$. For each point-interval pair $(\boldsymbol{t}, I) \in P$ satisfying $\boldsymbol{t} \in X_N$, we need to partition I in a special way. Let $\Phi_{I,0}([u_0, v_0]) := I$ and let

$$\mathcal{C}_r(I) := \left\{ \bigcap_{k=0}^{r} \Phi_{I,k}([u_k, v_k]) : [\boldsymbol{u}, \boldsymbol{v}] \in \mathcal{N}(I) \text{ and } \right. \\ \left. X_N \cap \bigcap_{k=0}^{r-1} \Phi_{I,k}([u_k, v_k]) \neq \emptyset = X_N \cap \bigcap_{k=0}^{r} \Phi_{I,k}([u_k, v_k]) \right\} \\ (r = 1, \ldots, m)$$

so that the following properties hold:

(d)

$$I = \bigcup_{\substack{[\boldsymbol{u},\boldsymbol{v}]\in\mathcal{N}(I)\\ [\boldsymbol{u},\boldsymbol{v}]\cap X_N\neq\emptyset}} [\boldsymbol{u}, \boldsymbol{v}] \cup \bigcup_{r=1}^{m} \bigcup_{U\in\mathcal{C}_r(I)} U;$$

(e) two elements $\prod_{i=1}^{m}[c_i, d_i]$ and $\prod_{i=1}^{m}[c_i', d_i']$ of $\mathcal{C}_1(I)$ are distinct if and only if $(c_1, d_1) \cap (c_1', d_1') = \emptyset$;

(f) for each $r \in \{2, \ldots, m\}$ two elements $\prod_{i=1}^{m}[\alpha_i, \beta_i]$ and $\prod_{i=1}^{m}[\alpha_i', \beta_i']$ of $\mathcal{C}_r(I)$ are distinct if $(\alpha_r, \beta_r) \cap (\alpha_r', \beta_r') = \emptyset$ and $[\alpha_i, \beta_i] = [\alpha_i', \beta_i']$ $(i = 1, \ldots, r-1)$.

Combining (d), the additivity of F, (e), (f) and the inequality $\inf_{x\in X_N}\delta(x) \geq \frac{1}{N}$, we get

$$S_5 \leq \sum_{\substack{(t,I)\in P\\ t\in X_N}} \sum_{r=1}^{m} \left| \sum_{U\in\mathcal{C}_r(I)} F(U) \right| < 2m\varepsilon_0.$$

□

We are now ready to state and prove the main result of this section.

Theorem 5.5.9. *Let $F : \mathcal{I}_m([a,b]) \longrightarrow \mathbb{R}$ be an additive interval function. The following conditions are equivalent.*

(i) *F is the indefinite Henstock-Kurzweil integral of some $f \in HK[a,b]$.*
(ii) $V_{\mathcal{HK}}F \ll \mu_m$.

Proof. The implication (i) $\Longrightarrow$ (ii) follows from Theorem 4.2.3.

Conversely, suppose that (ii) holds. By Theorem 5.4.11, Lemma 5.5.2 and Theorem 5.5.6, there exists an increasing sequence $(X_k)_{k=1}^{\infty}$ of pairwise disjoint of closed subsets of $[a,b]$ such that $\mu_m([a,b]\setminus\bigcup_{k=1}^{\infty} X_k) = 0$, F is derivable at each point of the set $\bigcup_{k=1}^{\infty} X_k$, and $\sup_{x\in X_k}|F'(x)| + V_{\mathcal{HK}}F(X_k)$ is finite for all $k \in \mathbb{N}$.

Define the function f on $[a,b]$ by setting

$$f(x) = \begin{cases} F'(x) & \text{if } x \in \bigcup_{k=1}^{\infty} X_k, \\ 0 & \text{otherwise.} \end{cases}$$

We claim that $f \in HK[a,b]$ and F is the indefinite Henstock-Kurzweil integral of f.

Let $\varepsilon > 0$ be given. Since $V_{\mathcal{HK}}F \ll \mu_m$ and $\mu_m([a,b]\setminus\bigcup_{k=1}^{\infty} X_k) = 0$, there exists a gauge δ_0 on $[a,b]\setminus\bigcup_{k=1}^{\infty} X_k$ such that

$$V(F, [a,b]\setminus\bigcup_{k=1}^{\infty} X_k, \delta_0) < \frac{\varepsilon}{2}.$$

For each $n \in \mathbb{N}$ we apply Lemma 5.5.8 to choose an upper semicontinuous gauge δ_n on X_n such that

$$\sum_{(t_n,I_n)\in P_n} |f(t_n)\mu_m(I_n) - F(I_n)| < \frac{\varepsilon}{2^{n+1}}$$

for each X-tagged δ_n-fine Perron subpartition P_n of $[a,b]$.

Let $X_0 = \emptyset$ and define a gauge δ on $[a,b]$ as follows:

$$\delta(x) = \begin{cases} \delta_0(x) & \text{if } x \in [a,b]\setminus\bigcup_{k=1}^{\infty} X_k, \\ \delta_n(x) & \text{if } x \in X_n\setminus X_{n-1} \text{ for some positive integer } n. \end{cases}$$

If P is a δ-fine Perron subpartitition of $[\boldsymbol{a}, \boldsymbol{b}]$, then

$$\begin{aligned}
&\sum_{(\boldsymbol{t},I)\in P} |f(\boldsymbol{t})\mu_m(I) - F(I)| \\
&\leq \sum_{\substack{(\boldsymbol{t},I)\in P \\ \boldsymbol{t}\in[\boldsymbol{a},\boldsymbol{b}]\setminus\bigcup_{k=1}^{\infty} X_k}} |F(I)| + \sum_{k=1}^{\infty} \sum_{\substack{(\boldsymbol{t},I)\in P \\ \boldsymbol{t}\in X_k\setminus X_{k-1}}} |f(\boldsymbol{t})\mu_m(I) - F(I)| \\
&< \varepsilon.
\end{aligned}$$

Since F is additive, we conclude that $f \in HK[\boldsymbol{a}, \boldsymbol{b}]$ and F is the indefinite Henstock-Kurzweil integral of f. The proof is complete. □

5.6 Product variational measures

In this section we let r and s be positive integers. For $k = r, s$, we fix an interval E_k in $\mathbb{R}^k$, and $\mathbb{R}^k$ will be equipped with the maximum norm $|||\cdot|||_k$. We begin with the following result.

Theorem 5.6.1 (Tonelli). *Let $f : E_r \times E_s \longrightarrow \mathbb{R}$ be a μ_{r+s}-measurable function. If one of the following iterated integrals*

$$\int_{E_r}\int_{E_s} |f(\xi,\eta)|\ d\mu_s(\eta)\ d\mu_r(\xi),\ \int_{E_s}\int_{E_r} |f(\xi,\eta)|\ d\mu_r(\xi)\ d\mu_s(\eta)$$

is finite, then $f \in L^1(E_r \times E_s)$.

Proof. By Theorem 3.5.13, there exists a sequence $(f_n)_{n=1}^{\infty}$ of μ_{r+s}-measurable simple functions on $E_r \times E_s$ with the following properties:

(i) $(|f_n|)_{n=1}^{\infty}$ is increasing;
(ii) $\sup_{n\in\mathbb{N}} |f_n(\xi,\eta)| \leq |f(\xi,\eta)|$ for all $(\xi,\eta) \in E_r \times E_s$;
(iii) $\lim_{n\to\infty} f_n(\xi,\eta) = f(\xi,\eta)$ for all $(\xi,\eta) \in E_r \times E_s$.

Using Fubini's Theorem and (ii), we get

$$\sup_{n\in\mathbb{N}} \int_{E_r\times E_s} |f_n(\xi,\eta)|\ d\mu_{r+s}(\xi,\eta) < \infty.$$

Consequently, (i), (iii), the Monotone Convergence Theorem and Theorem 3.7.1 yield the desired conclusion. □

If $f : E_r \longrightarrow \mathbb{R}$ and $g : E_s \longrightarrow \mathbb{R}$, we write $(f \otimes g)(\xi,\eta) = f(\xi)g(\eta)$. The following theorem is a consequence of Theorems 3.6.6 and 5.6.1.

Theorem 5.6.2. *If $f \in L^1(E_r)$ and $g \in L^1(E_s)$, then $f \otimes g \in L^1(E_r \times E_s)$ and*

$$\int_{E_r \times E_s} f \otimes g \, d\mu_{r+s} = \left\{ \int_{E_r} f \, d\mu_r \right\} \left\{ \int_{E_s} g \, d\mu_s \right\}.$$

Our next goal is to prove that

$$f \in HK(E_r) \text{ and } g \in L^1(E_s) \Longrightarrow f \otimes g \in HK(E_r \times E_s).$$

Since Theorem 5.6.1 does not hold for non-absolutely convergent Henstock-Kurzweil integrals, we need new ideas. We begin with the following crucial result.

Lemma 5.6.3. *If $I_1, I_2 \in \mathcal{I}_r(E_r)$ and $J_1, J_2, J_1 \cap J_2 \in \mathcal{I}_s(E_s)$ with $\mu_{r+s}((I_1 \times J_1) \cap (I_2 \times J_2)) = 0$, then $\mu_r(I_1 \cap I_2) = 0$.*

Proof. We have

$$0 \le \mu_r(I_1 \cap I_2)\mu_s(J_1 \cap J_2) \le \mu_{r+s}((I_1 \times J_1) \cap (I_2 \times J_2)) = 0. \qquad \square$$

Theorem 5.6.4. *If $f \in HK(E_r)$ and $g \in L^1(E_s)$, then $f \otimes g \in HK(E_r \times E_s)$ and*

$$(HK) \int_{E_r \times E_s} f \otimes g = \left\{ (HK) \int_{E_r} f \right\} \left\{ \int_{E_s} g \, d\mu_s \right\}.$$

Proof. Since $f \in HK(E_r)$, we apply Theorem 4.3.1 with $m = r$ to choose an increasing sequence $(X_k)_{k=1}^{\infty}$ of closed subsets of E_r such that $\bigcup_{k=1}^{\infty} X_k = E_r$, the sequence $(f\chi_{X_n})_{n=1}^{\infty}$ is in $L^1(E_r)$, and

$$\sup_{k \in \mathbb{N}} V_{\mathcal{HK}}(F_k - F)(X_k) = 0,$$

where F_k and F are the indefinite Henstock-Kurzweil integrals of $f\chi_{X_k}$ and f respectively.

Let G be the indefinite Lebesgue integral of g. A direct calculation shows that $(X_k \times E_s)_{k=1}^{\infty}$ is a sequence of closed subsets of $E_r \times E_s$ such that $\bigcup_{k=1}^{\infty} (X_k \times E_s) = E_r \times E_s$, and Theorem 5.6.2 implies that $(f\chi_{X_n} \otimes g)_{n=1}^{\infty}$ is a sequence in $L^1(E_r \times E_s)$. In view the $(r+s)$-dimensional version of Theorem 4.3.1, it remains to prove that

$$\sup_{k \in \mathbb{N}} V_{\mathcal{HK}}(F_k \otimes G - F \otimes G)(X_k \times E_s) = 0. \tag{5.6.1}$$

Let $k \in \mathbb{N}$ be fixed. Since $V_{\mathcal{HK}}(F_k - F)(X_k) = 0$, for each $\varepsilon > 0$ there exists a gauge Δ_k on X_k such that

$$V(F_k - F, X_k, \Delta_k) < \frac{\varepsilon}{1 + \|g\|_{L^1(E_s)}}.$$

Now we define a gauge δ_k on $X_k \times E_s$ by setting

$$\delta_k(\xi,\eta) = \Delta_k(\xi),$$

and let $\{((\xi_1,\eta_1), I_1 \times J_1), \ldots, ((\xi_p,\eta_p), I_p \times J_p)\}$ be an $(X_k \times E_s)$-tagged δ_k-fine Perron subpartition of $E_r \times E_s$. Next we choose non-overlapping subintervals $U_1, \ldots, U_\ell$ of E_s so that $\bigcup_{i=1}^p J_i = \bigcup_{j=1}^\ell U_j$, where each U_j is a subset of some interval J_i. Since

$$\max_{\xi\in I_i} |||\xi_i - \xi|||_r \leq \max_{(\xi,\eta)\in I_i\times J_i} |||(\xi_i,\eta_i) - (\xi,\eta)|||_{r+s} < \delta_k(\xi_i,\eta_i) = \Delta_k(\xi_i) \tag{5.6.2}$$

for all $i = 1, \ldots, p$, for each $j \in \{1, \ldots, \ell\}$ we infer from Lemma 5.6.3 that $\{(\xi_i, I_i) : U_j \subseteq J_i\}$ is an X_k-tagged Δ_k-fine Perron subpartition of E_r. Thus

$$\begin{aligned}
&\sum_{i=1}^p |F_k(I_i)G(J_i) - F(I_i)G(J_i)| \\
&\leq \sum_{i=1}^p \left| \left\{ \sum_{j=1}^\ell \int_{J_i\cap U_j} g\, d\mu_s \right\} (F_k(I_i) - F(I_i)) \right| \\
&\leq \sum_{j=1}^\ell \sum_{i=1}^p \left| \left\{ \int_{J_i\cap U_j} g\, d\mu_s \right\} (F_k(I_i) - F(I_i)) \right| \\
&= \sum_{j=1}^\ell \sum_{\substack{i=1 \\ U_j\subseteq J_i}}^p \left| \left\{ \int_{U_j} g\, d\mu_s \right\} (F_k(I_i) - F(I_i)) \right| \\
&< \sum_{j=1}^\ell \left| \left\{ \int_{U_j} g\, d\mu_s \right\} \right| \frac{\varepsilon}{1 + \|g\|_{L^1(E_s)}} \\
&< \varepsilon.
\end{aligned}$$

Since $\varepsilon > 0$ and $k \in \mathbb{N}$ are arbitrary, (5.6.1) holds. □

The following result generalizes a result of Henstock [57].

Theorem 5.6.5. *Let $[c,d]$ be a non-degenerate subinterval of $\mathbb{R}$. If $f \in HK(E_r)$ and $g \in HK[c,d]$, then $f \otimes g \in HK(E_r \times [c,d])$ and*

$$(HK)\int_{E_r\times[c,d]} f\otimes g = \left\{(HK)\int_{E_r} f\right\}\left\{(HK)\int_c^d g\right\}.$$

Proof. Since $f \in HK(E_r)$, we apply the r-dimensional version of Theorem 4.3.1 to choose an increasing sequence $(X_k)_{k=1}^\infty$ of closed subsets of E_r such that $\bigcup_{k=1}^\infty X_k = E_r$, the sequence $(f\chi_{X_n})_{n=1}^\infty$ belongs to $L^1(E_r)$, and

$$\sup_{k\in\mathbb{N}} V_{\mathcal{HK}}(F_k - F)(X_k) = 0,$$

where F_k and F are the indefinite Henstock-Kurzweil integrals of $f\chi_{X_k}$ and f respectively.

A direct calculation shows that $(X_n \times [c,d])_{n=1}^{\infty}$ is an increasing sequence of closed subsets of $E_r \times [c,d]$, and Theorem 5.6.4 implies that $(f\chi_{X_n} \otimes g)_{n=1}^{\infty}$ is a sequence in $HK(E_r \times E_s)$. Following the proof of the $(r+s)$-dimensional version of Theorem 4.3.1, it suffices to show that

$$\sup_{k \in \mathbb{N}} V_{\mathcal{HK}}(F_k \otimes G - F \otimes G)(X_k \times [c,d]) = 0,$$

where G denotes the indefinite Henstock-Kurzweil integral of g.

Let $k \in \mathbb{N}$ be given. According to monotonicity of $V_{\mathcal{HK}}(F_k \otimes G - F \otimes G)$ and Observation 4.4.1, it remains to prove that

$$V_{\mathcal{HK}}(F_k \otimes G - F \otimes G)(X_k \times Y) = 0,$$

where $Y \subseteq [c,d]$ is any closed set such that $\{c,d\} \subset Y$ and $V(G, Y, \kappa) < \infty$ for some $\kappa \in \mathbb{R}^+$. Let $\left((c_k, d_k)\right)_{k=1}^{\infty}$ be the sequence of pairwise disjoint open intervals such that $(a,b)\backslash Y = \bigcup_{k=1}^{\infty}(c_k, d_k)$.

Define a gauge δ_k on $X_k \times Y$ by setting

$$\delta_k(\xi, \eta) = \min\{\Delta_k(\xi), \kappa\},$$

and consider an arbitrary $(X_k \times Y)$-tagged δ_k-fine Perron subpartition $\left\{\left((\xi_1, \eta_1), I_1 \times J_1\right), \ldots, \left((\xi_p, \eta_p), I_p \times J_p\right)\right\}$ of $E_r \times E_s$. Since

$$\max_{\xi \in I_i} |||\xi_i - \xi|||_r \leq \max_{(\xi,\eta) \in I_i \times J_i} |||(\xi_i, \eta_i) - (\xi, \eta)|||_{r+s} < \delta_k(\xi_i, \eta_i) = \Delta_k(\xi_i) \tag{5.6.3}$$

for all $i = 1, \ldots, p$, we infer from Harnack extension and Lemma 5.6.3 that

$$\begin{aligned}
&\sum_{i=1}^{p} |F_k(I_i)G(J_i) - F(I_i)G(J_i)| \\
&\leq \sum_{i=1}^{p} |G(J_i)(F_k(I_i) - F(I_i))| \\
&\leq \sum_{i=1}^{p} |F_k(I_i) - F(I_i)| \int_{J_i} |g\chi_Y| \; d\mu_1 \\
&+ \sum_{k=1}^{\infty}\sum_{i=1}^{p} \left| \left\{ (HK)\int_{J_i} g\chi_{[c_k,d_k]} \right\} (F_k(I_i) - F(I_i)) \right| \\
&< \varepsilon \left\{ \|g\chi_Y\|_{L^1[c,d]} + 2\sum_{k=1}^{\infty} \|g\|_{HK[c_k,d_k]} \right\}.
\end{aligned}$$

Since $\varepsilon > 0$ is arbitrary, the proof is complete. □

Corollary 5.6.6. *If* $f_i \in HK[a_i, b_i]$ $(i = 1, \ldots, m)$, *then* $\otimes_{i=1}^m f_i \in HK[\boldsymbol{a}, \boldsymbol{b}]$ *and*

$$(HK) \int_{[\boldsymbol{a},\boldsymbol{b}]} \otimes_{i=1}^m f_i = \prod_{i=1}^m (HK) \int_{a_i}^{b_i} f_i.$$

On the other hand, the following conjecture appears to be still open.

Conjecture 5.6.7. *Let* $r, s \in \mathbb{N}\backslash\{1\}$. *If* $f \in HK(E_r)$ *and* $g \in HK(E_s)$, *then* $f \otimes g \in HK(E_r \times E_s)$.

5.7 Notes and Remarks

A different treatment of Lebesgue's outer measure can be found in Folland [42], Hewitt and Stromberg [59] or Royden [143].

Various examples of variational measures have been given by Bongiorno [9], Jurkat and Knizia [62], Faure [36], Di Piazza [30] and Pfeffer [137–140]. A general theory of variational measure can be found in Ostaszewski [134] or Thomson [152].

Lemma 5.3.1 is due to Lee [102]. The proof of Theorem 5.3.3 is similar to that of [30, Proposition 1]. A different approach to Theorem 5.3.4 can be found in [91].

Section 5.4 is based on the work of Lee [102]. For a one-dimensional version of Theorem 5.4.11, consult [15], [138] or [139]. Theorem 5.4.12 is a remarkable result of Bongiorno, Di Piazza and Skvortsov [15]. For more recent results, see [12, 13].

Lemmas 5.5.3–5.5.5 are due to Kurzweil and J. Jarník [78]. Lemmas 5.5.7, 5.5.8, and Theorem 5.5.9 are based on the paper [102]. A related result is also given by Lee and Ng [87].

Section 5.6 is based on the paper [92]. For other related results, see [25, 26, 56, 57, 99, 105].

Chapter 6

Multipliers for the Henstock-Kurzweil integral

A *multiplier* for a family X of functions on $[\boldsymbol{a}, \boldsymbol{b}]$ is a function on $[\boldsymbol{a}, \boldsymbol{b}]$ such that $fg \in X$ for each $f \in X$. According to Theorem 3.7.5, $L^\infty[\boldsymbol{a}, \boldsymbol{b}]$ is precisely the class of multipliers for $L^1[\boldsymbol{a}, \boldsymbol{b}]$. On the other hand, it follows from Example 2.5.8 that certain continuous functions are not multipliers for $HK([0,1]^2)$. The main goal of this chapter is to characterize the multipliers for $HK[\boldsymbol{a}, \boldsymbol{b}]$. We begin with the one-dimensional case.

6.1 One-dimensional integration by parts

We begin with the following one-dimensional integration by parts for the Lebesgue integral.

Lemma 6.1.1. *If $f \in L^1[a,b]$ and $\phi \in L^1[a,b]$, then the map $x \mapsto f(x)\int_a^x \phi\, d\mu_1$ belongs to $L^1[a,b]$ and*

$$\int_a^b f(x)\left\{\int_a^x \phi\, d\mu_1\right\} d\mu_1(x) = \int_a^b \left\{\int_t^b f\, d\mu_1\right\}\phi(t)\, d\mu_1(t).$$

Proof. By Theorems 3.11.9 and 3.7.3, the map $x \mapsto f(x)\int_a^x \phi\, d\mu_1$ belongs to $L^1[a,b]$. The conclusion follows from Theorems 5.6.2, 3.7.3 and 2.5.5:

$$\begin{aligned}\int_a^b f(x)\left\{\int_a^x \phi\, d\mu_1\right\} d\mu_1(x) &= \int_{[a,b]^2} (f\otimes\phi)(x,t)\chi_{[a,b]\times[a,x]}(x,t)\, d\mu_2(x,t)\\ &= \int_{[a,b]^2} (\phi\otimes f)(t,x)\chi_{[a,b]\times[t,b]}(t,x)\, d\mu_2(t,x)\\ &= \int_a^b \left\{\int_t^b f\, d\mu_1\right\}\phi(t)\, d\mu_1(t).\end{aligned}$$

□

The following lemma shows that an analogous version of Lemma 6.1.1 holds for the Henstock-Kurzweil integral.

Lemma 6.1.2. *If* $f \in HK[a,b]$ *and* $\phi \in L^1[a,b]$, *then the map* $x \mapsto f(x)\int_a^x \phi\, d\mu_1$ *belongs to* $HK[a,b]$ *and*

$$(HK)\int_a^b f(x)\left\{\int_a^x \phi\, d\mu_1\right\} dx = \int_a^b \left\{(HK)\int_x^b f\right\}\phi(x)\, d\mu_1(x).$$

Proof. Let $F(x) = (HK)\int_a^x f$ for each $x \in [a,b]$, and let $\varepsilon > 0$. According to the Saks-Henstock Lemma, there exists a gauge δ_1 on $[a,b]$ such that

$$\sum_{(t_1,[u_1,v_1])\in P_1} \left|f(t_1)(v_1-u_1) - (F(v_1)-F(u_1))\right| < \frac{\varepsilon}{2(1+\|\phi\|_{L^1[a,b]})}$$

for each δ_1-fine Perron subpartition P_1 of $[a,b]$. By Theorems 1.4.5 and 1.1.6, there exists a constant $\delta_2 > 0$ such that

$$|F(\beta) - F(\alpha)| < \frac{\varepsilon}{4(1+\|\phi\|_{L^1[a,b]})}$$

for every interval $[\alpha,\beta] \subseteq [a,b]$ with $0 < \beta - \alpha < \delta_2$.

Define a gauge δ on $[a,b]$ by setting $\delta(x) = \min\{\delta_1(x), \delta_2\}$, and let P be any δ-fine Perron partition of $[a,b]$. If $g(x) := \int_a^x \phi\, d\mu_1$ for each $x \in [a,b]$, we have $F(b)g(b) = \sum_{(t,[u,v])\in P} (F(v)g(v) - F(u)g(u))$ and so

$$\begin{aligned}
&\left|\sum_{(t,[u,v])\in P} f(t)g(t)(v-u) - \int_a^b \left\{(HK)\int_x^b f\right\}\phi(x)\, d\mu_1(x)\right| \\
&= \left|\sum_{(t,[u,v])\in P} \left\{f(t)g(t)(v-u) - (F(v)g(v) - F(u)g(u)) + \int_u^v F\phi\, d\mu_1\right\}\right| \\
&\leq \left|\sum_{(t,[u,v])\in P} \left\{f(t)g(t)(v-u) - F(v)g(t) + F(u)g(t)\right\}\right| \\
&\quad + \left|\sum_{(t,[u,v])\in P} \left\{(F(v)-F(u))\left(\int_v^t \phi\, d\mu_1\right) + \int_u^v (F - F(u))\phi\, d\mu_1\right\}\right| \\
&\leq \|\phi\|_{L^1[a,b]} \sum_{(t,[u,v])\in P} \left|f(t)(v-u) - (HK)\int_u^v f\right| + \frac{2\|\phi\|_{L^1[a,b]}\varepsilon}{4(1+\|\phi\|_{L^1[a,b]})} \\
&< \varepsilon.
\end{aligned}$$

Since $\varepsilon > 0$ is arbitrary, the lemma follows. □

Recall that $AC[a,b]$ is the space of absolutely continuous functions defined on $[a,b]$. Using the proof of Lemma 6.1.2 together with Theorem 4.4.8, we obtain the following generalization of Lemma 6.1.2.

Lemma 6.1.3. *Let $f \in HK[a,b]$ and let $(g_n)_{n=1}^{\infty}$ be a sequence in $AC[a,b]$ such that*

$$\sup_{n\in\mathbb{N}}\left\{|g_n(a)| + Var(g_n,[a,b])\right\} < \infty.$$

Then the sequence $(fg_n)_{n=1}^{\infty}$ is Henstock-Kurzweil equi-integrable on $[a,b]$. If, in addition, the sequence $(g_n)_{n=1}^{\infty}$ is pointwise convergent on $[a,b]$, then $\lim_{n\to\infty} fg_n$ is Henstock-Kurzweil integrable on $[a,b]$ and

$$(HK)\int_a^b \lim_{n\to\infty} fg_n = \lim_{n\to\infty}(HK)\int_a^b fg_n.$$

It is natural to ask for a useful description of the set

$$\Big\{g : \exists \text{ a sequence } (g_n)_{n=1}^{\infty} \text{ in } AC[a,b] \text{ such that}$$

$$g_n \to g \text{ pointwise on } [a,b] \text{ and } \sup_{n\in\mathbb{N}}\left\{|g_n(a)| + Var(g_n,[a,b])\right\} < \infty\Big\}.$$

The next theorem shows that this set is precisely the space $BV[a,b]$.

Theorem 6.1.4. *Let $g : [a,b] \longrightarrow \mathbb{R}$. The following conditions are equivalent.*

(i) $g \in BV[a,b]$.

(ii) *There exists a sequence $(\phi_n)_{n=1}^{\infty}$ of step functions on $[a,b]$ such that $\sup_{n\in\mathbb{N}} \|\phi_n\|_{L^1[a,b]}$ is finite and $\lim_{n\to\infty}\int_a^x \phi_n \, d\mu_1 = g(x) - g(a)$ for all $x \in [a,b]$.*

Proof. It is clear that (ii) implies (i).

Conversely, suppose that $g \in BV[a,b]$. By Lemma 3.11.2, $Var(g,[a,x])$ exists for every $x \in (a,b]$. Using the convention that $Var(g,\{a\}) = 0$, it follows from Lemma 3.11.2 that the function

$$x \mapsto Var(g,[a,x])$$

is non-decreasing on $[a,b]$. Hence, by Corollary 3.11.5, this function is continuous except at a countable set $\{d_\ell\}_{\ell=1}^{\infty} \subset [a,b]$.

We will next construct a sequence $(\phi_n)_{n=1}^{\infty}$ of step functions. For each $n \in \mathbb{N}$ we consider a fixed division

$$\mathcal{M}_n := \{[w_{n,k-1}, w_{n,k}] : k = 1,\ldots,j(n)\}$$

of $[a,b]$ satisfying the following conditions:

(A) $\max_{j=1,\dots,j(n)} (w_{n,j} - w_{n,j-1}) < \frac{b-a}{n}$;

(B) $\{d_\ell\}_{\ell=1}^{n} \subseteq \{w_{n,\ell}\}_{\ell=0}^{j(n)}$;

(C) $\{w_{n,\ell}\}_{\ell=0}^{j(n)} \subseteq \{w_{n+1,\ell}\}_{\ell=0}^{j(n+1)}$.

Let

$$\phi_n(x) := \begin{cases} \dfrac{g(v)-g(u)}{v_i - u_i} & \text{if } x \in (u_i, v_i) \text{ for some } [u,v] \in \mathcal{M}_n, \\ 0 & \text{otherwise.} \end{cases}$$

Then $(\phi_n)_{n=1}^{\infty}$ is a sequence of step functions defined on $[a,b]$.

We will prove that $(\phi_n)_{n=1}^{\infty}$ satisfies condition (ii). Let $g(a) = 0$ and let $g_n(x) := \int_a^x \phi_n \, d\mu_1$ for each $x \in (a,b]$. Then $\sup_{n \in \mathbb{N}} \|\phi_n\|_{L^1[a,b]}$ is finite since $g \in BV[a,b]$ and

$$\sup_{n\in\mathbb{N}} Var(g_n, [a,b]) = \sup_{n\in\mathbb{N}} \sum_{[u,v]\in\mathcal{M}_n} |g(v) - g(u)| \leq \mathrm{Var}(g, [a,b]). \quad (6.1.1)$$

It remains to prove that $\lim_{n\to\infty} g_n(x) = g(x) - g(a)$ for all $x \in [a,b]$. If $x = a$, then $g_n(x) = 0 = g(x) - g(a)$ for all $n \in \mathbb{N}$. On the other hand, for each $x \in (a,b]$ and $n \in \mathbb{N}$ it follows from our construction of $\mathcal{M}_n$ that there exists a unique $[\alpha_n, \beta_n] \in \mathcal{M}_n$ such that $x \in (\alpha_n, \beta_n]$. To this end, we consider two cases.

Case 1: There exists $N \in \mathbb{N}$ such that $x = \beta_N$.

For each integer $n \geq N$ condition (C) and the uniqueness of $[\alpha_n, \beta_n]$ imply that $x = \beta_n$ and so

$$g_n(x) = g_n(\beta_n) = g(\beta_n) - g(a) = g(x) - g(a).$$

Case 2: $x \neq \beta_n$ for all $n \in \mathbb{N}$.

In this case, we have $x \in (\alpha_n, \beta_n)$ for every $n \in \mathbb{N}$ and so conditions (A), (B) and (C) yield

$$\lim_{n\to\infty} Var(g, [\alpha_n, \beta_n]) = 0.$$

Combining this limit with $g_n(\beta_n) = g(\beta_n) - g(a)$ $(n \in \mathbb{N})$ and $Var(g_n, [\alpha_n, \beta_n]) \leq Var(g, [\alpha_n, \beta_n])$ $(n \in \mathbb{N})$, we get

$$\begin{aligned}
&\limsup_{n\to\infty} \big|g_n(x) - \big(g(x) - g(a)\big)\big| \\
&\leq \limsup_{n\to\infty} |g_n(\beta_n) - g_n(x)| + \limsup_{n\to\infty} |g(\beta_n) - g(a) - g(x) + g(a)| \\
&\leq \limsup_{n\to\infty} Var(g_n, [\alpha_n, \beta_n]) + \lim_{n\to\infty} Var(g, [\alpha_n, \beta_n]) \\
&\leq 2 \lim_{n\to\infty} Var(g, [\alpha_n, \beta_n]) \\
&= 0.
\end{aligned}$$

□

We are now ready to prove that if $g \in BV[a,b]$, then g is a multiplier for $HK[a,b]$.

Theorem 6.1.5. *If $f \in HK[a,b]$ and $g \in BV[a,b]$, then $fg \in HK[a,b]$.*

Proof. This follows from Theorem 6.1.4 and Lemma 6.1.3. □

Theorem 6.1.6. *If $f \in HK[a,b]$ and g is a positive increasing function on $[a,b]$, then there exists $\xi \in [a,b]$ such that*

$$(HK)\int_a^b fg = g(b)\left\{(HK)\int_\xi^b f\right\}.$$

Proof. We may assume that $g(a) = 0$. According to the proof of Theorem 6.1.4, there exists a sequence $(\phi_n)_{n=1}^\infty$ of non-negative steps functions such that $\sup_{n\in\mathbb{N}} \|\phi_n\|_{L^1[a,b]}$ is finite and $\lim_{n\to\infty} \int_a^x \phi \, d\mu_1 = g(x)$ for all $x \in [a,b]$. Thus

$$\begin{aligned} g_n(b) \min_{x\in[a,b]} (HK)\int_x^b f &\leq \int_a^b \left\{(HK)\int_x^b f\right\} \phi_n(x)\, d\mu_1(x) \\ &\leq g_n(b) \max_{x\in[a,b]} (HK)\int_x^b f \qquad (6.1.2) \end{aligned}$$

for all $n \in \mathbb{N}$. Next we combine (6.1.2), Lemmas 6.1.2 and 6.1.3 to get

$$\begin{aligned} g(b) \min_{x\in[a,b]} (HK)\int_x^b f &\leq (HK)\int_a^b fg \\ &\leq g(b) \max_{x\in[a,b]} (HK)\int_x^b f. \qquad (6.1.3) \end{aligned}$$

Finally, since $g(b) > 0$, the desired conclusion follows from (6.1.3) and the Intermediate Value Theorem. □

A similar reasoning gives the following result.

Theorem 6.1.7. *If $f \in HK[a,b]$ and g is a positive decreasing function on $[a,b]$, then there exists $\xi \in [a,b]$ such that*

$$(HK)\int_a^b fg = g(a)\left\{(HK)\int_a^\xi f\right\}.$$

Theorem 6.1.8. *If $f \in HK[a,b]$ and $g \in BV[a,b]$, then $fg \in HK[a,b]$ and*

$$\left|(HK)\int_a^b fg\right| \leq \|f\|_{HK[a,b]}\left\{|g(a)| + Var(g,[a,b])\right\}.$$

Proof. Let $(g_n)_{n=1}^{\infty}$ be given as in the proof of Theorem 6.1.4. Since Lemma 6.1.2 yields

$$\left|(HK)\int_a^b fg_n\right| \leq \|f\|_{HK[a,b]}Var(g_n,[a,b])$$

for all $n \in \mathbb{N}$, the result follows from Lemma 6.1.3 and (6.1.1). □

The following theorem shows that Theorem 6.1.8 is, in some sense, sharp.

Theorem 6.1.9. *Let $g : [a,b] \longrightarrow \mathbb{R}$. If $fg \in HK[a,b]$ for each $f \in HK[a,b]$, then there exists $g_0 \in BV[a,b]$ such that $g_0 = g$ μ_1-almost everywhere on $[a,b]$.*

Example 6.1.10. For each $s \in \mathbb{R}$, we define

$$f_s(x) := \begin{cases} \frac{1}{x^s}\sin\frac{1}{x} & \text{if } x \in (0,1], \\ 0 & \text{otherwise.} \end{cases}$$

Then the following assertions hold.

(i) $f_s \in HK[0,1]$ if and only if $s < 2$.
(ii) $f_s \in L^1[0,1]$ if and only if $s < 1$.

Proof. (i) If $s < 0$, then $f_s \in L^1[0,1] \subset HK[0,1]$.

On the other hand, for each $s \geq 0$ we define

$$g_s(x) := \begin{cases} x^s & \text{if } x \in (0,1], \\ 0 & \text{otherwise.} \end{cases}$$

If $0 \leq s < 2$, then $g_{2-s} \in BV[0,1]$ and so $g_{2-s} \in BV[\alpha,1]$ for all $\alpha \in (0,1)$. Since an integration by parts yields

$$\int_\alpha^1 \frac{1}{x^s}\sin\frac{1}{x}\,dx = \left[x^{2-s}\cos\frac{1}{x}\right]_\alpha^1 - \int_\alpha^1 (2-s)x^{1-s}\cos\frac{1}{x}\,dx$$

for all $\alpha \in (0,1)$, an application of Cauchy extension shows that $f_s \in HK[0,1]$.

If $s = 2$, a similar argument shows that $f_2 \notin HK[0,1]$.

Now suppose that $s > 2$. Proceeding towards a contradiction, suppose that $f_s \in HK[0,1]$. Since $g_{s-2} \in BV[0,1]$, it follows from Theorem 6.1.8 that $f_2 = f_s g_{s-2} \in HK[0,1]$, a contradiction. Thus, $f_s \notin HK[0,1]$.

Combining the above cases yields (i).

(ii) If $s < 1$, then a simple computation reveals that $f_s \in L^1[0,1]$.

We claim that $f_1 \notin L^1[0,1]$. Since (i) implies that $f_1 \in HK[0,1]$, it is enough to show that $F_1 \notin BV[0,1]$, where $F_1(x) = (HK)\int_0^x f_1$ $(x \in (0,1])$ and $F_1(0) = 0$. Using integration by parts again, we get

$$F_1(x) = x\cos\frac{1}{x} - \int_0^x \cos\frac{1}{t}\,dt \tag{6.1.4}$$

for all $x \in (0,1]$. Since the function $F : [0,1] \to \mathbb{R}$ defined by

$$F(x) := \begin{cases} x\cos\frac{1}{x} & \text{if } x \in (0,1], \\ 0 & \text{if } x = 0 \end{cases}$$

is not of bounded variation on $[0,1]$, we infer from (6.1.4) that $F_1 \notin BV[0,1]$. Thus, $f_1 \notin L^1[0,1]$.

If $s > 1$, then $f_1 = f_s g_{s-1}$. Since $f_1 \notin L^1[0,1]$ and g_{s-1} is continuous (and hence bounded) on $[0,1]$, we conclude that $f_s \notin L^1[0,1]$.

Combining the above cases yields (ii) to be proved. □

6.2 On functions of bounded variation in the sense of Vitali

The aim of this section is to prove a useful multidimensional analogue of Theorem 6.1.4. We begin the following m-dimensional analogue of Definition 3.11.1.

Definition 6.2.1. Let $g : [\boldsymbol{a},\boldsymbol{b}] \longrightarrow \mathbb{R}$. The total variation of g over $[\boldsymbol{a},\boldsymbol{b}]$ is given by

$$Var(g,[\boldsymbol{a},\boldsymbol{b}]) := \sup\left\{\sum_{[\boldsymbol{u},\boldsymbol{v}]\in D} |\Delta_g([\boldsymbol{u},\boldsymbol{v}])| : D \text{ is a division of } [\boldsymbol{a},\boldsymbol{b}]\right\},$$

where

$$\Delta_g([\boldsymbol{u},\boldsymbol{v}]) := \sum_{\boldsymbol{t}\in\mathcal{V}[\boldsymbol{u},\boldsymbol{v}]} g(\boldsymbol{t}) \prod_{k=1}^m (-1)^{\chi_{\{u_k\}}(t_k)} \quad ([\boldsymbol{u},\boldsymbol{v}] \in \mathcal{I}_m([\boldsymbol{a},\boldsymbol{b}])). \tag{6.2.1}$$

Example 6.2.2. When $m = 2$, (6.2.1) becomes

$$\Delta_g([u_1,v_1]\times[u_2,v_2]) = g(u_2,v_2) - g(u_1,v_2) + g(u_1,u_2) - g(v_1,u_2).$$

Definition 6.2.3. A function $g : [\boldsymbol{a}, \boldsymbol{b}] \longrightarrow \mathbb{R}$ is said to be of bounded variation (in the sense of Vitali) on $[\boldsymbol{a}, \boldsymbol{b}]$ if $\mathrm{Var}(g, [\boldsymbol{a}, \boldsymbol{b}])$ is finite.

The space of functions of bounded variation (in the sense of Vitali) on $[\boldsymbol{a}, \boldsymbol{b}]$ is denoted by $BV[\boldsymbol{a}, \boldsymbol{b}]$. The next theorem is an immediate consequence of Definition 6.2.3.

Theorem 6.2.4. *If* $g \in BV[\boldsymbol{a}, \boldsymbol{b}]$ *and* $[\boldsymbol{u}, \boldsymbol{v}] \in \mathcal{I}_m([\boldsymbol{a}, \boldsymbol{b}])$, *then* $g|_{[\boldsymbol{u},\boldsymbol{v}]} \in BV[\boldsymbol{u}, \boldsymbol{v}]$.

Remark 6.2.5. Let $g \in BV[\boldsymbol{a}, \boldsymbol{b}]$. We set $Var(g, J) := 0$ whenever J is a degenerate subinterval of $[\boldsymbol{a}, \boldsymbol{b}]$.

On the other hand, when $m \geq 2$ the space $BV[\boldsymbol{a}, \boldsymbol{b}]$ has certain undesirable properties; for example, it contains unbounded functions. Therefore we are interested in the following space

$$BV_0[\boldsymbol{a}, \boldsymbol{b}] := \{g \in BV[\boldsymbol{a}, \boldsymbol{b}] : g(\boldsymbol{x}) = 0 \text{ whenever } \boldsymbol{x} \in [\boldsymbol{a}, \boldsymbol{b}] \backslash (\boldsymbol{a}, \boldsymbol{b}]\},$$

where $(\boldsymbol{a}, \boldsymbol{b}] := \prod_{i=1}^{m} (a_i, b_i]$.

We have the following result.

Theorem 6.2.6 (Jordan Decomposition Theorem). *If* $g \in BV_0[\boldsymbol{a}, \boldsymbol{b}]$, *then there exist* $g_1, g_2 \in BV_0[\boldsymbol{a}, \boldsymbol{b}]$ *such that* $g = g_1 - g_2$ *and*

$$\inf_{I \in \mathcal{I}_m([\boldsymbol{a},\boldsymbol{b}])} \min_{i=1,2} \Delta_{g_i}(I) \geq 0.$$

Proof. For each $\boldsymbol{x} \in [\boldsymbol{a}, \boldsymbol{b}]$, we let $g_1(\boldsymbol{x}) = \frac{1}{2}(\mathrm{Var}(g, [\boldsymbol{a}, \boldsymbol{x}]) + g(\boldsymbol{x}))$ and $g_2(\boldsymbol{x}) = \frac{1}{2}(\mathrm{Var}(g, [\boldsymbol{a}, \boldsymbol{x}]) - g(\boldsymbol{x}))$. Then g_1 and g_2 have the desired properties. □

In order to state and prove a multidimensional analogue of Theorem 6.1.4, we follow Section 5.4 to write $\Phi_{[\boldsymbol{a},\boldsymbol{b}],k}(X_k) := \prod_{i=1}^{m} W_i$, where $W_k = X_k$ and $W_i = [a_i, b_i]$ for every $i \in \{1, \ldots, m\} \backslash \{k\}$.

Lemma 6.2.7. *Let* $g \in BV_0[\boldsymbol{a}, \boldsymbol{b}]$. *If* $\boldsymbol{x}, \boldsymbol{t} \in [\boldsymbol{a}, \boldsymbol{b}]$ *and* $x_i \leq t_i$ $(i = 1, \ldots, m)$, *then*

$$|g(\boldsymbol{x}) - g(\boldsymbol{t})| \leq \sum_{k=1}^{m} Var(g, \Phi_{[\boldsymbol{a},\boldsymbol{b}],k}([x_k, t_k])).$$

Proof. By the triangle inequality,

$$\begin{aligned}&|g(\boldsymbol{x}) - g(\boldsymbol{t})|\\ &\le |g(x_1, \dots, x_m) - g(t_1, x_2, \dots, x_m)|\\ &+ \sum_{k=2}^{m-1} |g(t_1, \dots, t_{k-1}, x_k, \dots, x_m) - g(t_1, \dots, t_k, x_{k+1}, \dots, x_m)|\\ &+ |g(t_1, \dots, t_{m-1}, x_m) - g(t_1, \dots, t_m)| .\end{aligned}$$

It remains to prove that the last sum is less or equal to

$$\sum_{k=1}^{m} \mathrm{Var}(g, E([x_k, t_k])),$$

where $E([x_k, t_k])$ denotes the cartesian product

$$[a_1, t_1] \times \cdots \times [a_{k-1}, t_{k-1}] \times [x_k, t_k] \times [a_{k+1}, x_{k+1}] \times \cdots \times [a_m, x_m].$$

The proof is now complete because

$$\begin{aligned}&|g(t_1, \dots, t_{k-1}, x_k, \dots, x_m) - g(t_1, \dots, t_k, x_{k+1}, \dots, x_m)|\\ &= |\Delta_g(E([x_k, t_k]))|\\ &\le \mathrm{Var}(g, E([x_k, t_k]))\end{aligned}$$

for $k = 1, \dots, m$. □

In order to state and prove the main result of this section, we consider the following space

$$AC_0[\boldsymbol{a}, \boldsymbol{b}] := \left\{ F : \exists f \in L^1[\boldsymbol{a}, \boldsymbol{b}] \text{ such that } F(\boldsymbol{x}) = \int_{[\boldsymbol{a}, \boldsymbol{x}]} f(\boldsymbol{t})\, d\mu_m(\boldsymbol{t})\ \forall\, \boldsymbol{x} \in [\boldsymbol{a}, \boldsymbol{b}] \right\}.$$

The reader can check that $AC_0[\boldsymbol{a}, \boldsymbol{b}] \subset BV_0[\boldsymbol{a}, \boldsymbol{b}]$. The following theorem is a multidimensional analogue of Theorem 6.1.4.

Theorem 6.2.8. *Let $g : [\boldsymbol{a}, \boldsymbol{b}] \longrightarrow \mathbb{R}$. The following conditions are equivalent.*

(i) $g \in BV_0[\boldsymbol{a}, \boldsymbol{b}]$.
(ii) *There exists a sequence $(g_n)_{n=1}^{\infty}$ in $AC_0[\boldsymbol{a}, \boldsymbol{b}]$ such that $\sup_{n\in\mathbb{N}} Var(g_n, [\boldsymbol{a}, \boldsymbol{b}])$ is finite and $\lim_{n\to\infty} g_n(\boldsymbol{x}) = g(\boldsymbol{x})$ for all $\boldsymbol{x} \in [\boldsymbol{a}, \boldsymbol{b}]$.*

Proof. It is clear that (ii) implies (i).

For the converse, we suppose that $g \in BV_0[\boldsymbol{a}, \boldsymbol{b}]$. For each $k \in \{1, \dots, m\}$, it follows from Theorem 6.2.4 and Remark 6.2.5 that $g \in BV(\Phi_{[\boldsymbol{a},\boldsymbol{b}],k}([a_k, x_k]))$ for every $x_k \in [a_k, b_k]$. Since the function

$$x_k \mapsto Var(g, \Phi_{[\boldsymbol{a},\boldsymbol{b}],k}([a_k, x_k]))$$

is non-decreasing on $[a_k, b_k]$. Hence, by Corollary 3.11.5, the function

$$x_k \mapsto Var(g, \Phi_{[\boldsymbol{a},\boldsymbol{b}],k}([a_k, x_k]))$$

is continuous except at a countable set $\{d_{k,\ell}\}_{\ell=1}^{\infty}$.

We will next construct the sequence $(g_n)_{n=1}^{\infty}$ of functions. For each $k \in \{1, \dots, m\}$ and $n \in \mathbb{N}$ we let

$$D_{k,n} = \{[w_{k,n,\ell-1}, w_{k,n,\ell}] : \ell = 1, \dots, j(k,n)\}$$

be a fixed division of $[a_k, b_k]$ satisfying the following conditions:

(A) $\max\limits_{\ell=1,\dots,j(k,n)} (w_{k,n,\ell} - w_{k,n,\ell-1}) < \dfrac{b_k - a_k}{n}$;

(B) $\{d_{k,\ell}\}_{\ell=1}^{n} \subseteq \{w_{k,n,\ell}\}_{\ell=0}^{j(k,n)}$;

(C) $\{w_{k,n,\ell}\}_{\ell=0}^{j(k,n)} \subseteq \{w_{k,n+1,\ell}\}_{\ell=0}^{j(k,n+1)}$.

Next we set

$$\mathcal{M}_n := \left\{ \prod_{k=1}^{m} [w_{k,n,\ell(k)-1}, w_{k,n,\ell(k)}] : [w_{k,n,\ell(k)-1}, w_{k,n,\ell(k)}] \in D_{k,n} \, (k = 1, \dots, m) \right\},$$

$$\phi_n(\boldsymbol{x}) := \begin{cases} \dfrac{\Delta_g([\boldsymbol{u}, \boldsymbol{v}])}{\prod_{i=1}^{m}(v_i - u_i)} & \text{if } \boldsymbol{x} \in \prod_{i=1}^{m}(u_i, v_i) \text{ for some } [\boldsymbol{u}, \boldsymbol{v}] \in \mathcal{M}_n, \\ 0 & \text{otherwise.} \end{cases}$$

and $g_n(\boldsymbol{x}) := \int_{[\boldsymbol{a},\boldsymbol{x}]} \phi_n(\boldsymbol{t}) \, d\mu_m(\boldsymbol{t})$ $(\boldsymbol{x} \in [\boldsymbol{a}, \boldsymbol{b}])$. From the above construction of the sequence $(g_n)_{n=1}^{\infty}$ it is clear that $\sup_{n\in\mathbb{N}} Var(g_n, [\boldsymbol{a}, \boldsymbol{b}])$ is finite, since $g \in BV_0[\boldsymbol{a}, \boldsymbol{b}]$ and

$$\sup_{n\in\mathbb{N}} Var(g_n, [\boldsymbol{a}, \boldsymbol{b}]) = \sup_{n\in\mathbb{N}} \sum_{I\in\mathcal{M}_n} |\Delta_g(I)| \le \mathrm{Var}(g, [\boldsymbol{a}, \boldsymbol{b}]). \tag{6.2.2}$$

Now we prove that $\lim_{n\to\infty} g_n(\boldsymbol{x}) = g(\boldsymbol{x})$ for all $\boldsymbol{x} \in [\boldsymbol{a}, \boldsymbol{b}]$. If $\boldsymbol{x} \in [\boldsymbol{a}, \boldsymbol{b}] \setminus (\boldsymbol{a}, \boldsymbol{b}]$, then $g_n(\boldsymbol{x}) = g(\boldsymbol{x}) = 0$ for all $n \in \mathbb{N}$. On the other hand, we select any $\boldsymbol{x} \in (\boldsymbol{a}, \boldsymbol{b}]$. Then for each $n \in \mathbb{N}$ it follows from our construction of $\mathcal{M}_n$ that there exists a unique $[\boldsymbol{\alpha}_n, \boldsymbol{\beta}_n] \in \mathcal{M}_n$ such that $\boldsymbol{x} \in (\boldsymbol{\alpha}_n, \boldsymbol{\beta}_n]$. To this end, we consider two cases.

Case 1: There exists $N \in \mathbb{N}$ such that $\boldsymbol{x} = \boldsymbol{\beta}_N$.

For each integer $n \geq N$ we deduce from condition (C) and the uniqueness of $[\boldsymbol{\alpha}_n, \boldsymbol{\beta}_n]$ that $\boldsymbol{x} = \boldsymbol{\beta}_n$ and so $g_n(\boldsymbol{x}) = g_n(\boldsymbol{\beta}_n) = g(\boldsymbol{\beta}_n) = g(\boldsymbol{x})$.

Case 2: $\boldsymbol{x} \neq \boldsymbol{\beta}_n$ for every $n \in \mathbb{N}$.

In this case, there exists $k \in \{1, \ldots, m\}$ such that $x_k \in (\alpha_{k,n}, \beta_{k,n})$ for infinitely many $n \in \mathbb{N}$. This assertion, together with condition (C), implies that $x_k \in (\alpha_{k,n}, \beta_{k,n})$ and

$$
\begin{aligned}
&|g_n(\boldsymbol{x}) - g(\boldsymbol{x})| \\
&\leq |g_n(\boldsymbol{\beta}_n) - g_n(\boldsymbol{x})| + |g(\boldsymbol{\beta}_n) - g(\boldsymbol{x})| \\
&\leq \sum_{k=1}^{m} \mathrm{Var}(g_n, \Phi_{[\boldsymbol{a},\boldsymbol{b}],k}([x_k, \beta_{k,n}])) + \sum_{k=1}^{m} \mathrm{Var}(g, \Phi_{[\boldsymbol{a},\boldsymbol{b}],k}([x_k, \beta_{k,n}])) \\
&\qquad\qquad\qquad\qquad\qquad\qquad\qquad\qquad \text{(by Lemma 6.2.7)} \\
&\leq \sum_{\substack{k=1 \\ x_k \in (\alpha_{k,n}, \beta_{k,n})}}^{m} \Big(\mathrm{Var}(g_n, \Phi_{[\boldsymbol{a},\boldsymbol{b}],k}([\alpha_{k,n}, \beta_{k,n}])) + \mathrm{Var}(g, \Phi_{[\boldsymbol{a},\boldsymbol{b}],k}([\alpha_{k,n}, \beta_{k,n}]))\Big) \\
&\leq 2 \sum_{\substack{k=1 \\ x_k \in (\alpha_{k,n}, \beta_{k,n})}}^{m} \mathrm{Var}(g, \Phi_{[\boldsymbol{a},\boldsymbol{b}],k}([\alpha_{k,n}, \beta_{k,n}])) \qquad \text{(by our choice of } g_n\text{)}
\end{aligned}
$$

for every integers n.

It remains to prove that if $k \in \{1, \ldots, m\}$ and $\alpha_{k,n} < x_k < \beta_{k,n}$ for all $n \in \mathbb{N}$, then

$$\mathrm{Var}(g, \Phi_{[\boldsymbol{a},\boldsymbol{b}],k}([\alpha_{k,n}, \beta_{k,n}])) \to 0 \text{ as } n \to \infty.$$

But it is clear, since $x_k \neq d_{k,n}$ ($k \in \{1, \ldots, m\}; n \in \mathbb{N}$), conditions (A), (B) and (C) imply that

$$\lim_{n \to \infty} \max_{k=1,\ldots,m} (\beta_{k,n} - \alpha_{k,n}) = 0.$$

□

Remark 6.2.9. While the proof of Theorem 6.2.8 is similar to that of [117, Theorem 1], we observe that our proof is even more general. In fact, the proof does not depend on the Jordon Decomposition Theorem (Theorem 6.2.6).

A modification of the proof of Theorem 6.2.8 yields the following results.

Theorem 6.2.10. *Let* $(x_1, \ldots, x_m) \in [\boldsymbol{a}, \boldsymbol{b}]$ *and let* $\big((x_{1,n}, \ldots, x_{m,n})\big)_{n=1}^{\infty}$ *be a sequence in* $[\boldsymbol{a}, \boldsymbol{b}]$ *such that* $sgn(x_{i,k} - x_k) = sgn(x_{j,k} - x_k)$ *for all* $i, j \in \mathbb{N}$ *and* $k \in \{1, \ldots, m\}$. *If* $g \in BV_0[\boldsymbol{a}, \boldsymbol{b}]$ *and* $\lim_{n\to\infty}(x_{1,n}, \ldots, x_{m,n}) = (x_1, \ldots, x_m)$, *then* $\lim_{n\to\infty} g(x_{1,n}, \ldots, x_{m,n})$ *exists.*

Theorem 6.2.11. *If* $g \in BV_0[\boldsymbol{a}, \boldsymbol{b}]$, *then* g *is continuous everywhere on* $[\boldsymbol{a}, \boldsymbol{b}]$ *except for a countable number of hyperplanes parallel to the coordinate axes.*

6.3 The m-dimensional Riemann-Stieltjes integral

In this section we give a generalization of the m-dimensional Riemann integral. As a result, we obtain a refinement of Lemma 6.1.1.

Definition 6.3.1. Let F and H be two real-valued functions defined on $[\boldsymbol{a}, \boldsymbol{b}]$. F is said to be Riemann-Stieltjes integrable with respect to H on $[\boldsymbol{a}, \boldsymbol{b}]$ if there exists $A \in \mathbb{R}$ with the following property: for each $\varepsilon > 0$ there exists $\delta > 0$ such that

$$\left| \sum_{(\boldsymbol{t},I)\in P} F(\boldsymbol{t})\Delta_H(I) - A \right| < \varepsilon \tag{6.3.1}$$

for each δ-fine Perron partition P of $[\boldsymbol{a}, \boldsymbol{b}]$.

Theorem 6.3.2. *There is at most one number* A *satisfying (6.3.1).*

Proof. Exercise. □

Let A be given as in Definition 6.3.1. We write A as $(RS)\int_{[\boldsymbol{a},\boldsymbol{b}]} F\, dH$ or $(RS)\int_{[\boldsymbol{a},\boldsymbol{b}]} F(\boldsymbol{x})\, dH(\boldsymbol{x})$; in this case, we say that the Riemann-Stieltjes integral $(RS)\int_{[\boldsymbol{a},\boldsymbol{b}]} F\, dH$ exists.

We have the following Cauchy criterion for Riemann-Stieltjes integrability.

Theorem 6.3.3. *Let* F *and* H *be real-valued functions defined on* $[\boldsymbol{a}, \boldsymbol{b}]$. *Then the Riemann-Stieltjes integral* $(RS)\int_{[\boldsymbol{a},\boldsymbol{b}]} F\, dH$ *exists if and only if for each* $\varepsilon > 0$ *there exists* $\delta > 0$ *such that*

$$\left| \sum_{(\boldsymbol{t},I)\in P} F(\boldsymbol{t})\Delta_H(I) - \sum_{(\boldsymbol{x},J)\in Q} F(\boldsymbol{x})\Delta_H(J) \right| < \varepsilon$$

whenever P *and* Q *are* δ*-fine Perron partitions of* $[\boldsymbol{a}, \boldsymbol{b}]$.

Proof. The proof is similar to that of Theorem 2.3.4. □

Recall that $C[a,b]$ is the space of continuous functions on $[a,b]$. According to Corollary 2.2.4, $\|f\|_{C[a,b]} := \sup_{x\in[a,b]} |f(x)|$ is a non-negative real number whenever $f \in C[a,b]$. The following theorem is an immediate consequence of Theorem 6.3.3.

Theorem 6.3.4. *If $F \in C[a,b]$ and $g \in BV[a,b]$, then the Riemann-Stieltjes integral $(RS)\int_{[a,b]} F\,dg$ exists.*

Proof. In view of Theorem 2.2.3, the proof is similar to that of Theorem 2.3.5. □

Theorem 6.3.5. *If $F \in C[a,b]$ and $g \in BV[a,b]$, then*

$$\left|(RS)\int_{[a,b]} F\,dg\right| \le \|F\|_{C[a,b]} Var(g,[a,b]).$$

Proof. For each $\varepsilon > 0$ there exists a positive constant δ such that

$$\left|\sum_{(t,I)\in P} F(t)\Delta_g(I) - (RS)\int_{[a,b]} F\,dg\right| < \varepsilon$$

for each δ-fine Perron partition P of $[a,b]$. Hence, for any δ-fine partition P of $[a,b]$, we have

$$\left|(RS)\int_{[a,b]} F\,dg\right| \le \left|\sum_{(t,I)\in P} F(t)\Delta_g(I)\right| + \varepsilon \le \|F\|_{C[a,b]} Var(g,[a,b]) + \varepsilon.$$

Since $\varepsilon > 0$ is arbitrary, the theorem is proved. □

Theorem 6.3.6. *If $F \in C[\,a,b]$, $g \in BV[a,b]$, and D is a division of $[a,b]$, then $(RS)\int_I F\,dg$ exists for all $I \in D$, and*

$$(RS)\int_{[a,b]} F\,dg = \sum_{I\in D} (RS)\int_I F\,dg.$$

Proof. Exercise. □

Theorem 6.3.7. *If $F \in C[a,b]$, $h \in L^1[a,b]$ and $H(x) = \int_{[a,x]} h\,d\mu_m$ for all $x \in [a,b]$, then the Riemann-Stieltjes integral $(RS)\int_{[a,b]} F\,dH$ exists, $Fh \in L^1[a,b]$ and*

$$(RS)\int_{[a,b]} F\,dH = \int_{[a,b]} Fh\,d\mu_m. \tag{6.3.2}$$

Proof. In view of Theorem 6.3.4, Corollary 2.2.4 and Theorem 3.7.3, it suffices to prove that (6.3.2) holds. For each $\varepsilon > 0$ we apply Theorem 2.2.3 to select a sufficient small $\eta > 0$ so that

$$\omega(F, I) < \frac{\varepsilon}{2(1 + \|h\|_{L^1[\boldsymbol{a},\boldsymbol{b}]})}$$

for each $I \in \mathcal{I}_m([\boldsymbol{a}, \boldsymbol{b}])$ with $\text{diam}(I) < \eta$. If P is any η-fine Perron partition of $[\boldsymbol{a}, \boldsymbol{b}]$, it follows from Theorems 6.3.6, 2.3.10 and 6.3.5 that

$$\begin{aligned}
&\left|(RS)\int_{[\boldsymbol{a},\boldsymbol{b}]} F\, dH - \int_{[\boldsymbol{a},\boldsymbol{b}]} Fh\, d\mu_m\right| \\
&\leq \sum_{(\boldsymbol{t},I)\in P} \left|(RS)\int_I F\, dH - \int_I Fh\, d\mu_m\right| \\
&= \sum_{(\boldsymbol{t},I)\in P} \left|(RS)\int_I (F - F(\boldsymbol{t}))\, dH - \int_I (F - F(\boldsymbol{t}))h\, d\mu_m\right| \\
&\leq \sum_{(\boldsymbol{t},I)\in P} \omega(F, I)(Var(H, I) + \|h\|_{L^1[\boldsymbol{a},\boldsymbol{b}]}) \\
&< \varepsilon.
\end{aligned}$$

Since $\varepsilon > 0$ is arbitrary, the theorem is proved. □

Theorem 6.3.8. *Let $F \in C[\boldsymbol{a}, \boldsymbol{b}]$ and suppose that the following conditions are satisfied:*

(i) *$(g_n)_{n=1}^{\infty}$ is a sequence in $BV[\boldsymbol{a}, \boldsymbol{b}]$ with $\sup_{n\in\mathbb{N}} Var(g_n, [\boldsymbol{a}, \boldsymbol{b}]) < \infty$;*
(ii) *$g_n \to g$ pointwise on $[\boldsymbol{a}, \boldsymbol{b}]$.*

Then the Riemann-Stieltjes integral $(RS)\int_{[\boldsymbol{a},\boldsymbol{b}]} F\, dg$ exists. Moreover, the limit $\lim_{n\to\infty}(RS)\int_{[\boldsymbol{a},\boldsymbol{b}]} F\, dg_n$ exists and

$$\lim_{n\to\infty} (RS)\int_{[\boldsymbol{a},\boldsymbol{b}]} F\, dg_n = (RS)\int_{[\boldsymbol{a},\boldsymbol{b}]} F\, dg.$$

Proof. For each $\varepsilon > 0$ we apply Theorem 2.2.3 to select a sufficient small $\eta > 0$ so that

$$\omega(F, I) < \frac{\varepsilon}{4(1 + \sup_{n\in\mathbb{N}} Var(g_n, [\boldsymbol{a}, \boldsymbol{b}]))}$$

whenever $I \in \mathcal{I}_m([\boldsymbol{a}, \boldsymbol{b}])$ with $\mathrm{diam}(I) < \eta$. If P is any η-fine Perron partition of $[\boldsymbol{a}, \boldsymbol{b}]$, we have

$$\begin{aligned}
&\left| \sum_{(\boldsymbol{t},I)\in P} F(\boldsymbol{t})\Delta_{g_n}(I) - (RS)\int_{[\boldsymbol{a},\boldsymbol{b}]} F\, dg_n \right| \\
&\le \sum_{(\boldsymbol{t},I)\in P} \left| F(\boldsymbol{t})\Delta_{g_n}(I) - (RS)\int_I F\, dg_n \right| \\
&= \sum_{(\boldsymbol{t},I)\in P} \left| (RS)\int_I (F(\boldsymbol{t}) - F)\, dg_n \right| \\
&\le \sum_{(\boldsymbol{t},I)\in P} \omega(F, I) Var(g_n, I) \\
&< \frac{\varepsilon}{2}.
\end{aligned} \tag{6.3.3}$$

Combining (6.3.3) and (ii) shows that $\left((RS)\int_{[\boldsymbol{a},\boldsymbol{b}]} F\ dg_n\right)_{n=1}^{\infty}$ is a Cauchy sequence of real numbers; therefore this sequence converges to some real number A. Finally, we let $n \to \infty$ in (6.3.3) to complete the proof. □

The following theorem is a refinement of Lemma 6.1.1.

Theorem 6.3.9. *If $f \in L^1[\boldsymbol{a}, \boldsymbol{b}]$ and $g \in BV_0[\boldsymbol{a}, \boldsymbol{b}]$, then $fg \in L^1[\boldsymbol{a}, \boldsymbol{b}]$ and*

$$\int_{[\boldsymbol{a},\boldsymbol{b}]} fg\, d\mu_m = (RS)\int_{[\boldsymbol{a},\boldsymbol{b}]} \left\{ \int_{[\boldsymbol{x},\boldsymbol{b}]} f\, d\mu_m \right\} dg(\boldsymbol{x}).$$

Proof. Let $(\phi_n)_{n=1}^{\infty}$ be given as in the proof of Theorem 6.2.8. For each $n \in \mathbb{N}$ we observe that the proof of Lemma 6.1.1 yields

$$\begin{aligned}
&\int_{[\boldsymbol{a},\boldsymbol{b}]} f(\boldsymbol{x}) \left\{ \int_{[\boldsymbol{a},\boldsymbol{x}]} \phi_n(\boldsymbol{t})\, d\mu_m(\boldsymbol{t}) \right\} d\mu_m(\boldsymbol{x}) \\
&= \int_{[\boldsymbol{a},\boldsymbol{b}]} \left\{ \int_{[\boldsymbol{t},\boldsymbol{b}]} f(\boldsymbol{x})\, d\mu_m(\boldsymbol{x}) \right\} \phi_n(\boldsymbol{t})\, d\mu_m(\boldsymbol{t}).
\end{aligned}$$

Next we let $g_n(\boldsymbol{t}) := \int_{[\boldsymbol{a},\boldsymbol{t}]} \phi_n\, d\mu_m$ $(\boldsymbol{t} \in [\boldsymbol{a}, \boldsymbol{b}])$ and apply Theorem 6.3.7 to obtain

$$\begin{aligned}
&\int_{[\boldsymbol{a},\boldsymbol{b}]} \left\{ \int_{[\boldsymbol{t},\boldsymbol{b}]} f(\boldsymbol{x})\, d\mu_m(\boldsymbol{x}) \right\} \phi_n(\boldsymbol{t})\, d\mu_m(\boldsymbol{t}) \\
&= \int_{[\boldsymbol{a},\boldsymbol{b}]} \left\{ \int_{[\boldsymbol{t},\boldsymbol{b}]} f(\boldsymbol{x})\, d\mu_m(\boldsymbol{x}) \right\} dg_n(\boldsymbol{t}).
\end{aligned}$$

The desired conclusion follows from the above equalities, Lebesgue's Dominated Convergence Theorem and Theorem 6.3.8. □

It is now easy to deduce the following Mean-Value Theorem from Theorem 6.3.9. For other versions of such theorem, consult for instance [162].

Theorem 6.3.10. *Let $f \in L^1[\boldsymbol{a}, \boldsymbol{b}]$ and let $g \in BV_0[\boldsymbol{a}, \boldsymbol{b}]$. If $\Delta_g(I) \geq 0$ for every $I \in \mathcal{I}_m([\boldsymbol{a}, \boldsymbol{b}])$, then there exists $\boldsymbol{c} \in [\boldsymbol{a}, \boldsymbol{b}]$ such that*

$$\int_{[\boldsymbol{a},\boldsymbol{b}]} fg \, d\mu_m = g(\boldsymbol{b}) \int_{[\boldsymbol{c},\boldsymbol{b}]} f \, d\mu_m.$$

Proof. This is an immediate consequence of Theorem 6.3.9, the following inequalities

$$g(\boldsymbol{b}) \min_{\boldsymbol{x}\in[\boldsymbol{a},\boldsymbol{b}]} \left\{ \int_{[\boldsymbol{x},\boldsymbol{b}]} f \, d\mu_m \right\} \leq \int_{[\boldsymbol{a},\boldsymbol{b}]} \left\{ \int_{[\boldsymbol{x},\boldsymbol{b}]} f \, d\mu_m \right\} dg(\boldsymbol{x})$$
$$\leq g(\boldsymbol{b}) \max_{\boldsymbol{x}\in[\boldsymbol{a},\boldsymbol{b}]} \left\{ \int_{[\boldsymbol{x},\boldsymbol{b}]} f \, d\mu_m \right\}$$

and the Intermediate Value Theorem. □

6.4 A multiple integration by parts for the Henstock-Kurzweil integral

The aim of this section is to show that an analogous version of Theorem 6.3.9 holds for the Henstock-Kurzweil integral.

Theorem 6.4.1. *If $f \in HK[\boldsymbol{a}, \boldsymbol{b}]$ and $g \in BV_0[\boldsymbol{a}, \boldsymbol{b}]$, then $fg \in HK[\boldsymbol{a}, \boldsymbol{b}]$ and*

$$(HK) \int_{[\boldsymbol{a},\boldsymbol{b}]} fg = (RS) \int_{[\boldsymbol{a},\boldsymbol{b}]} \left\{ (HK) \int_{[\boldsymbol{x},\boldsymbol{b}]} f \right\} dg(\boldsymbol{x}). \qquad (6.4.1)$$

Proof. Let $\varepsilon > 0$ and use Theorem 2.4.7 to choose a gauge δ_1 on $[\boldsymbol{a}, \boldsymbol{b}]$ so that

$$\sum_{(\boldsymbol{t}',I')\in P'} \|f(\boldsymbol{t}') - f\|_{HK(I')} < \frac{\varepsilon}{10(1 + \mathrm{Var}(g, [\boldsymbol{a}, \boldsymbol{b}]))} \qquad (6.4.2)$$

for each δ_1-fine Perron subpartition P' of $[\boldsymbol{a}, \boldsymbol{b}]$.

Since $g \in BV_0[\boldsymbol{a}, \boldsymbol{b}]$, it follows from Theorem 6.2.11 that there exists a set $Z \subseteq [\boldsymbol{a}, \boldsymbol{b}]$ such that $\mu_m(Z) = 0$ and g is continuous at each point of $[\boldsymbol{a}, \boldsymbol{b}]\backslash Z$. Hence there exists a gauge δ_2 on $[\boldsymbol{a}, \boldsymbol{b}]\backslash Z$ such that

$$|g(\boldsymbol{x}) - g(\boldsymbol{y})| < \frac{\varepsilon}{4(1 + |f(\boldsymbol{x})|)(\mu_m([\boldsymbol{a}, \boldsymbol{b}]))}$$

whenever $\boldsymbol{x} \in [\boldsymbol{a}, \boldsymbol{b}]\backslash Z$ and $\boldsymbol{y} \in B(\boldsymbol{x}, \delta_2(\boldsymbol{x})) \cap [\boldsymbol{a}, \boldsymbol{b}]$.

Since $Z \subset [\boldsymbol{a}, \boldsymbol{b}]$ has μ_m-measure zero and f is real-valued, it follows from (6.4.2) that there exists a gauge δ_3 on Z such that

$$\sum_{(\boldsymbol{t}'',I'')\in P''} \|f\|_{HK(I'')} < \frac{\varepsilon}{8(1+\mathrm{Var}(g,[\boldsymbol{a},\boldsymbol{b}]))}$$

for each Z-tagged δ_3-fine Perron subpartition P'' of $[\boldsymbol{a}, \boldsymbol{b}]$.

Now we define a gauge δ on $[\boldsymbol{a}, \boldsymbol{b}]$ by setting

$$\delta(\boldsymbol{x}) = \begin{cases} \min\{\delta_1(\boldsymbol{x}), \delta_2(\boldsymbol{x})\} \text{ if } \boldsymbol{x} \in [\boldsymbol{a},\boldsymbol{b}]\backslash Z, \\ \\ \min\{\delta_1(\boldsymbol{x}), \delta_3(\boldsymbol{x})\} \text{ if } \boldsymbol{x} \in Z, \end{cases}$$

and consider any δ-fine Perron partition P of $[\boldsymbol{a}, \boldsymbol{b}]$. Then

$$\begin{aligned}
&\left|\sum_{(\boldsymbol{t},I)\in P} f(\boldsymbol{t})g(\boldsymbol{t})\mu_m(I) - (RS)\int_{[\boldsymbol{a},\boldsymbol{b}]}\left\{(HK)\int_{[\boldsymbol{x},\boldsymbol{b}]} f\right\} dg(\boldsymbol{x})\right| \\
&\le \sum_{\substack{(\boldsymbol{t},I)\in P\\ \boldsymbol{t}\in[\boldsymbol{a},\boldsymbol{b}]\backslash Z}} \left|f(\boldsymbol{t})g(\boldsymbol{t})\mu_m(I) - (RS)\int_{[\boldsymbol{a},\boldsymbol{b}]}\left\{(HK)\int_{[\boldsymbol{x},\boldsymbol{b}]} f\chi_I\right\} dg(\boldsymbol{x})\right| \\
&+ \sum_{\substack{(\boldsymbol{t},I)\in P\\ \boldsymbol{t}\in Z}} \left|f(\boldsymbol{t})g(\boldsymbol{t})\mu_m(I) - (RS)\int_{[\boldsymbol{a},\boldsymbol{b}]}\left\{(HK)\int_{[\boldsymbol{x},\boldsymbol{b}]} f\chi_I\right\} dg(\boldsymbol{x})\right| \\
&\le \sum_{\substack{(\boldsymbol{t},I)\in P\\ \boldsymbol{t}\in[\boldsymbol{a},\boldsymbol{b}]\backslash Z}} \left|f(\boldsymbol{t})\left\{g(\boldsymbol{t})\mu_m(I) - \int_I g\, d\mu_m\right\}\right| \\
&+ \sum_{\substack{(\boldsymbol{t},I)\in P\\ \boldsymbol{t}\in[\boldsymbol{a},\boldsymbol{b}]\backslash Z}} \left|f(\boldsymbol{t})\int_I g\, d\mu_m - (RS)\int_{[\boldsymbol{a},\boldsymbol{b}]}\left\{(HK)\int_{[\boldsymbol{x},\boldsymbol{b}]} f\chi_I\right\} dg(\boldsymbol{x})\right| \\
&+ \sum_{\substack{(\boldsymbol{t},I)\in P\\ \boldsymbol{t}\in Z}} \left|f(\boldsymbol{t})g(\boldsymbol{t})\mu_m(I) - (RS)\int_{[\boldsymbol{a},\boldsymbol{b}]}\left\{(HK)\int_{[\boldsymbol{x},\boldsymbol{b}]} f\chi_I\right\} dg(\boldsymbol{x})\right| \\
&= S_1 + S_2 + S_3,
\end{aligned}$$

say.

According to our choice of δ_2, we get

$$S_1 = \sum_{\substack{(\boldsymbol{t},I)\in P\\ \boldsymbol{t}\in[\boldsymbol{a},\boldsymbol{b}]\backslash Z}} \left|f(\boldsymbol{t})\left\{\int_I (g(\boldsymbol{t}) - g)\, d\mu_m\right\}\right| < \frac{\varepsilon}{4}.$$

Theorems 6.3.9, 6.3.5 and (6.4.2) yield an upper bound for S_2:

$$\begin{aligned}
&S_2\\
&= \sum_{\substack{(\boldsymbol{t},I)\in P\\ \boldsymbol{t}\in[\boldsymbol{a},\boldsymbol{b}]\setminus Z}} \left| f(\boldsymbol{t})\int_I g\, d\mu_m - (RS)\int_{[\boldsymbol{a},\boldsymbol{b}]} \left\{ (HK)\int_{[\boldsymbol{x},\boldsymbol{b}]} f\chi_I \right\} dg(\boldsymbol{x}) \right|\\
&= \sum_{\substack{(\boldsymbol{t},I)\in P\\ \boldsymbol{t}\in[\boldsymbol{a},\boldsymbol{b}]\setminus Z}} \left| (RS)\int_{[\boldsymbol{a},\boldsymbol{b}]} \left\{ (HK)\int_{[\boldsymbol{x},\boldsymbol{b}]} (f(\boldsymbol{t})-f)\chi_I \right\} dg(\boldsymbol{x}) \right|\\
&< \frac{\varepsilon}{10}.
\end{aligned}$$

We also have

$$\begin{aligned}
&S_3\\
&= \sum_{\substack{(\boldsymbol{t},I)\in P\\ \boldsymbol{t}\in Z}} \left| f(\boldsymbol{t})g(\boldsymbol{t})\mu_m(I) - (RS)\int_{[\boldsymbol{a},\boldsymbol{b}]} \left\{ (HK)\int_{[\boldsymbol{x},\boldsymbol{b}]} f\chi_I \right\} dg(\boldsymbol{x}) \right|\\
&\le \left\{ \sup_{\boldsymbol{t}\in[\boldsymbol{a},\boldsymbol{b}]} |g(\boldsymbol{t})| \right\} \sum_{\substack{(\boldsymbol{t},I)\in P\\ \boldsymbol{t}\in Z}} \left\{ \left| f(\boldsymbol{t})\mu_m(I) - (HK)\int_I f \right| + \left| (HK)\int_I f \right| \right\}\\
&+ \sum_{\substack{(\boldsymbol{t},I)\in P\\ \boldsymbol{t}\in Z}} \left| (RS)\int_{[\boldsymbol{a},\boldsymbol{b}]} \left\{ (HK)\int_{[\boldsymbol{x},\boldsymbol{b}]} f\chi_I \right\} dg(\boldsymbol{x}) \right|\\
&\le \left\{ \sup_{\boldsymbol{t}\in[\boldsymbol{a},\boldsymbol{b}]} |g(\boldsymbol{t})| \right\} \sum_{\substack{(\boldsymbol{t},I)\in P\\ \boldsymbol{t}\in Z}} \left\{ \left| f(\boldsymbol{t})\mu_m(I) - (HK)\int_I f \right| + \left| (HK)\int_I f \right| \right\}\\
&+\mathrm{Var}(g,[\boldsymbol{a},\boldsymbol{b}]) \sum_{\substack{(\boldsymbol{t},I)\in P\\ \boldsymbol{t}\in Z}} \|f\|_{HK(I)} \qquad \text{(by Theorem 6.3.5)}\\
&\le 2\,\mathrm{Var}(g,[\boldsymbol{a},\boldsymbol{b}]) \left\{ \sum_{\substack{(\boldsymbol{t},I)\in P\\ \boldsymbol{t}\in Z}} \|f(\boldsymbol{t})-f\|_{HK(I)} + \sum_{\substack{(\boldsymbol{t},I)\in P\\ \boldsymbol{t}\in Z}} \|f\|_{HK(I)} \right\}\\
&\qquad \text{(since } g\in BV_0[\boldsymbol{a},\boldsymbol{b}]\text{)}\\
&< \frac{\varepsilon}{2}.
\end{aligned}$$

Combining the above estimates completes the proof. □

Following the argument of Theorem 6.3.10, we obtain the following result.

Theorem 6.4.2. *Let $f \in HK[\boldsymbol{a}, \boldsymbol{b}]$ and let $g \in BV_0[\boldsymbol{a}, \boldsymbol{b}]$. If $\Delta_g(I) \geq 0$ for every $I \in \mathcal{I}_m([\boldsymbol{a}, \boldsymbol{b}])$, then there exists $\boldsymbol{c} \in [\boldsymbol{a}, \boldsymbol{b}]$ such that*

$$(HK) \int_{[\boldsymbol{a},\boldsymbol{b}]} fg = g(\boldsymbol{b}) \Big\{ (HK) \int_{[\boldsymbol{c},\boldsymbol{b}]} f \Big\}.$$

The following theorem is a consequence of Theorems 6.3.5 and 6.4.1.

Theorem 6.4.3. *If $f \in HK[\boldsymbol{a}, \boldsymbol{b}]$ and $g \in BV_0[\boldsymbol{a}, \boldsymbol{b}]$, then $fg \in HK[\boldsymbol{a}, \boldsymbol{b}]$ and*

$$\left| (HK) \int_{[\boldsymbol{a},\boldsymbol{b}]} fg \right| \leq \|f\|_{HK[\boldsymbol{a},\boldsymbol{b}]} Var(g, [\boldsymbol{a}, \boldsymbol{b}]). \tag{6.4.3}$$

6.5 Kurzweil's multiple integration by parts formula for the Henstock-Kurzweil integral

The aim of this section is to establish Kurzweil's multiple integration by parts formula for the Henstock-Kurzweil integral. We begin with the following definition.

Definition 6.5.1. A function $g : [a_1, b_1] \times [a_2, b_2] \longrightarrow \mathbb{R}$ is said to be of bounded variation in the sense of Hardy and Krause if the following conditions are satisfied:

(i) the function $x \mapsto g(x, a_2)$ is of bounded variation on $[a_1, b_1]$;
(ii) the function $y \mapsto g(a_1, y)$ is of bounded variation on $[a_2, b_2]$;
(iii) $g \in BV([a_1, b_1] \times [a_2, b_2])$.

Recall that $\Phi_{[\boldsymbol{a},\boldsymbol{b}],k}(X_k) := \prod_{i=1}^m W_i$, where $W_k = X_k$ and $W_i = [a_i, b_i]$ for all $i \in \{1, \ldots, m\} \backslash \{k\}$. The following definition is an m-dimensional version of Definition 6.5.1.

Definition 6.5.2. A function $g : [\boldsymbol{a}, \boldsymbol{b}] \longrightarrow \mathbb{R}$ is said to be of bounded variation in the sense of Hardy and Krause on $[\boldsymbol{a}, \boldsymbol{b}]$ if the following conditions are satisfied:

(i) if $\Gamma \subset \{1, \ldots, m\}$ is non-empty, then

$$g \Big|_{\bigcap_{\substack{k=1 \\ k \notin \Gamma}}^{m} \Phi_{[\boldsymbol{a},\boldsymbol{b}],k}(\{a_k\})} \in BV\Big(\prod_{\substack{k=1 \\ k \in \Gamma}}^{m} [a_k, b_k] \Big);$$

(ii) $g \in BV[\boldsymbol{a}, \boldsymbol{b}]$.

The class of functions of bounded variation in the sense of Hardy and Krause on $[\boldsymbol{a}, \boldsymbol{b}]$ will be denoted by $BV_{HK}[\boldsymbol{a}, \boldsymbol{b}]$. The following theorem is an immediate consequence of Definition 6.5.2.

Theorem 6.5.3. $BV_0[\boldsymbol{a}, \boldsymbol{b}] \subset BV_{HK}[\boldsymbol{a}, \boldsymbol{b}]$.

We need the following lemmas in order to prove Theorem 6.5.6 below.

Lemma 6.5.4. *Let $g \in BV_{HK}[\boldsymbol{a}, \boldsymbol{b}]$. If $\mathcal{T} \subset \{1, \ldots, m\}$ is non-empty and $c_k \in \{a_k, b_k\}$ for all $k \in \{1, \ldots, m\} \setminus \mathcal{T}$, then*

$$g \Bigg|_{\bigcap\limits_{\substack{k=1\\k \notin \mathcal{T}}}^{m} \Phi_{[\boldsymbol{a},\boldsymbol{b}],k}(\{c_k\})} \in BV\Bigg(\prod_{\substack{k=1\\k \in \mathcal{T}}}^{m} [a_k, b_k]\Bigg).$$

Proof. This is an immediate consequence of Definition 6.5.2. □

In order to proceed further, let

$$\mathcal{P}_m := \left\{\prod_{k=1}^{m} Y_k : Y_k \in \{\{a_k\}, \{b_k\}, [a_k, b_k]\} \text{ for each } k \in \{1, \ldots, m\}\right\}$$

and for $\prod_{k=1}^{m} Y_k \in \mathcal{P}_m$, let

$$\Gamma\left(\prod_{k=1}^{m} Y_k\right) := \{i \in \{1, \ldots, m\} : Y_i = [a_i, b_i]\}.$$

Lemma 6.5.5. *If $g \in BV_{HK}[\boldsymbol{a}, \boldsymbol{b}]$ and $Y \in \mathcal{P}_m$, then $g\chi_Y \in BV[\boldsymbol{a}, \boldsymbol{b}]$.*

Proof. If $Y \in \mathcal{P}_m$ with $\text{card}(\Gamma(Y)) = m$, then $Y = [\boldsymbol{a}, \boldsymbol{b}]$ and hence $g\chi_Y \in BV[\boldsymbol{a}, \boldsymbol{b}]$. On the other hand, for any $Y \in \mathcal{P}_m$ satisfying $\text{card}(\Gamma(Y)) < m$, Lemma 6.5.4 implies that $g\chi_Y \in BV[\boldsymbol{a}, \boldsymbol{b}]$. □

Theorem 6.5.6. *If $g \in BV_{HK}[\boldsymbol{a}, \boldsymbol{b}]$, then $g\chi_{(\boldsymbol{a},\boldsymbol{b})} \in BV_0[\boldsymbol{a}, \boldsymbol{b}]$ and*

$$g\chi_{(\boldsymbol{a},\boldsymbol{b})} = \sum_{Y \in \mathcal{P}_m} (-1)^{m - card(\Gamma(Y))} g\chi_Y. \tag{6.5.1}$$

Proof. It is clear that (6.5.1) holds for any real-valued function g defined on $[\boldsymbol{a}, \boldsymbol{b}]$. Since (6.5.1) and Lemma 6.5.5 imply $g\chi_{(\boldsymbol{a},\boldsymbol{b})} \in BV[\boldsymbol{a}, \boldsymbol{b}]$, we conclude that $g\chi_{(\boldsymbol{a},\boldsymbol{b})} \in BV_0[\boldsymbol{a}, \boldsymbol{b}]$. □

For each $k \in \{1,\dots,m\}$ and $\boldsymbol{c} \in [\boldsymbol{a},\boldsymbol{b}]$, we define the function $\pi_{k,\boldsymbol{c}} : [\boldsymbol{a},\boldsymbol{b}] \longrightarrow [\boldsymbol{a},\boldsymbol{b}]$ by setting $\pi_{k,\boldsymbol{c}}(\boldsymbol{x}) = \boldsymbol{z}$, where $z_k = c_k$ and $z_j = x_j$ for all $j \in \{1,\dots,m\}\backslash\{k\}$. Using Theorems 6.2.8, 6.3.7 and 6.3.8, we obtain the following result.

Theorem 6.5.7. *Let $F \in C[\boldsymbol{a},\boldsymbol{b}]$, let $g \in BV[\boldsymbol{a},\boldsymbol{b}]$, and suppose that $g(\boldsymbol{x}) = 0$ for all $\boldsymbol{x} \in [\boldsymbol{a},\boldsymbol{b}]\backslash(\boldsymbol{a},\boldsymbol{b})$. If there exists $\boldsymbol{c} \in [\boldsymbol{a},\boldsymbol{b}]$ such that $F(\boldsymbol{x}) = (F \circ \pi_{k,\boldsymbol{c}})(\boldsymbol{x})$ for all $\boldsymbol{x} \in [\boldsymbol{a},\boldsymbol{b}]$, then*

$$(RS)\int_{[\boldsymbol{a},\boldsymbol{b}]} F\, dg = 0.$$

Proof. According to Theorem 6.2.8, there exists a sequence $(\phi_n)_{n=1}^{\infty}$ in $L^1[\boldsymbol{a},\boldsymbol{b}]$ such that

$$\sup_{n\in\mathbb{N}} ||\phi_n||_{L^1[\boldsymbol{a},\boldsymbol{b}]} < \infty \tag{6.5.2}$$

and

$$\lim_{n\to\infty}\int_{[\boldsymbol{a},\boldsymbol{x}]} \phi_n\, d\mu_m = g(\boldsymbol{x}) \ \text{ for all } \boldsymbol{x} \in [\boldsymbol{a},\boldsymbol{b}]. \tag{6.5.3}$$

Consequently, we infer from (6.5.2), (6.5.3), Theorems 6.3.7 and 6.3.8 that

$$(RS)\int_{[\boldsymbol{a},\boldsymbol{b}]} F\, dg = \lim_{n\to\infty}\int_{[\boldsymbol{a},\boldsymbol{b}]} F\phi_n\, d\mu_m. \tag{6.5.4}$$

Using the assumption $F = F \circ \pi_{k,\boldsymbol{c}}$ and Fubini's Theorem, we get

$$\int_{[\boldsymbol{a},\boldsymbol{b}]} F\phi_n\, d\mu_m = \int_{\prod_{\substack{i=1\\ i\neq k}}^m [a_i,b_i]} (F\circ\pi_{k,\boldsymbol{c}})\left\{\int_{[a_k,b_k]} \phi_n\, d\mu_1\right\} d\mu_{m-1} \tag{6.5.5}$$

for all $n \in \mathbb{N}$.

In view of (6.5.4) and (6.5.5), it remains to prove that

$$\lim_{n\to\infty}\int_{\prod_{\substack{i=1\\ i\neq k}}^m [a_i,b_i]} (F\circ\pi_{k,\boldsymbol{c}})\left\{\int_{[a_k,b_k]} \phi_n\, d\mu_1\right\} d\mu_{m-1} = 0. \tag{6.5.6}$$

To prove (6.5.6), we first observe that (6.5.2) yields

$$\sup_{n\in\mathbb{N}}\int_{\prod_{\substack{i=1\\ i\neq k}}^m [a_i,b_i]} \left|\int_{[a_k,b_k]} \phi_n\, d\mu_1\right| d\mu_{m-1} \leq \sup_{n\in\mathbb{N}} ||\phi_n||_{L^1[\boldsymbol{a},\boldsymbol{b}]} < \infty. \tag{6.5.7}$$

Next, for each $\boldsymbol{x} \in \Phi_{[\boldsymbol{a},\boldsymbol{b}],k}(\{a_k\})$ Fubini's Theorem, (6.5.3) and our choice of g yield

$$\lim_{n\to\infty} \int_{\prod_{\substack{i=1\\i\neq k}}^{m} [a_i,x_i]} \left\{ \int_{[a_k,b_k]} \phi_n \, d\mu_1 \right\} d\mu_{m-1} = 0. \tag{6.5.8}$$

Finally, an application of an $(m-1)$-dimensional analogue of Theorem 6.3.8 yields (6.5.6) to be proved. □

Theorem 6.5.8. *Let $f \in HK[\boldsymbol{a},\boldsymbol{b}]$ and let $\widetilde{F} : [\boldsymbol{a},\boldsymbol{b}] \longrightarrow \mathbb{R}$ be any function such that $\Delta_{\widetilde{F}}(I) = (HK)\int_I f$ for every $I \in \mathcal{I}_m([\boldsymbol{a},\boldsymbol{b}])$. If $g \in BV_{HK}[\boldsymbol{a},\boldsymbol{b}]$, then $fg \in HK[\boldsymbol{a},\boldsymbol{b}]$ and*

$$(HK)\int_{[\boldsymbol{a},\boldsymbol{b}]} fg = \sum_{Y\in\mathcal{P}_m} (-1)^{card(\Gamma(Y))} \left\{ (RS)\int_{[\boldsymbol{a},\boldsymbol{b}]} \widetilde{F} \, d(g\chi_Y) \right\}. \tag{6.5.9}$$

Proof. Let $g_0 = g\chi_{(\boldsymbol{a},\boldsymbol{b})}$. By Theorems 6.5.6 and 6.4.1, $fg_0 \in HK[\boldsymbol{a},\boldsymbol{b}]$ and

$$(HK)\int_{[\boldsymbol{a},\boldsymbol{b}]} fg_0 = (RS)\int_{[\boldsymbol{a},\boldsymbol{b}]} \left\{ (HK)\int_{[\boldsymbol{x},\boldsymbol{b}]} f \right\} dg_0(\boldsymbol{x}).$$

Since $g = g_0$ μ_m-almost everywhere on $[\boldsymbol{a},\boldsymbol{b}]$, we conclude that $fg \in HK[\boldsymbol{a},\boldsymbol{b}]$ and

$$(HK)\int_{[\boldsymbol{a},\boldsymbol{b}]} fg = (HK)\int_{[\boldsymbol{a},\boldsymbol{b}]} fg_0.$$

It remains to prove that (6.5.9) holds. Using the additivity of the indefinite HK-integral of f over $[\boldsymbol{a},\boldsymbol{b}]$, we see that

$$(HK)\int_{[\boldsymbol{x},\boldsymbol{b}]} f = \Delta_{\widetilde{F}}([\boldsymbol{x},\boldsymbol{b}])$$

for each $\boldsymbol{x} \in [\boldsymbol{a},\boldsymbol{b}]$. Therefore the linearity of the Riemann-Stieltjes integral and Theorem 6.5.7 yield

$$(RS)\int_{[\boldsymbol{a},\boldsymbol{b}]} \Delta_{\widetilde{F}}([\boldsymbol{x},\boldsymbol{b}]) \, dg_0(\boldsymbol{x}) = (RS)\int_{[\boldsymbol{a},\boldsymbol{b}]} (-1)^m \widetilde{F} \, dg_0.$$

Finally, since the linearity of the Riemann-Stieltjes integral and Theorem 6.5.6 yield

$$(RS)\int_{[\boldsymbol{a},\boldsymbol{b}]} (-1)^m \widetilde{F} \, dg_0 = \sum_{Y\in\mathcal{P}_m} (-1)^{\mathrm{card}(\Gamma(Y))} \left\{ (RS)\int_{[\boldsymbol{a},\boldsymbol{b}]} \widetilde{F} \, d(g\chi_Y) \right\},$$

the above equalities yield (6.5.9). □

For $\prod_{k=1}^{m} Y_k \in \mathcal{P}_m$, set $\sigma(\prod_{k=1}^{m} Y_k) := \{i \in \{1,\ldots,m\} : Y_i = \{a_i\}\}$. It remains to show that Theorem 6.5.8 is equivalent to the following Kurzweil's multidimensional integration by parts formula [73, Theorem 2.10].

Theorem 6.5.9. *Let $f \in HK[\boldsymbol{a},\boldsymbol{b}]$ and let $\widetilde{F} : [\boldsymbol{a},\boldsymbol{b}] \longrightarrow \mathbb{R}$ be any function such that $\Delta_{\widetilde{F}}(I) = (HK)\int_I f$ for every $I \in \mathcal{I}_m([\boldsymbol{a},\boldsymbol{b}])$. If $g \in BV_{HK}[\boldsymbol{a},\boldsymbol{b}]$, then $fg \in HK[\boldsymbol{a},\boldsymbol{b}]$ and*

$$\begin{aligned}
&(HK)\int_{[\boldsymbol{a},\boldsymbol{b}]} fg \\
&= \Delta_{\widetilde{F}g}([\boldsymbol{a},\boldsymbol{b}]) \\
&+\sum_{k=1}^{m} \sum_{\substack{Y \in \mathcal{P}_m \\ card(\Gamma(Y))=k}} (-1)^k \, (RS)\int_{\prod\limits_{\substack{j=1 \\ j\in\Gamma(Y)}}^{m} [a_j,b_j]} (-1)^{card(\sigma(Y))} \widetilde{F}\big|_Y \, d(g\big|_Y).
\end{aligned} \tag{6.5.10}$$

Proof. If $\boldsymbol{v} \in \mathcal{V}[\boldsymbol{a},\boldsymbol{b}]$, then it follows from the definition of the Riemann-Stieltjes integral that

$$(RS)\int_{[\boldsymbol{a},\boldsymbol{b}]} \widetilde{F} \, d(g\chi_{\{\boldsymbol{v}\}}) = \widetilde{F}(\boldsymbol{v})g(\boldsymbol{v})\prod_{i=1}^{m}(-1)^{\chi_{\{a_i\}}(v_i)}.$$

Similarly, for each $Y \in \mathcal{P}_m$ satisfying $0 < \text{card}(\Gamma(Y)) \leq m$, we write $Y = \prod_{i=1}^{m} Y_i$ to get

$$\begin{aligned}
&(RS)\int_{[\boldsymbol{a},\boldsymbol{b}]} \widetilde{F} \, d(g\chi_Y) \\
&= \Big\{ \prod_{i\in\{1,\ldots,m\}\setminus\Gamma(Y)} (-1)^{\chi_{Y_i}(a_i)} \Big\} \Big\{ (RS)\int_{\prod\limits_{\substack{j=1 \\ j\in\Gamma(Y)}}^{m} [a_j,b_j]} \widetilde{F}\big|_Y \, d(g\big|_Y) \Big\} \\
&= (-1)^{\text{card}(\sigma(Y))} \Big\{ (RS)\int_{\prod\limits_{\substack{j=1 \\ j\in\Gamma(Y)}}^{m} [a_j,b_j]} \widetilde{F}\big|_Y \, d(g\big|_Y) \Big\}.
\end{aligned}$$

Combining the above equalities together with Theorem 6.5.8 yields (6.5.10) to be proved. □

Corollary 6.5.10. Let $m = 1$, let $f \in HK[a,b]$, and let $\widetilde{F} : [a,b] \longrightarrow \mathbb{R}$ be any function such that $\Delta_{\widetilde{F}}([u,v]) = (HK)\int_u^v f$ for every $[u,v] \in \mathcal{I}_1([a,b])$. If $g \in BV[a,b]$, then the Riemann-Stieltjes integral $(RS)\int_a^b \widetilde{F}\, dg$ exists and

$$(HK)\int_a^b fg = \widetilde{F}(b)g(b) - \widetilde{F}(a)g(a) - (RS)\int_a^b \widetilde{F}\, dg.$$

Example 6.5.11. If $m = 2$, then (6.5.10) becomes

$$
\begin{aligned}
&(HK)\int_{[a_1,b_1]\times[a_2,b_2]} f(x_1,x_2)g(x_1,x_2)\, d(x_1,x_2)\\
&= \widetilde{F}(b_1,b_2)g(b_1,b_2) - \widetilde{F}(a_1,b_2)g(a_1,b_2)\\
&+\widetilde{F}(a_1,a_2)g(a_1,a_2) - \widetilde{F}(a_2,b_1)g(a_2,b_1)\\
&-\left\{(RS)\int_{a_1}^{b_1} \widetilde{F}(x_1,b_2)\, dg(x_1,b_2) - (RS)\int_{a_1}^{b_1} \widetilde{F}(x_1,a_2)\, dg(x_1,a_2)\right\}\\
&-\left\{(RS)\int_{a_2}^{b_2} \widetilde{F}(b_1,x_2)\, dg(b_1,x_2) - (RS)\int_{a_2}^{b_2} \widetilde{F}(a_1,x_2)\, dg(a_1,x_2)\right\}\\
&+(RS)\int_{[a_1,b_1]\times[a_2,b_2]} \widetilde{F}(x,y)\, dg(x,y).
\end{aligned}
$$

Theorem 6.5.12. *If $f \in HK[\boldsymbol{a},\boldsymbol{b}]$ and $g \in BV_{HK}[\boldsymbol{a},\boldsymbol{b}]$, then there exists $g_0 \in BV_0[\boldsymbol{a},\boldsymbol{b}]$ such that $g = g_0$ μ_m-almost everywhere on $[\boldsymbol{a},\boldsymbol{b}]$ and*

$$\left|(HK)\int_{[\boldsymbol{a},\boldsymbol{b}]} fg\right| \leq \|f\|_{HK[\boldsymbol{a},\boldsymbol{b}]} Var(g_0,[\boldsymbol{a},\boldsymbol{b}]).$$

Proof. Exercise. □

We are now ready to give a generalization of Theorem 6.1.6.

Theorem 6.5.13. *Let $f \in HK[\boldsymbol{a},\boldsymbol{b}]$. For each $i \in \{1,\ldots,m\}$, let g_i be a positive increasing function defined on $[a_i,b_i]$. Then there exists $\mathbf{c} \in [\boldsymbol{a},\boldsymbol{b}]$ such that*

$$(HK)\int_{[\boldsymbol{a},\boldsymbol{b}]} f(\otimes_{i=1}^m g_i) = (\otimes_{i=1}^m g_i)(\boldsymbol{b})\left\{(HK)\int_{[\mathbf{c},\boldsymbol{b}]} f\right\}.$$

Proof. Exercise. □

Example 6.5.14. Define the function $f : [0,1]^2 \longrightarrow \mathbb{R}$ by setting

$$f(x_1,x_2) = \begin{cases} \frac{1}{x_1x_2}\sin\frac{1}{x_1x_2} & \text{if } (x_1,x_2) \in (0,1]^2,\\ 0 & \text{if } (x_1,x_2) \in [0,1]^2\backslash(0,1]^2. \end{cases}$$

Then $f \in HK([0,1]^2)\backslash L^1([0,1]^2)$.

Proof. Suppose that $f \in L^1([0,1]^2)$. Then it follows from Fubini's Theorem that the map $x \mapsto f(x,y)$ belongs to $L^1[0,1]$ for μ_1-almost all $y \in [0,1]$, a contradiction. Thus, $f \notin L^1([0,1]^2)$.

We will next use Theorem 5.5.9 to show that $f \in HK([0,1]^2)$. For each $x_2 \in (0,1]$ and $0 < \alpha < \beta \leq 1$, an integration by parts gives

$$\int_\alpha^\beta \frac{1}{x_1x_2} \sin \frac{1}{x_1x_2}\, dx_1 = \left[x_1 \cos \frac{1}{x_1x_2}\right]_\alpha^\beta - \int_\alpha^\beta \cos \frac{1}{x_1x_2}\, dx_1 \tag{6.5.11}$$

and hence

$$(HK)\int_0^\beta \frac{1}{x_1x_2} \sin \frac{1}{x_1x_2}\, dx_1 = \beta \cos \frac{1}{\beta x_2} - \int_0^\beta \cos \frac{1}{x_1x_2}\, dx_1. \tag{6.5.12}$$

Furthermore, it follows from (6.5.11) and (6.5.12) that if $[\alpha,\beta] \times [\gamma,\delta] \in \mathcal{I}_2([0,1]^2)$ with $0 < \beta \leq 1$ and $0 < \gamma < \delta \leq 1$, then the iterated integral

$$\int_\gamma^\delta \left\{(HK)\int_0^\beta \frac{1}{x_1x_2} \sin \frac{1}{x_1x_2}\, dx_1\right\} d\mu_1(x_2) \text{ exists}$$

and

$$\left|\int_\gamma^\delta \left\{(HK)\int_0^\beta \frac{1}{x_1x_2} \sin \frac{1}{x_1x_2}\, dx_1\right\} d\mu_1(x_2)\right| \leq 2\beta(\delta-\gamma). \tag{6.5.13}$$

We will next prove that the inequality

$$\left|\int_0^\delta \left\{\int_\alpha^\beta \frac{1}{x_1x_2} \sin \frac{1}{x_1x_2}\, d\mu_1(x_1)\right\} d\mu_1(x_2)\right| \leq 2\delta(\beta-\alpha) \tag{6.5.14}$$

holds whenever $0 < \delta < 1$ and $0 < \alpha < \beta \leq 1$. Indeed,

$$\begin{aligned}
&\left|\int_0^\delta \left\{\int_\alpha^\beta \frac{1}{x_1x_2} \sin \frac{1}{x_1x_2}\, d\mu_1(x_1)\right\} d\mu_1(x_2)\right| \\
&= \left|\lim_{\gamma\to 0+}\int_\gamma^\delta \left\{\int_\alpha^\beta \frac{1}{x_1x_2} \sin \frac{1}{x_1x_2}\, d\mu_1(x_1)\right\} d\mu_1(x_2)\right| \\
&= \left|\lim_{\gamma\to 0+}\int_{[\alpha,\beta]\times[\gamma,\delta]} \frac{1}{x_1x_2} \sin \frac{1}{x_1x_2}\, d\mu_2(x_1,x_2)\right| \\
&= \left|\lim_{\gamma\to 0+}\int_\alpha^\beta \left\{\int_\gamma^\delta \frac{1}{x_1x_2} \sin \frac{1}{x_1x_2}\, d\mu_1(x_2)\right\} d\mu_1(x_1)\right| \\
&= \left|\lim_{\gamma\to 0+}\int_\alpha^\beta \left\{\delta \cos \frac{1}{\delta x_1} - \gamma \cos \frac{1}{\gamma x_1} - \int_\gamma^\delta \cos \frac{1}{x_1x_2}\, d\mu_1(x_2)\right\} d\mu_1(x_1)\right| \\
&= \left|\int_\alpha^\beta \left\{\delta \cos \frac{1}{\delta x_1} - \int_0^\delta \cos \frac{1}{x_1x_2}\, d\mu_1(x_2)\right\} d\mu_1(x_1)\right|
\end{aligned}$$

(by Lebesgue's Dominated Convergence Theorem)

$\leq 2\delta(\beta-\alpha)$.

Finally, we prove that $f \in HK([0,1]^2)$. To do this, we define the interval function $F : \mathcal{I}_2([0,1]^2) \longrightarrow \mathbb{R}$ by setting

$$F([\alpha,\beta]\times[\gamma,\delta]) = (HK)\int_\gamma^\delta \left\{(HK)\int_\alpha^\beta \frac{1}{x_1x_2}\sin\frac{1}{x_1x_2}\,dx_1\right\}dx_2.$$

Clearly, F is an additive interval function. Since f is continuous on each interval $I \subset (0,1]^2$, we need to prove that

$$V_{\mathcal{HK}}F((\{0\}\times[0,1])\cup([0,1]\times\{0\})) = 0.$$

But this is an easy consequence of (6.5.13) and (6.5.14). □

6.6 Riesz Representation Theorems

The aim of this section is to prove a generalization of Theorem 6.4.1, which will be used to characterize the multipliers for $HK[\boldsymbol{a},\boldsymbol{b}]$. We begin with the following definition.

Definition 6.6.1. A norm on a linear space X is a function $\|\cdot\|_X : X \longrightarrow [0,\infty)$ satisfying the following conditions:

(i) $\|x\|_X = 0$ if and only if $x = 0_X$, where 0_X denotes the zero vector of X;
(ii) $\|\alpha x\|_X = |\alpha|\,\|x\|_X$ for all $\alpha \in \mathbb{R}$ and $x \in X$;
(iii) (Triangle inequality) $\|x+y\|_X \le \|x\|_X + \|y\|_X$ for all $x, y \in X$.

The normed linear space just defined is denoted by $(X, \|\cdot\|_X)$, or simply X.

Example 6.6.2. The space $C[0,1]$ is a normed linear space.

Definition 6.6.3. Let $(X,\|\cdot\|_X)$ and $(Y,\|\cdot\|_Y)$ be normed spaces, and let $T : (X,\|\cdot\|_X) \longrightarrow (Y,\|\cdot\|_Y)$. If

$$T(\alpha x_1 + \beta x_2) = \alpha T(x_1) + \beta T(x_2)$$

for all $\alpha, \beta \in \mathbb{R}$ and $x_1, x_2 \in (X,\|\cdot\|_X)$, we say that T is a *linear operator*. An operator taking real values is known as a *functional*.

Example 6.6.4. The function $S : C[0,1] \longrightarrow \mathbb{R}$ is defined by

$$S(f) = \int_0^1 f\,d\mu_1.$$

Then S is a functional on $C[0,1]$, and $|S(f)| \le \|f\|_{C[0,1]}$ for all $f \in C[0,1]$.

Definition 6.6.5. Let $(X, \|\cdot\|_X)$ and $(Y, \|\cdot\|_Y)$ be normed spaces. A linear operator $T : (X, \|\cdot\|_X) \longrightarrow (Y, \|\cdot\|_Y)$ is said to be *bounded* if there exists $C \in [0, \infty)$ such that $\|T(x)\|_Y \leq C\|x\|_X$ for all $x \in X$. The *operator norm* for a bounded linear operator T is defined as

$$\|T\| = \inf \left\{C : \|T(x)\|_Y \leq C\|x\|_X \text{ for all } x \in X\right\}.$$

Theorem 6.6.6. *Let $(X, \|\cdot\|_X)$ and $(Y, \|\cdot\|_Y)$ be normed linear spaces, and let $\mathcal{B}(X, Y)$ be the set of all bounded linear operators from X to Y. Then $(\mathcal{B}(X, Y), \|\cdot\|)$ is a normed space.*

Definition 6.6.7. Let $(X, \|\cdot\|_X)$ be a normed linear space. The space $(\mathcal{B}(X, \mathbb{R}), \|\cdot\|)$ is usually denoted by X^*, which is known as the *dual* of $(X, \|\cdot\|_X)$.

The following theorem is an immediate consequence of Theorem 6.3.5.

Theorem 6.6.8. *If $g \in BV_0[\boldsymbol{a}, \boldsymbol{b}]$, then the map $F \mapsto (RS) \int_{[\boldsymbol{a},\boldsymbol{b}]} F\, dg$ is a bounded linear functional on $C[\boldsymbol{a}, \boldsymbol{b}]$.*

At this moment, it is natural to ask whether every bounded linear functional on $C[\boldsymbol{a}, \boldsymbol{b}]$ can be represented in the form $F \mapsto (RS) \int_{[\boldsymbol{a},\boldsymbol{b}]} F\, dg_0$ for some $g_0 \in BV_0[\boldsymbol{a}, \boldsymbol{b}]$. In order to answer this question, we need the following useful result.

Theorem 6.6.9 (Hahn-Banach). *Let $(X, \|\cdot\|_X)$ be a normed linear space, $(Y, \|\cdot\|_X)$ a linear subspace of $(X, \|\cdot\|_X)$, and $y^* \in Y^*$. Then there exists $x^* \in X^*$ such that $\|x^*\| = \|y^*\|$ and $x^*(y) = y^*(y)$ for all $y \in (Y, \|\cdot\|_X)$.*

We are now ready to state and prove the following result concerning the dual space of $C[\boldsymbol{a}, \boldsymbol{b}]$.

Theorem 6.6.10. *Let $T : C[\boldsymbol{a}, \boldsymbol{b}] \longrightarrow \mathbb{R}$ be a bounded linear functional. Then there exists $g_0 \in BV_0[\boldsymbol{a}, \boldsymbol{b}]$ such that*

$$T(F) = (RS) \int_{[\boldsymbol{a},\boldsymbol{b}]} F\, dg_0$$

for all $F \in C[\boldsymbol{a}, \boldsymbol{b}]$. Moreover, $\|T\| = Var(g_0, [\boldsymbol{a}, \boldsymbol{b}])$.

Proof. Let $B[\boldsymbol{a},\boldsymbol{b}]$ be the space of bounded functions on $[\boldsymbol{a},\boldsymbol{b}]$, and we equip the space $B[\boldsymbol{a},\boldsymbol{b}]$ with the supremum norm $\|\cdot\|_\infty$. According to the Hahn-Banach Theorem, T has an extension $\widetilde{T}$ such that $\widetilde{T} \in B[\boldsymbol{a},\boldsymbol{b}]^*$ and $\|T\| = \|\widetilde{T}\|$.

Define a function $g_0 : [\boldsymbol{a},\boldsymbol{b}] \longrightarrow \mathbb{R}$ by setting

$$g_0(\boldsymbol{x}) := \begin{cases} \widetilde{T}(\chi_{[\boldsymbol{a},\boldsymbol{x}]}) & \text{if } \boldsymbol{x} \in (\boldsymbol{a},\boldsymbol{b}], \\ 0 & \text{otherwise.} \end{cases}$$

Then we have

$$\Delta_{g_0}([\boldsymbol{u},\boldsymbol{v}]) = \widetilde{T}\Big(\prod_{i=1}^m \Big(\chi_{(u_i,v_i]} + \chi_{\{u_i\}\cap\{a_i\}}\Big)\Big) \quad ([\boldsymbol{u},\boldsymbol{v}] \in \mathcal{I}_m([\boldsymbol{a},\boldsymbol{b}])).$$

We claim $g_0 \in BV[\boldsymbol{a},\boldsymbol{b}]$ and $\mathrm{Var}(g_0,[\boldsymbol{a},\boldsymbol{b}]) \leq \|T\|$. Indeed, for any net D of $[\boldsymbol{a},\boldsymbol{b}]$, we have

$$\begin{aligned}
&\sum_{[\boldsymbol{u},\boldsymbol{v}]\in D} |\Delta_{g_0}([\boldsymbol{u},\boldsymbol{v}])| \\
&= \sum_{[\boldsymbol{u},\boldsymbol{v}]\in D} \left|\widetilde{T}\Big(\prod_{i=1}^m \kappa_i([u_i,v_i])\Big)\right| \\
&\qquad\qquad (\text{where } \kappa_i([u_i,v_i]) := \chi_{(u_i,v_i]} + \chi_{\{u_i\}\cap\{a_i\}} \quad (i=1,\ldots,m)) \\
&= \sum_{[\boldsymbol{u},\boldsymbol{v}]\in D} \widetilde{T}\Big(\prod_{i=1}^m \kappa_i([u_i,v_i])\Big)\, \mathrm{sgn}\Big(\widetilde{T}\Big(\prod_{i=1}^m \kappa_i([u_i,v_i])\Big)\Big) \\
&= \widetilde{T}\Big(\sum_{[\boldsymbol{u},\boldsymbol{v}]\in D} \Big(\prod_{i=1}^m \kappa_i([u_i,v_i])\Big)\, \mathrm{sgn}\Big(\widetilde{T}\Big(\prod_{i=1}^m \kappa_i([u_i,v_i])\Big)\Big)\Big) \\
&\leq \|\widetilde{T}\| \left\|\sum_{[\boldsymbol{u},\boldsymbol{v}]\in D} \Big(\prod_{i=1}^m \kappa_i([u_i,v_i])\Big)\, \mathrm{sgn}\Big(\widetilde{T}\Big(\prod_{i=1}^m \kappa_i([u_i,v_i])\Big)\Big)\right\|_\infty \\
&\leq \|T\|.
\end{aligned}$$

We will next prove that $T(F) = (RS)\int_{[\boldsymbol{a},\boldsymbol{b}]} F\,dg$ for each $F \in C[\boldsymbol{a},\boldsymbol{b}]$. To prove this, we let $F \in C[\boldsymbol{a},\boldsymbol{b}]$ and let $(D_n)_{n=1}^\infty$ be a sequence of divisions of $[\boldsymbol{a},\boldsymbol{b}]$ such that $\lim_{n\to\infty} \max_{[\boldsymbol{s},\boldsymbol{t}]\in D_n} |||\boldsymbol{t}-\boldsymbol{s}||| = 0$. By Theorem 2.2.3, F is uniformly continuous on $[\boldsymbol{a},\boldsymbol{b}]$ and so

$$\lim_{n\to\infty} \left\| \sum_{[\boldsymbol{s},\boldsymbol{t}]\in D_n} F(\boldsymbol{t})\Big(\prod_{i=1}^m (\chi_{(s_i,t_i]} + \chi_{\{s_i\}\cap\{a_i\}})\Big) - F \right\|_\infty = 0.$$

Thus,

$$\begin{aligned}
&\left|\sum_{[s,t]\in D_n} F(t)\Delta_{g_0}([s,t]) - \widetilde{T}(F)\right| \\
&= \left|\widetilde{T}\left(\sum_{[s,t]\in D_n} F(t)\left(\prod_{i=1}^{m}(\chi_{(s_i,t_i]} + \chi_{\{s_i\}\cap\{a_i\}})\right) - F\right)\right| \\
&\le \|T\| \left\|\sum_{[s,t]\in D_n} F(t)\left(\prod_{i=1}^{m}(\chi_{(s_i,t_i]} + \chi_{\{s_i\}\cap\{a_i\}})\right) - F\right\|_{\infty} \\
&\to 0 \text{ as } n \to \infty.
\end{aligned}$$

Consequently, $T(F) = \widetilde{T}(F) = (RS)\int_{[a,b]} F\, dg_0$. As $F \in C[a,b]$ is arbitrary, we get

$$\|T\| \le \mathrm{Var}(g_0, [a,b]).$$

It is now clear that g_0 has the desired properties. □

We now turn to the normed linear space $HK[a,b]$. Here we write $f = g$ if $f = g$ μ_m-almost everywhere on $[a,b]$. The following theorem is an immediate consequence of Theorem 6.4.3.

Theorem 6.6.11. *If $g \in BV_0[a,b]$, then the map $f \mapsto (HK)\int_{[a,b]} fg$ is a bounded linear functional on $HK[a,b]$.*

Our next goal is to prove that every bounded linear functional on $HK[a,b]$ is of the form $f \mapsto (HK)\int_{[a,b]} fg_0$ for some $g_0 \in BV_0[a,b]$.

Theorem 6.6.12. *If T is a $\|\cdot\|$-bounded linear functional on $HK[a,b]$, then there exists $g \in BV_0[a,b]$ such that*

$$T(f) = (HK)\int_{[a,b]} fg$$

for all $f \in HK[a,b]$. Moreover, $\|T\| = \mathrm{Var}(g,[a,b])$.

Proof. For each $f \in HK[a,b]$ it follows from Theorem 2.4.8 that the function $x \mapsto (HK)\int_{[x,b]} f$ belongs to $C[a,b]$. A simple computation shows that

$$\begin{aligned}
&C_{HK}[a,b] \\
&:= \left\{F : \exists f \in HK[a,b] \text{ such that } F(x) = (HK)\int_{[x,b]} f \quad (x \in [a,b])\right\}
\end{aligned}$$

is a linear subspace of $C[\boldsymbol{a}, \boldsymbol{b}]$.

Define $T_0 : C_{HK}[\boldsymbol{a}, \boldsymbol{b}] \longrightarrow \mathbb{R}$ by setting

$$T_0(F) = T(f),$$

where $f \in HK[\boldsymbol{a}, \boldsymbol{b}]$ and $F(\boldsymbol{x}) = (HK)\int_{[\boldsymbol{x},\boldsymbol{b}]} f \quad (\boldsymbol{x} \in [\boldsymbol{a}, \boldsymbol{b}])$. Since T_0 is a bounded linear functional on $C_{HK}[\boldsymbol{a}, \boldsymbol{b}]$, we can apply Hahn-Banach Theorem and Theorem 6.6.10 to choose $g \in BV_0[\boldsymbol{a}, \boldsymbol{b}]$ so that

$$T_0(F) = (RS)\int_{[\boldsymbol{a},\boldsymbol{b}]} F\, dg$$

for all $F \in C_{HK}[\boldsymbol{a}, \boldsymbol{b}]$. An appeal to Theorem 6.4.1 completes the proof. □

6.7 Characterization of multipliers for the Henstock-Kurzweil integral

The aim of this section is to give a characterization of multipliers for the Henstock-Kurzweil integral. We begin with the following theorem, which asserts that each multiplier for $HK[\boldsymbol{a}, \boldsymbol{b}]$ induces a bounded linear functional on $HK[\boldsymbol{a}, \boldsymbol{b}]$.

Theorem 6.7.1. *If $g : [\boldsymbol{a}, \boldsymbol{b}] \longrightarrow \mathbb{R}$ is a multiplier for $HK[\boldsymbol{a}, \boldsymbol{b}]$, then the linear functional*

$$f \mapsto (HK)\int_{[\boldsymbol{a},\boldsymbol{b}]} fg \;:\; HK[\boldsymbol{a}, \boldsymbol{b}] \longrightarrow \mathbb{R}$$

is bounded.

Proof. Suppose that the linear functional

$$f \mapsto (HK)\int_{[\boldsymbol{a},\boldsymbol{b}]} fg : HK[\boldsymbol{a}, \boldsymbol{b}] \longrightarrow \mathbb{R}$$

is not bounded. Then there exists $[\boldsymbol{a}_1, \boldsymbol{b}_1] \in \mathcal{I}_m([\boldsymbol{a}, \boldsymbol{b}])$ such that $|||\boldsymbol{b}_1 - \boldsymbol{a}_1||| < \frac{1}{2}|||\boldsymbol{b} - \boldsymbol{a}|||$ and the linear functional

$$f \mapsto (HK)\int_{[\boldsymbol{a}_1,\boldsymbol{b}_1]} fg : HK[\boldsymbol{a}_1, \boldsymbol{b}_1] \longrightarrow \mathbb{R}$$

is not bounded. Thus, we can select an $f_1 \in HK[\boldsymbol{a}_1, \boldsymbol{b}_1]$ such that $\|f_1\|_{HK[\boldsymbol{a}_1,\boldsymbol{b}_1]} < 1$ and $\left|(HK)\int_{[\boldsymbol{a}_1,\boldsymbol{b}_1]} f_1 g\right| > 4$.

Since the linear functional

$$f \mapsto (HK)\int_{[\boldsymbol{a}_1,\boldsymbol{b}_1]} fg : HK[\boldsymbol{a}_1, \boldsymbol{b}_1] \longrightarrow \mathbb{R}$$

is not bounded, there exists $[\boldsymbol{a}_2, \boldsymbol{b}_2] \in \mathcal{I}_m([\boldsymbol{a}_1, \boldsymbol{b}_1])$ such that $|||\boldsymbol{b}_2 - \boldsymbol{a}_2||| < \frac{1}{2}|||\boldsymbol{b}_1 - \boldsymbol{a}_1|||$, $\left|(HK)\int_{[\boldsymbol{a}_2,\boldsymbol{b}_2]} f_1 g\right| < 2$ and the linear functional

$$f \mapsto (HK)\int_{[\boldsymbol{a}_2,\boldsymbol{b}_2]} fg : HK[\boldsymbol{a}_2, \boldsymbol{b}_2] \longrightarrow \mathbb{R}$$

is not bounded. Proceeding inductively, we construct a sequence $\left([\boldsymbol{a}_n, \boldsymbol{b}_n]\right)_{n=1}^{\infty}$ of intervals so that the following properties hold for every $n \in \mathbb{N}$:

(i) $[\boldsymbol{a}_{n+1}, \boldsymbol{b}_{n+1}] \subset [\boldsymbol{a}_n, \boldsymbol{b}_n]$;
(ii) $|||\boldsymbol{b}_{n+1} - \boldsymbol{a}_{n+1}||| < \frac{1}{2}|||\boldsymbol{b}_n - \boldsymbol{a}_n|||$;
(iii) there exists $f_n \in HK[\boldsymbol{a}_n, \boldsymbol{b}_n]$ such that $\|f_n\|_{HK[\boldsymbol{a}_n,\boldsymbol{b}_n]} < 1$ and $\left|(HK)\int_{[\boldsymbol{a}_{n+1},\boldsymbol{b}_{n+1}]} f_n g\right| < 2^n$;
(iv) $\left|(HK)\int_{[\boldsymbol{a}_n,\boldsymbol{b}_n]} f_n g\right| > 4^n$.

Since (i) and (ii) hold for all $n \in \mathbb{N}$, it follows from the Nested Interval Theorem that $\bigcap_{n=1}^{\infty}[\boldsymbol{a}_n, \boldsymbol{b}_n] = \{\boldsymbol{c}\}$ for some $\boldsymbol{c} \in [\boldsymbol{a}, \boldsymbol{b}]$. Now we define the function $f : [\boldsymbol{a}, \boldsymbol{b}] \longrightarrow \mathbb{R}$ by setting

$$f := \sum_{k=1}^{\infty} \frac{f_k}{2^k} \chi_{[\boldsymbol{a}_k,\boldsymbol{b}_k]\setminus(\boldsymbol{a}_{k+1},\boldsymbol{b}_{k+1})}.$$

Since the set $\{\boldsymbol{c}\}$ is μ_m-negligible, it follows from (i), (ii), (iii) and Theorem 5.5.9 that $f \in HK[\boldsymbol{a}, \boldsymbol{b}]$ and

$$(HK)\int_I f = \sum_{k=1}^{\infty} (HK)\int_I \frac{f_k}{2^k} \chi_{[\boldsymbol{a}_k,\boldsymbol{b}_k]\setminus(\boldsymbol{a}_{k+1},\boldsymbol{b}_{k+1})}$$

for all $I \in \mathcal{I}_m([\boldsymbol{a}, \boldsymbol{b}])$. On the other hand, it follows from (iv) and (iii) that

$$\begin{aligned}
&\left|(HK)\int_{[\boldsymbol{a}_n,\boldsymbol{b}_n]} fg - (HK)\int_{[\boldsymbol{a}_{n+1},\boldsymbol{b}_{n+1}]} fg\right| \\
&= \left|(HK)\int_{[\boldsymbol{a},\boldsymbol{b}]} fg\chi_{[\boldsymbol{a}_n,\boldsymbol{b}_n]\setminus(\boldsymbol{a}_{n+1},\boldsymbol{b}_{n+1})}\right| \\
&= \left|(HK)\int_{[\boldsymbol{a},\boldsymbol{b}]} \frac{f_n g}{2^n} \chi_{[\boldsymbol{a}_n,\boldsymbol{b}_n]\setminus(\boldsymbol{a}_{n+1},\boldsymbol{b}_{n+1})}\right| \\
&\geq \frac{1}{2^n}\left|(HK)\int_{[\boldsymbol{a}_n,\boldsymbol{b}_n]} f_n g\right| - \frac{1}{2^n}\left|(HK)\int_{[\boldsymbol{a}_{n+1},\boldsymbol{b}_{n+1}]} f_n g\right| \\
&\geq 2^n - 1 \\
&\geq 1
\end{aligned}$$

for all $n \in \mathbb{N}$, contradicting Theorem 2.4.9. This contradiction proves the theorem. $\square$

The following theorem gives the main result of this chapter.

Theorem 6.7.2. *A function $g : [\boldsymbol{a}, \boldsymbol{b}] \longrightarrow \mathbb{R}$ is a multiplier for $HK[\boldsymbol{a}, \boldsymbol{b}]$ if and only if there exists $g_0 \in BV_0[\boldsymbol{a}, \boldsymbol{b}]$ such that $g = g_0$ μ_m-almost everywhere on $[\boldsymbol{a}, \boldsymbol{b}]$.*

Proof. This is a consequence of Theorems 6.4.1, 6.7.1 and 6.6.12. $\square$

6.8 A Banach-Steinhaus Theorem for the space of Henstock-Kurzweil integrable functions

In this section we modify the proof of Theorem 6.7.1 to obtain a Banach-Steinhaus Theorem for $HK[\boldsymbol{a}, \boldsymbol{b}]$. We begin with the following lemma.

Lemma 6.8.1. *Let Y be a normed space, let $[\boldsymbol{u}, \boldsymbol{v}] \in \mathcal{I}_m([\boldsymbol{a}, \boldsymbol{b}])$, and let $\mathcal{T} \subset \mathcal{B}(HK[\boldsymbol{u}, \boldsymbol{v}], Y)$. If $\sup\left\{\|T\| : T \in \mathcal{T}\right\} = \infty$, then there exists $J \in \mathcal{I}_m([\boldsymbol{u}, \boldsymbol{v}])$ such that $diam(J) \leq \frac{1}{2}|||\boldsymbol{v} - \boldsymbol{u}|||$ and*

$$\sup\left\{\left\|T\Big|_{HK(J)}\right\| : T \in \mathcal{T}\right\} = \infty.$$

Proof. Exercise. $\square$

Theorem 6.8.2 (Banach-Steinhaus Theorem). *Let Y be a normed space and suppose that $\mathcal{T} \subseteq \mathcal{B}(HK[\boldsymbol{a}, \boldsymbol{b}], Y)$. If $M(f) := \sup_{T \in \mathcal{T}} \|T(f)\|_Y$ is finite for each $f \in HK[\boldsymbol{a}, \boldsymbol{b}]$, then $\sup_{T \in \mathcal{T}} \|T\|$ is finite.*

Proof. Proceeding towards a contradiction, suppose that $\sup_{T \in \mathcal{T}} \|T\| = \infty$. By Lemma 6.8.1, there exists $[\boldsymbol{a}_1, \boldsymbol{b}_1] \in \mathcal{I}_m([\boldsymbol{a}, \boldsymbol{b}])$ such that $|||\boldsymbol{b}_1 - \boldsymbol{a}_1||| < \frac{1}{2}|||\boldsymbol{b} - \boldsymbol{a}|||$ and

$$\sup_{T \in \mathcal{T}} \left\|T\Big|_{HK[\boldsymbol{a}_1, \boldsymbol{b}_1]}\right\| = \infty. \tag{6.8.1}$$

Hence there exist $T_1 \in \mathcal{T}$ and $f_1 \in HK[\boldsymbol{a}_1, \boldsymbol{b}_1]$ such that $\|f_1\|_{HK[\boldsymbol{a}_1, \boldsymbol{b}_1]} < 1$ and $\|T_1(f_1)\|_Y > 3$.

Proceeding inductively, we construct a decreasing sequence $\left([\boldsymbol{a}_n, \boldsymbol{b}_n]\right)_{n=1}^{\infty}$ of intervals such that the following conditions hold for every $n \in \mathbb{N}$:

(i) $[\boldsymbol{a}_{n+1}, \boldsymbol{b}_{n+1}] \in \mathcal{I}_m([\boldsymbol{a}_n, \boldsymbol{b}_n])$ and $|||\boldsymbol{b}_{n+1} - \boldsymbol{a}_{n+1}||| < \frac{1}{2}|||\boldsymbol{b}_n - \boldsymbol{a}_n|||$;
(ii) we have

$$\sup_{T \in \mathcal{T}} \left\| T \Big|_{HK[\boldsymbol{a}_n, \boldsymbol{b}_n]} \right\| = \infty;$$

(iii) there exist $f_n \in HK[\boldsymbol{a}_n, \boldsymbol{b}_n]$ and $T_n \in \mathcal{T}$ such that

$$||f_n||_{HK[\boldsymbol{a}_n, \boldsymbol{b}_n]} < \frac{1}{2^{n-1}(1 + \sum_{k=1}^{n-1} ||T_k||)}, \ ||T_n(f_n \chi_{[\boldsymbol{a}_{n+1}, \boldsymbol{b}_{n+1}]})||_Y < 2$$

and

$$||T_n(f_n)||_Y > n + 2 + \sum_{k=1}^{n-1} \left(M(f_k \chi_{[\boldsymbol{a}_k, \boldsymbol{b}_k]}) + M(f_k \chi_{[\boldsymbol{a}_{k+1}, \boldsymbol{b}_{k+1}]}) \right).$$

Set

$$f = \sum_{k=1}^{\infty} f_k \chi_{[\boldsymbol{a}_k, \boldsymbol{b}_k] \setminus (\boldsymbol{a}_{k+1}, \boldsymbol{b}_{k+1})}.$$

Since (i), (iii) and Theorem 5.5.9 imply that $f \in HK[\boldsymbol{a}, \boldsymbol{b}]$ and

$$\lim_{n \to \infty} \left\| \sum_{k=1}^{n} f_k \chi_{[\boldsymbol{a}_k, \boldsymbol{b}_k] \setminus (\boldsymbol{a}_{k+1}, \boldsymbol{b}_{k+1})} - f \right\|_{HK[\boldsymbol{a}, \boldsymbol{b}]} = 0,$$

for each $n \in \mathbb{N}$ the continuity of the linear operator T_{n+1} yields

$$T_{n+1}(f) = \sum_{k=1}^{\infty} T_{n+1}(f_k \chi_{[\boldsymbol{a}_k, \boldsymbol{b}_k] \setminus (\boldsymbol{a}_{k+1}, \boldsymbol{b}_{k+1})})$$

so that

$$\begin{aligned}
&\left\| T_{n+1}(f) \right\|_Y \\
&\geq \left\| T_{n+1}(f_{n+1} \chi_{[\boldsymbol{a}_{n+1}, \boldsymbol{b}_{n+1}]}) \right\|_Y - \left\| T_{n+1}(f_{n+1} \chi_{[\boldsymbol{a}_{n+2}, \boldsymbol{b}_{n+2}]}) \right\|_Y \\
&- \left\| \sum_{k=1}^{n} T_{n+1}(f_k \chi_{[\boldsymbol{a}_k, \boldsymbol{b}_k] \setminus (\boldsymbol{a}_{k+1}, \boldsymbol{b}_{k+1})}) \right\|_Y \\
&- \left\| \sum_{k=n+2}^{\infty} T_{n+1}(f_k \chi_{[\boldsymbol{a}_k, \boldsymbol{b}_k] \setminus (\boldsymbol{a}_{k+1}, \boldsymbol{b}_{k+1})}) \right\|_Y
\end{aligned}$$

and

$$\begin{aligned}
&\left\| T_{n+1}(f) \right\|_Y \\
&\geq n+3+\sum_{k=1}^{n}\left(M(f_k\chi_{[\boldsymbol{a}_k,\boldsymbol{b}_k]})+M(f_k\chi_{[\boldsymbol{a}_{k+1},\boldsymbol{b}_{k+1}]})\right)-2 \\
&-\sum_{k=1}^{n}\left(M(f_k\chi_{[\boldsymbol{a}_k,\boldsymbol{b}_k]})+M(f_k\chi_{[\boldsymbol{a}_{k+1},\boldsymbol{b}_{k+1}]})\right) \\
&-\sum_{k=n+2}^{\infty}\|T_{n+1}\|\frac{2}{2^{k-1}(1+\sum_{j=1}^{k-1}\|T_j\|)} \\
&\geq n.
\end{aligned}$$

As $n \in \mathbb{N}$ is arbitrary, we obtain $M(f) = \infty$, which contradicts the assumption that $M(f)$ is finite. This contradiction proves the theorem. □

Theorem 6.8.3. *Let Y be a normed space and assume that $(T_n)_{n=1}^{\infty}$ is a sequence in $\mathcal{B}(HK[\boldsymbol{a},\boldsymbol{b}],Y)$. If $\lim_{n\to\infty} T_n(f)$ exists for each $f \in HK[\boldsymbol{a},\boldsymbol{b}]$, then there exists $T \in \mathcal{B}(HK[\boldsymbol{a},\boldsymbol{b}],Y)$ such that $\lim_{n\to\infty} T_n(f) = T(f)$ for each $f \in HK[\boldsymbol{a},\boldsymbol{b}]$.*

Proof. Define $T : HK[\boldsymbol{a},\boldsymbol{b}] \longrightarrow Y$ by setting

$$T(f) = \lim_{n\to\infty} T_n(f).$$

Then it is easy to check that T is linear. It remains to prove that T is bounded. For each $f \in HK[\boldsymbol{a},\boldsymbol{b}]$, the existence of the limit $\lim_{n\to\infty} T_n(f)$ implies that $\sup_{n\in\mathbb{N}} \|T_n(f)\|_Y$ is finite. Since $f \in HK[\boldsymbol{a},\boldsymbol{b}]$ is arbitrary, we infer from Theorem 6.8.2 that $\sup_{n\in\mathbb{N}} \|T_n\|$ is finite. It follows that $T \in \mathcal{B}(HK[\boldsymbol{a},\boldsymbol{b}],Y)$:

$$\|T(f)\|_Y = \lim_{n\to\infty} \|T_n(f)\|_Y \leq \limsup_{n\to\infty} \|T_n\|\|f\|_{HK[\boldsymbol{a},\boldsymbol{b}]} \leq \sup_{n\in\mathbb{N}} \|T_n\|\|f\|_{HK[\boldsymbol{a},\boldsymbol{b}]}$$

for each $f \in HK[\boldsymbol{a},\boldsymbol{b}]$. □

6.9 Notes and Remarks

There are many excellent books on integration by parts for one-dimensional Henstock-Kurzweil integrals; see, for example, Gordon [44], Pfeffer [137], and Lee and Výborný [88]. Theorem 6.1.4 is due to Jeffery [67], but the present proof is similar to that of [117, Theorem 1]. Sargent [146] used improper Lebesgue integrals to prove Theorem 6.1.9 in the context of

Denjoy-Perron integration. For other results concerning multipliers of one-dimensional non-absolutely convergent integrals; see, for example, Bruckner, J. Mařík and C.E. Weil [21], Bongiorno [9], Fleissner [39–41], Lee [100], and Mařík and Weil [119].

Section 6.2 is largely based on Macphail [117] and Lee [104]. Theorem 6.2.11 is due to W.H. Young and G.C. Young [164].

The presentation of Section 6.3 is similar to that of Lee [104]. Several variations of Theorem 6.3.10 can be found in W.H. Young [162].

Sections 6.4 and 6.5 are based on the paper [104] too. W.H. Young [162] used a different method to establish Theorems 6.5.9 and 6.5.13 for Lebesgue integrable functions. For an application of multipliers to first order partial differential equations, see [28]. For other results involving multipliers of non-absolutely convergent multiple integrals, see, for example, De Pauw and Pfeffer [29], Kurzweil [73], Lee [95, 105], Liu [113], Mikusiński and Ostaszewski [132], Ostaszewski [135], and Pfeffer [140]. Example 6.5.14 is a special case of [92, Example 4.8].

A two-dimensional analogue of Theorem 6.6.10 can be found in Hildebrandt and Schoenberg [61]. Theorem 6.6.12 is due to Lee [105].

Sections 6.7 and 6.8 are based on [105, Theorem 3.9] and [89, Section 3] respectively. For other related results, see Ostaszewski [135] and Lee [105].

[illegible] interpolation. For other results concerning multipliers [illegible] [illegible] [illegible] [illegible] [illegible], see, for example, [illegible] [illegible], E. Weil [21], Bourgeois [illegible] [illegible] [illegible], [illegible] and [illegible].

[illegible] is largely based on [illegible] [illegible] and [illegible]. Theorem [illegible] is due to W.H. Young and G.C. Young [illegible].

The presentation of Section [illegible] [illegible] that of [illegible] [illegible] several [illegible]. Theorem 6.3.10 can be found in W.H. Young [illegible].

Section [illegible] [illegible] [illegible] [illegible] [illegible] [illegible] [illegible] [illegible] [illegible]. [illegible] a different [illegible] [illegible] Theorems [illegible] and 6.5.11 [illegible] integrable functions. For [illegible] of multipliers [illegible] partial differential equations see [2]. [illegible] other results involving multipliers of [illegible] [illegible] multiple integrals [illegible], for example, [illegible] [illegible] [illegible] [illegible] [illegible] [illegible] [illegible] [illegible] [illegible] [illegible] and [illegible] [illegible] [illegible] [illegible] of [62] [illegible].

[illegible] analogue of Theorem [illegible] [illegible] [illegible] [illegible] [illegible] [61]. Theorem [illegible] [illegible] [illegible].

Section 6.7 [illegible] [illegible] [illegible] [illegible] [illegible] [illegible]. For other related results, [illegible] [illegible] [illegible] [illegible] [illegible].

Chapter 7

Some selected topics in trigonometric series

The main aim of this chapter is to give some recent applications of the Henstock-Kurzweil integral to the single trigonometric series

$$\frac{a_0}{2} + \sum_{k=1}^{\infty}(a_k \cos kt + b_k \sin kt), \tag{7.0.1}$$

where $a_0, a_1, b_1, a_2, b_2, \ldots$ are real numbers independent of t. Our starting point depends on a new generalized Dirichlet test.

7.1 A generalized Dirichlet test

Let $\mathbb{N}_0 := \mathbb{N} \cup \{0\}$. We begin with the following summation by parts.

Theorem 7.1.1 (Summation By Parts). *Let $(u_k)_{k=0}^{\infty}$ and $(v_k)_{k=0}^{\infty}$ be two sequences of real numbers. If $n \in \mathbb{N}_0$, then*

$$\sum_{k=0}^{n} u_k v_k = v_n \sum_{k=0}^{n} u_k + \sum_{k=0}^{n-1}(v_k - v_{k+1}) \sum_{j=0}^{k} u_j. \tag{7.1.1}$$

Proof. For each $k \in \{0, \ldots, n\}$, we have

$$\begin{aligned}
& u_k v_k \\
&= v_k \left(\sum_{j=0}^{k} u_j - \sum_{j=0}^{k-1} u_j \right) \text{ (where an empty sum of real numbers is zero)} \\
&= v_k \sum_{j=0}^{k} u_j - v_k \sum_{j=0}^{k-1} u_j \\
&= \left(v_k \sum_{j=0}^{k} u_j - v_{k-1} \sum_{j=0}^{k-1} u_j \right) + \left(v_{k-1} \sum_{j=0}^{k-1} u_j - v_k \sum_{j=0}^{k-1} u_j \right)
\end{aligned}$$

and hence

$$\begin{aligned}\sum_{k=0}^{n} u_k v_k &= \sum_{k=0}^{n}\left\{\left(v_k\sum_{j=0}^{k}u_j - v_{k-1}\sum_{j=0}^{k-1}u_j\right) + \left(v_{k-1}\sum_{j=0}^{k-1}u_j - v_k\sum_{j=0}^{k-1}u_j\right)\right\}\\ &= \sum_{k=0}^{n}\left\{v_k\sum_{j=0}^{k}u_j - v_{k-1}\sum_{j=0}^{k-1}u_j\right\} + \sum_{k=0}^{n}\left\{v_{k-1}\sum_{j=0}^{k-1}u_j - v_k\sum_{j=0}^{k-1}u_j\right\}\\ &= v_n\sum_{j=0}^{n}u_j + \sum_{k=0}^{n-1}\left\{(v_k - v_{k+1})\sum_{j=0}^{k}u_j\right\}.\end{aligned}$$

□

We have the following generalized Dirichlet test for infinite series.

Theorem 7.1.2. *Let* $(u_k)_{k=0}^{\infty}$ *and* $(v_k)_{k=0}^{\infty}$ *be two sequences of real numbers such that* $\lim_{n\to\infty} v_n = 0$ *and*

$$\sum_{k=0}^{\infty} |\, v_k - v_{k+1} \,| \max_{r=0,\ldots,k}\left|\sum_{j=0}^{r} u_j\right| < \infty. \tag{7.1.2}$$

Then the following assertions hold.

(i) *The series* $\sum_{k=0}^{\infty}(v_k - v_{k+1})\sum_{j=0}^{k} u_j$ *converges absolutely.*

(ii) *If* $n \in \mathbb{N}_0$, *then*

$$\left|\sum_{k=0}^{n} u_k v_k - \sum_{k=0}^{\infty}(v_k - v_{k+1})\sum_{j=0}^{k} u_j\right| \leq 2\sum_{k=n}^{\infty} |\, v_k - v_{k+1} \,| \max_{r=0,\ldots,k}\left|\sum_{j=0}^{r} u_j\right|.$$

(iii) *The series* $\sum_{k=0}^{\infty} u_k v_k$ *converges and*

$$\sum_{k=0}^{\infty} u_k v_k = \sum_{k=0}^{\infty}(v_k - v_{k+1})\sum_{j=0}^{k} u_j.$$

Proof. It is clear that assertion (i) holds. To prove assertion (ii), we let $n \in \mathbb{N}_0$ be arbitrary and consider two cases.

Case 1: Suppose that

$$\lim_{n\to\infty}\max_{r=0,\ldots,n}\left|\sum_{j=0}^{r} u_j\right| = 0.$$

In this case, it is clear that assertion (ii) holds.

Case 2: Suppose that

$$\lim_{n\to\infty}\max_{r=0,\ldots,n}\left|\sum_{j=0}^{r} u_j\right| > 0.$$

In this case, it follows from (7.1.2) that the series $\sum_{k=0}^{\infty} | v_k - v_{k+1} |$ converges. Thus

$$\left| \sum_{k=0}^{n} u_k v_k - \sum_{k=0}^{\infty} (v_k - v_{k+1}) \sum_{j=0}^{k} u_j \right|$$
$$= \left| v_n \sum_{k=0}^{n} u_k + \sum_{k=0}^{n-1} (v_k - v_{k+1}) \sum_{j=0}^{k} u_j - \sum_{k=0}^{\infty} (v_k - v_{k+1}) \sum_{j=0}^{k} u_j \right|$$
$$\leq \left| v_n \sum_{k=0}^{n} u_k \right| + \sum_{k=n}^{\infty} \left| (v_k - v_{k+1}) \sum_{j=0}^{k} u_j \right|$$
$$\leq \left| \sum_{k=0}^{n} u_k \right| \left\{ \sum_{k=n}^{\infty} | v_k - v_{k+1} | \right\} + \sum_{k=n}^{\infty} \left| (v_k - v_{k+1}) \sum_{j=0}^{k} u_j \right|$$
$$\left(\text{since } \sum_{k=0}^{\infty} |v_k - v_{k+1}| \text{ converges and } \lim_{n\to\infty} v_n = 0 \right)$$
$$\leq 2 \sum_{k=n}^{\infty} | v_k - v_{k+1} | \max_{r=0,\ldots,k} \left| \sum_{j=0}^{r} u_j \right| .$$

Since assertion (iii) follows from assertion (ii) and (7.1.2), the theorem is proved. □

We remark that Theorem 7.1.2 is a proper generalization of the following classical Dirichlet test.

Corollary 7.1.3 (Dirichlet). *Let $(u_k)_{k=0}^{\infty}$ and $(v_k)_{k=0}^{\infty}$ be two sequences of real numbers. If $(v_k)_{k=0}^{\infty}$ is decreasing, $\lim_{n\to\infty} v_n = 0$, and $\sup_{n\in\mathbb{N}} | \sum_{k=0}^{n} u_k |$ is finite, then $\sum_{k=0}^{\infty} u_k v_k$ converges.*

Before we present some useful consequences of Theorem 7.1.2, we need the following simple lemma.

Lemma 7.1.4. *If a real number t is not an integral multiple of π, then*

$$\sum_{k=1}^{n} \sin kt = \frac{1}{2}\sin nt + \frac{1-\cos nt}{2\tan\frac{t}{2}} \tag{7.1.3}$$

and

$$\sum_{k=1}^{n} \cos kt = -\frac{1-\cos nt}{2} + \frac{\sin nt}{2\tan\frac{t}{2}} \tag{7.1.4}$$

for all $n \in \mathbb{N}$.

Proof. Exercise. □

The next lemma is a consequence of Theorem 7.1.2.

Lemma 7.1.5. *The series $\sum_{k=1}^{\infty} \frac{\sin kt}{k}$ converges for every $t \in \mathbb{R}$. Moreover, this series converges uniformly on $[-\pi,-\delta] \cup [\delta,\pi]$ whenever $0 < \delta < \pi$.*

Proof. It is easy to see that the series $\sum_{k=1}^{\infty} \frac{\sin kt}{k}$ converges if t is an integral multiple of π. On the other hand, if t is not an integral multiple of π, then (7.1.3) yields

$$\sum_{k=1}^{\infty}\left(\frac{1}{k}-\frac{1}{k+1}\right)\max_{r=1,\ldots,k}\left|\sum_{j=1}^{r}\sin jt\right| \leq \frac{1}{2}+\frac{1}{\left|\tan\frac{t}{2}\right|} < \infty;$$

hence Theorem 7.1.2(iii) implies that the series $\sum_{k=1}^{\infty} \frac{\sin kt}{k}$ converges.

Finally, for any $0 < \delta < \pi$, we infer from (7.1.3) and Theorem 7.1.2(ii) that $\sum_{k=1}^{\infty} \frac{\sin kt}{k}$ converges uniformly on $[-\pi,-\delta] \cup [\delta,\pi]$. □

The following lemmas play a crucial role in Chapters 7 and 8.

Lemma 7.1.6. *We have*

$$\sum_{k=1}^{\infty} \frac{\sin kx}{k} = \begin{cases} \frac{\pi-x}{2} & \textit{if } x \in (0,2\pi), \\ \frac{-\pi-x}{2} & \textit{if } x \in (-2\pi,0), \\ 0 & \textit{if } x \in \{-2\pi,0,2\pi\}. \end{cases} \tag{7.1.5}$$

Proof. Since the function $x \mapsto \sin x$ is odd, and $\sin k\pi = 0$ for every $k \in \mathbb{N}$, it is enough to consider the case $x \in (0, 2\pi)$. Then we have

$$\begin{aligned}
&\sum_{k=1}^{\infty} \frac{\sin kx}{k} \\
&= \lim_{n\to\infty} \int_x^{\pi} \sum_{k=1}^{n} \cos kt \, dt \\
&= \lim_{n\to\infty} \int_x^{\pi} \left(\frac{1}{2} - \frac{\cos nt}{2} + \frac{\sin nt \cos \frac{t}{2}}{2 \sin \frac{t}{2}} \right) dt && (\text{by } (7.1.4)) \\
&= \frac{\pi - x}{2} + \lim_{n\to\infty} \int_x^{\pi} \frac{\sin nt \cos \frac{t}{2}}{2 \sin \frac{t}{2}} \, dt && \left(\text{since } \lim_{n\to\infty} \frac{\sin nx}{2n} = 0 \right) \\
&= \frac{\pi - x}{2} + \lim_{n\to\infty} \left\{ \left[- \frac{\cos nt \cot \frac{t}{2}}{2n} \right]_x^{\pi} - \int_x^{\pi} \frac{\cos nt}{4n \sin^2 \frac{t}{2}} \, dt \right\} \\
& && (\text{by integration by parts, since } 0 < x < 2\pi) \\
&= \frac{\pi - x}{2}.
\end{aligned}$$

□

Lemma 7.1.7. $\sup_{n\in\mathbb{N}} \sup_{t\in\mathbb{R}} \left| \sum_{k=1}^{n} \frac{\sin kt}{k} \right| \le 3\pi.$

Proof. Let $n \in \mathbb{N}$ be given. Since the function $x \mapsto \sum_{k=1}^{n} \frac{\sin kx}{k}$ is odd and 2π-periodic, it suffices to prove the lemma for $t \in [0, \pi]$. For each $t \in [0, \pi]$ we apply Lemma 7.1.6, Theorem 7.1.2(iii) and (7.1.3) to obtain

$$\begin{aligned}
&\left| \sum_{k=1}^{n} \frac{\sin kt}{k} \right| \\
&\le \left| \sum_{k=1}^{n} \frac{\sin kt - \sin k(t + \frac{1}{n})}{k} \right| + \left| \sum_{k=n+1}^{\infty} \frac{\sin k(t + \frac{1}{n})}{k} \right| + \frac{\pi - (t + \frac{1}{n})}{2} \\
&\le \sum_{k=1}^{n} \left| \frac{2 \sin \frac{k}{2n}}{k} \right| + \left| \sum_{k=n+1}^{\infty} \frac{1}{k(k+1)} \sum_{j=n+1}^{k} \sin j(t + \frac{1}{n}) \right| + \frac{\pi}{2} \\
&= \sum_{k=1}^{n} \left| \frac{2 \sin \frac{k}{2n}}{k} \right| + \frac{1 + \left| \cot \frac{1}{2}(t + \frac{1}{n}) \right|}{n+1} + \frac{\pi}{2} \\
&\le 3\pi && \left(\text{since } 2n \sin \frac{1}{2n} \ge \frac{2}{\pi} \text{ for all } n \in \mathbb{N} \right).
\end{aligned}$$

□

The following theorems are immediate consequences of Theorem 7.1.2(ii) and Lemma 7.1.7.

Theorem 7.1.8. *Let $(a_k)_{k=1}^{\infty}$ be a sequence of real numbers such that $\sum_{k=1}^{\infty} |a_k - a_{k+1}|$ converges and $\lim_{n\to\infty} a_n = 0$. Then $\sum_{k=1}^{\infty} \frac{a_k \sin kt}{k}$ converges uniformly on $[-\pi, \pi]$.*

Theorem 7.1.9. *Let $(a_k)_{k=0}^{\infty}$ be a sequence of real numbers such that $\sum_{k=0}^{\infty} |a_k - a_{k+1}|$ converges and $\lim_{n\to\infty} a_n = 0$. Then $\frac{a_0}{2} + \sum_{k=1}^{\infty} a_k \cos kt$ converges uniformly on $[\delta, \pi]$ whenever $0 < \delta < \pi$.*

7.2 Fourier series

A 2π-periodic function $f : \mathbb{R} \longrightarrow \mathbb{R}$ is said to be in $L^1(\mathbb{T})$ if $f \in L^1[-\pi, \pi]$.

Definition 7.2.1. Let $f \in L^1(\mathbb{T})$. The *Fourier series* of f is the trigonometric series

$$\frac{a_0}{2} + \sum_{k=1}^{\infty} (a_k \cos kt + b_k \sin kt),$$

where the a_n's and b_n's are given by the formulas

$$a_n = \frac{1}{\pi} \int_{-\pi}^{\pi} f(t) \cos nt \, d\mu_1(t) \quad (n = 0, 1, 2, \dots)$$

and

$$b_n = \frac{1}{\pi} \int_{-\pi}^{\pi} f(t) \sin nt \, d\mu_1(t) \quad (n = 1, 2, \dots).$$

In this case, we write

$$f(t) \sim \frac{a_0}{2} + \sum_{k=1}^{\infty} (a_k \cos kt + b_k \sin kt).$$

Here a_n and b_n are known as the *Fourier coefficients* of f.

The above sign "$\sim$" can be replaced by the "$=$" sign only if we can prove that the series converges and that its sum is equal to $f(t)$.

Theorem 7.2.2 (Riemann-Lebesgue Lemma). *Let $f \in L^1(\mathbb{T})$ and assume that*

$$f(t) \sim \frac{a_0}{2} + \sum_{k=1}^{\infty} (a_k \cos kt + b_k \sin kt).$$

Then $\lim_{n\to\infty} a_n = \lim_{n\to\infty} b_n = 0$.

Proof. It is easy to check that the theorem is true if f is a step function. Now suppose that $f \in L^1(\mathbb{T})$. For each $\varepsilon > 0$ we use Exercise 3.7.6 to choose a step function ϕ on $[-\pi, \pi]$ such that

$$\int_{-\pi}^{\pi} |f - \phi| < \varepsilon.$$

Then

$$\limsup_{n\to\infty} |a_n| \leq \|f - \phi\|_{L^1[-\pi,\pi]} + \lim_{n\to\infty} \frac{1}{\pi} \left| \int_{-\pi}^{\pi} \phi(t) \cos nt \, d\mu_1(t) \right| \leq \varepsilon.$$

Since a similar reasoning shows that $\limsup_{n\to\infty} |b_n| \leq \varepsilon$, the theorem follows. □

At this point it is unclear whether the converse of Theorem 7.2.2 is true. In order to prove that this converse is false, we need the following classical result.

Theorem 7.2.3. *Let $f \in L^1(\mathbb{T})$ and assume that*

$$f(t) \sim \sum_{k=1}^{\infty} (a_k \cos kt + b_k \sin kt). \tag{7.2.1}$$

Then the series $\sum_{k=1}^{\infty} \frac{b_k}{k}$ converges and

$$\lim_{n\to\infty} \max_{x\in[-\pi,\pi]} \left| \int_{-\pi}^{x} \left\{ \sum_{k=1}^{n} (a_k \cos kt + b_k \sin kt) - f(t) \right\} d\mu_1(t) \right| = 0.$$

Proof. We will first prove that

$$\sup_{x\in[-\pi,\pi]} \left| \sum_{k=p}^{q} \left(\frac{a_k}{k} \sin kx - \frac{b_k}{k} \cos kx \right) \right| \to 0 \text{ as } p, q \to \infty. \tag{7.2.2}$$

Let $\varepsilon > 0$ be given. Since $f \in L^1[-\pi, \pi]$ there exists $\eta > 0$ so that $\eta < \frac{\pi}{4}$ and

$$\sup_{x\in[-\pi,\pi]} \int_{-\pi}^{\pi} |f| \, \chi_{U(x)} \, d\mu_1 < \frac{\varepsilon}{12},$$

where

$$U(x) := [-\pi, -\pi + \eta) \cup ((x - \eta, x + \eta) \cap [-\pi, \pi]) \cup (\pi - \eta, \pi]$$

whenever $x \in [-\pi, \pi]$. According to Lemma 7.1.5 and our choice of η, we select a positive integer N so that

$$\sup_{x\in[-\pi,\pi]} \max_{t\in[-\pi,\pi]\setminus U(x)} \left| \sum_{k=r}^{s} \frac{\sin k(t-x)}{k} \right| < \frac{\varepsilon}{2(1 + \|f\|_{L^1[-\pi,\pi]})}$$

for all integers $s \geq r \geq N$. Hence for any $x \in [-\pi, \pi]$ it follows from Lemma 7.1.7 that

$$\begin{aligned}
&\left| \sum_{k=p}^{q} \left(\frac{a_k}{k} \sin kx - \frac{b_k}{k} \cos kx \right) \right| \\
&= \frac{1}{\pi} \left| \int_{-\pi}^{\pi} f(t) \Big\{ \sum_{k=p}^{q} \frac{\sin k(t-x)}{k} \Big\} \, d\mu_1(t) \right| \\
&\leq \int_{-\pi}^{\pi} |f(t)| \left\{ 6\, \chi_{U(x)}(t) + \left| \sum_{k=p}^{q} \frac{\sin k(t-x)}{k\pi} \right| \chi_{[-\pi,\pi]\setminus U(x)}(t) \right\} \, d\mu_1(t) \\
&< \varepsilon
\end{aligned}$$

for all integers $q \geq p \geq N$. It follows that (7.2.2) holds; in particular, the series $\sum_{k=1}^{\infty} \frac{b_k}{k}$ converges.

Next we observe that (7.2.1) implies that the function $t \mapsto \int_{-\pi}^{t} f \, d\mu_1$ belongs to $L^1(\mathbb{T})$. In view of (7.2.2), it remains to prove that

$$\int_{-\pi}^{t} f \, d\mu_1 \sim \sum_{k=1}^{\infty} \frac{b_k}{k} \cos k\pi + \sum_{k=1}^{\infty} \left\{ \frac{a_k \sin kt}{k} - \frac{b_k \cos kt}{k} \right\}. \tag{7.2.3}$$

By integration by parts, we get

$$\frac{1}{\pi} \int_{-\pi}^{\pi} \left(\int_{-\pi}^{t} f \, d\mu_1 \right) \cos kt \, d\mu_1(t) = -\frac{b_k}{k}$$

and

$$\frac{1}{\pi} \int_{-\pi}^{\pi} \left(\int_{-\pi}^{t} f \, d\mu_1 \right) \sin kt \, d\mu_1(t) = \frac{a_k}{k}$$

for all $k \in \mathbb{N}$. Thus

$$\int_{-\pi}^{t} f \, d\mu_1 \sim C + \sum_{k=1}^{\infty} \left\{ \frac{a_k \sin kt}{k} - \frac{b_k \cos kt}{k} \right\} \tag{7.2.4}$$

for some constant C to be determined. Finally, since (7.2.1) implies that $\int_{-\pi}^{\pi} f \, d\mu_1 = 0$, we infer from (7.2.2) and (7.2.4) that $C = \sum_{k=1}^{\infty} \frac{b_k}{k} \cos k\pi$. □

The following example is an immediate consequence of Theorem 7.2.3.

Example 7.2.4 (Fatou). The trigonometric series $\sum_{k=2}^{\infty} \frac{\sin kt}{\ln k}$ is not the Fourier series of a function in $L^1(\mathbb{T})$.

Example 7.2.5 (Perron). Let $(b_k)_{k=0}^{\infty}$ be a sequence of real numbers such that $b_0 = 1$ and

$$b_{n+1} = \sum_{k=0}^{n} \left(\frac{1}{n+1-k} - \frac{1}{n+2-k} \right) b_k$$

for $n \in \mathbb{N}_0$. Then $(b_n)_{n=0}^{\infty}$ is a decreasing sequence, $\lim_{n\to\infty} b_n = 0$ and

$$\sum_{k=1}^{\infty} b_k \sin kt = \frac{-(\cos\frac{t}{2})\ln(2\sin\frac{t}{2}) - \frac{t-\pi}{2}\sin\frac{t}{2}}{2\sin\frac{t}{2}\left[(\ln(2\sin\frac{t}{2}))^2 + (\frac{t-\pi}{2})^2\right]} \qquad (t \in (0,\pi)).$$

Since the function

$$t \mapsto \frac{-(\cos\frac{t}{2})\ln(2\sin\frac{t}{2}) - \frac{t-\pi}{2}\sin\frac{t}{2}}{2\sin\frac{t}{2}\left[(\ln(2\sin\frac{t}{2}))^2 + (\frac{t-\pi}{2})^2\right]}$$

is *not* Lebesgue integrable on $[0,\pi]$, the sine series $\sum_{k=1}^{\infty} b_k \sin kt$ *cannot be* the Fourier series of a function in $L^1(\mathbb{T})$.

7.3 Some examples of Fourier series

According to Example 7.2.4, not every trigonometric series is the Fourier series of a function in $L^1(\mathbb{T})$. In this section we give some simple sufficient conditions for a trigonometric series to be the Fourier series of a function in $L^1(\mathbb{T})$. We begin with following modification of Theorem 7.1.2.

Theorem 7.3.1. *Let $(c_k)_{k=1}^{\infty}$ be a sequence of real numbers with $\lim_{n\to\infty} c_n = 0$. If $(f_k)_{k=1}^{\infty}$ is a sequence in $L^1[a,b]$ satisfying*

$$\sum_{k=1}^{\infty} |c_k - c_{k+1}| \max_{r=1,\dots,k} \left\| \sum_{j=1}^{r} f_j \right\|_{L^1[a,b]} < \infty, \tag{7.3.1}$$

then there exists $f \in L^1[a,b]$ such that

$$\lim_{n\to\infty} \left\| \sum_{k=1}^{n} c_k f_k - f \right\|_{L^1[a,b]} = 0.$$

Proof. Exercise. □

We need the following two lemmas to prove Theorem 7.3.4 below.

Lemma 7.3.2. *Let $(u_k)_{k=0}^{\infty}$ be a decreasing sequence of real numbers with $\lim_{n\to\infty} u_n = 0$. If $(v_k)_{k=0}^{\infty}$ is a sequence of non-negative numbers, then $\sum_{k=0}^{\infty} u_k v_k$ converges if and only if $\sum_{k=0}^{\infty}(u_k - u_{k+1})\sum_{j=0}^{k} v_j$ converges.*

Proof. This follows from Theorems 7.1.1 and 7.1.2(iii). □

Lemma 7.3.3. $\sup_{n\in\mathbb{N}} \frac{1}{\ln(n+1)} \int_0^\pi \left|\sum_{k=1}^n \sin kt\right| \, dt \le 4.$

Proof. Clearly, $\int_0^\pi \sin t \, dt = 2 \le 4\ln 2$. On the other hand, for any positive integer $n \ge 2$, we have

$$\begin{aligned}
\int_0^\pi \left|\sum_{k=1}^n \sin kt\right| \, dt &= \int_0^\pi \left|\frac{1-\cos nt}{2\tan\frac{t}{2}} + \frac{1}{2}\sin nt\right| \, dt \qquad \text{(by (7.1.3))} \\
&\le \int_0^\pi \frac{1}{2\sin\frac{t}{2}} \sum_{k=0}^{n-1} \Big(\cos kt - \cos(k+1)t\Big) \, dt + \frac{\pi}{2} \\
&= \sum_{k=0}^{n-1} \frac{2}{2k+1} + \frac{\pi}{2} \\
&\le \ln(n+1)^2 + 2\ln 3 \\
&\le 4\ln(n+1).
\end{aligned}$$

□

Theorem 7.3.4. *Let $(b_k)_{k=1}^\infty$ be a decreasing sequence of real numbers with $\lim_{n\to\infty} b_n = 0$. Then the series $\sum_{k=1}^\infty \frac{b_k}{k}$ converges if and only if there exists $g \in L^1(\mathbb{T})$ such that $g(t) = \sum_{k=1}^\infty b_k \sin kt$ for every $t \in [-\pi, \pi]$. In this case,*

$$\lim_{n\to\infty} \int_{-\pi}^{\pi} \left|\sum_{k=1}^n b_k \sin kt - g(t)\right| \, d\mu_1(t) = 0.$$

Proof. ($\Longleftarrow$) This follows from Theorem 7.2.3.

($\Longrightarrow$) Since $(b_k)_{k=1}^\infty$ is a decreasing sequence such that $\lim_{n\to\infty} b_n = 0$ and $\sum_{k=1}^\infty \frac{b_k}{k}$ converges, it follows from Lemma 7.3.2 that the series $\sum_{k=1}^\infty |b_k - b_{k+1}| \ln k$ converges. Hence the series

$$\sum_{k=1}^\infty |\, b_k - b_{k+1} \,| \max_{r=1,\ldots,k} \int_0^\pi \left|\sum_{j=1}^r \frac{\sin jt}{t}\right| \, dt$$

converges by Lemma 7.3.3. An application of Theorem 7.3.1 yields the result. □

Our next aim is to establish a Lebesgue integrability theorem for single cosine series via Theorem 7.3.1. To do this, we need the following modification of Lemma 7.3.3.

Lemma 7.3.5. $\sup_{n\in\mathbb{N}} \frac{1}{\ln(n+1)} \int_0^\pi \left|\frac{1}{2} + \sum_{k=1}^n \cos kt\right| \, dt \le 4.$

Proof. First we select any integer $n \geq 3$. Then

$$\begin{aligned}
&\int_0^\pi \left| \frac{1}{2} + \sum_{k=1}^{n} \cos kt \right| dt \\
&\leq \int_0^\pi \left| \frac{1}{2} \sin nt \cot \frac{t}{2} \right| dt + \frac{1}{2} \int_0^\pi |\cos nt|\ dt \qquad \text{(by (7.1.4))} \\
&\leq \int_0^\pi \left| \frac{\sin nt}{t} \right| dt + \int_0^\pi \left(\frac{1}{t} - \frac{1}{2} \cot \frac{t}{2} \right) dt + \frac{1}{2} \int_0^\pi |\cos nt|\ dt \\
&\leq \sum_{k=1}^{n} \int_{\frac{(k-1)\pi}{n}}^{\frac{k\pi}{n}} \left| \frac{\sin nt}{t} \right| dt + 2 \\
&= \sum_{k=1}^{n} \int_0^\pi \frac{\sin \eta}{\eta + (k-1)\pi}\, d\eta + 2 \\
&\leq 4 + \ln(n+1) \\
&\leq 4 \ln(n+1)
\end{aligned}$$

because $3\ln(n+1) \geq 3\ln 4 = 6\ln 2 > \frac{12}{3} = 4$ for every integer $n \geq 3$. Since we also have

$$\int_0^\pi \left| \frac{1}{2} + \cos t \right| dt = \frac{\pi}{6} + \sqrt{3} < \frac{8}{3} < 4\ln 2$$

and

$$\begin{aligned}
\int_0^\pi \left| \frac{1}{2} + \sum_{k=1}^{2} \cos kt \right| dt &\leq \int_0^\pi (\cos^2 t + \cos t)\ dt + \frac{1}{2} \int_0^\pi |\cos 2t|\ dt \\
&= \frac{\pi}{2} + 1 \\
&\leq 4 \ln 3,
\end{aligned}$$

the lemma is proved. □

In view of Lemma 7.3.5, the proof of the following theorem is similar to that of Theorem 7.3.4.

Theorem 7.3.6. *Let $(a_k)_{k=0}^{\infty}$ be a decreasing sequence of real numbers such that $\lim_{n\to\infty} a_n = 0$. If the series $\sum_{k=1}^{\infty} \frac{a_k}{k}$ converges, then there exists $f \in L^1(\mathbb{T})$ such that $f(t) = \frac{a_0}{2} + \sum_{k=1}^{\infty} a_k \cos kt$ for all $t \in [-\pi, \pi] \setminus \{0\}$ and*

$$\lim_{n\to\infty} \int_{-\pi}^{\pi} \left| \frac{a_0}{2} + \sum_{k=1}^{n} a_k \cos kt - f(t) \right| d\mu_1(t) = 0.$$

Now we weaken the assumptions of Theorem 7.3.6 to obtain the following theorem.

Theorem 7.3.7. *Let $(a_k)_{k=0}^{\infty}$ be a sequence of real numbers such that $\sum_{k=1}^{\infty} |a_k - a_{k+1}|$ converges and $\lim_{n\to\infty} a_n = 0$. Then there exists a 2π-periodic function $f : \mathbb{R} \longrightarrow \mathbb{R}$ such that*

$$\lim_{n\to\infty} \sup_{x\in(0,\pi]} \left| \lim_{\delta\to 0+} \int_\delta^x \left\{ \frac{a_0}{2} + \sum_{k=1}^{n} a_k \cos kt - f(t) \right\} dt \right| = 0. \quad (7.3.2)$$

In particular,

$$\frac{2}{\pi}\left\{ \lim_{\delta\to 0+} \int_\delta^\pi f(t) \cos nt \, dt \right\} = a_n$$

for all $n \in \mathbb{N}_0$.

Proof. By Theorem 7.1.9, the series

$$\frac{a_0}{2} + \sum_{k=1}^{\infty} a_k \cos kt$$

converges uniformly on $[\delta, \pi]$ for every $0 < \delta < \pi$. Define the function $f : \mathbb{R} \longrightarrow \mathbb{R}$ by setting

$$f(t) = \begin{cases} \frac{a_0}{2} + \sum_{k=1}^{\infty} a_k \cos kt & \text{if } t \in [-\pi, 0) \cup (0, \pi], \\ 0 & \text{if } t = 0, \\ f(t + 2\pi) & \text{if } t \in \mathbb{R} \end{cases}$$

so that f is 2π-periodic. By Theorems 7.1.9 and 7.1.8,

$$\begin{aligned} &\limsup_{n\to\infty} \sup_{x\in(0,\pi]} \left| \lim_{\delta\to 0+} \int_\delta^x \left\{ \frac{a_0}{2} + \sum_{k=1}^{n} a_k \cos kt - f(t) \right\} dt \right| \\ &= \limsup_{n\to\infty} \max_{x\in[0,\pi]} \left| \sum_{k=n+1}^{\infty} \frac{a_k \sin kx}{k} \right| \\ &= 0. \end{aligned}$$

□

We say that a sequence $(a_k)_{k=0}^{\infty}$ of real numbers is *convex* if $a_n - 2a_{n+1} + a_{n+2} \geq 0$ for all $n \in \mathbb{N}_0$.

Corollary 7.3.8. *Let $(a_k)_{k=0}^{\infty}$ be a convex sequence of real numbers such that $\lim_{n\to\infty} a_n = 0$. Then there exists a non-negative function $f \in L^1(\mathbb{T})$ such that $f(t) = \frac{a_0}{2} + \sum_{k=1}^{\infty} a_k \cos kt$ for all $t \in (0, \pi]$, and $\frac{a_0}{2} + \sum_{k=1}^{\infty} a_k \cos kt$ is the Fourier series of f.*

Proof. Since our assumptions imply that the series $\sum_{k=0}^{\infty} |a_k - a_{k+1}|$ converges and $\lim_{n\to\infty} a_n = 0$, we let f be given as in Theorem 7.3.7. Then $f(t) \geq 0$ for every $t \in (0, \pi]$. Indeed, for any $t \in (0, \pi]$, we have

$$\begin{aligned} f(t) &= \sum_{k=0}^{\infty} (a_k - a_{k+1}) \left(\frac{1}{2} + \sum_{j=1}^{k} \cos jt \right) \\ &= \sum_{k=0}^{\infty} (u_k - u_{k+1}) \frac{\sin(k + \frac{1}{2})t}{2 \sin \frac{t}{2}} \\ &= \sum_{k=0}^{\infty} (a_k - 2a_{k+1} + a_{k+2}) \sum_{j=0}^{k} \frac{\sin(j + \frac{1}{2})t}{2 \sin \frac{t}{2}} \\ &= \sum_{k=0}^{\infty} (a_k - 2a_{k+1} + a_{k+2}) \frac{\sin^2(\frac{k+1}{2})t}{2 \sin^2 \frac{t}{2}} \\ &\geq 0. \end{aligned}$$

□

The following example, which is a consequence of Corollary 7.3.8, shows that the converse of Theorem 7.3.6 is false.

Example 7.3.9. The trigonometric series $\sum_{k=2}^{\infty} \frac{\cos kt}{\ln k}$ is a Fourier series, but the series $\sum_{k=2}^{\infty} \frac{1}{k \ln k}$ diverges.

Example 7.3.10 (cf. [149, p.524]). Let $g(t) = \sum_{k=2}^{\infty} \frac{2 \sin \frac{k\pi}{3}}{\ln k} \sin kt$ if the series converges, and $g(t) = 0$ otherwise. Then $g \in L^1(\mathbb{T})$ but $\sum_{k=2}^{\infty} \left| \frac{2 \sin \frac{k\pi}{3}}{k \ln k} \right| = \infty$.

Proof. Since we have

$$g(t) = \sum_{k=2}^{\infty} \frac{\cos k(t - \frac{\pi}{3})}{\ln k} - \sum_{k=2}^{\infty} \frac{\cos k(t + \frac{\pi}{3})}{\ln k}$$

for μ_1-almost all $t \in [0, \pi]$, we can modify the proof of Corollary 7.3.8 to conclude that $g \in L^1(\mathbb{T})$. On the other hand, it follows from the Cauchy Condensation Test that $\sum_{k=2}^{\infty} \left| \frac{2 \sin \frac{k\pi}{3}}{k \ln k} \right| = \infty$. □

7.4 Some Lebesgue integrability theorems for trigonometric series

Let $(a_k)_{k=0}^{\infty}$ and f be given as in Theorem 7.3.7. At this point, it is not clear whether $f \in L^1(\mathbb{T})$. One of the aims of this section is to show that f need not belong to $L^1(\mathbb{T})$. We begin with the following lemmas.

Lemma 7.4.1. *The following statements are true.*

(i) *If $j, k, \ell \in \mathbb{N}$ with $j \leq k < \ell$ or $j < k \leq \ell$, then*

$$\int_{\frac{\pi}{2}}^{\pi} \sin 2^j t \sin^2 2^k t \sin 2^\ell t \, dt = \int_{\frac{\pi}{2}}^{\pi} \sin 2^j t \sin 2^k t \sin^2 2^\ell t \, dt = 0.$$

(ii) *If $j, k, \ell \in \mathbb{N}$ with $\ell > k+1 > j+1$ or $\ell > k > j+1$, then*

$$\int_{\frac{\pi}{2}}^{\pi} \sin^2 2^j t \sin 2^k t \sin 2^\ell t \, dt = 0.$$

(iii) *If $j \in \mathbb{N}$, then*

$$\int_{\frac{\pi}{2}}^{\pi} \sin^2 2^j t \sin 2^{j+1} t \sin 2^{j+2} t \, dt = -\frac{\pi}{16}.$$

(iv) *If $j, k, \ell, p \in \mathbb{N}$ with $j < k < \ell < p$, then*

$$\int_{\frac{\pi}{2}}^{\pi} \sin 2^j t \sin 2^k t \sin 2^\ell t \sin 2^p t \, dt = 0.$$

Proof. Exercise. □

Lemma 7.4.2. *If $b_1, \ldots, b_n$ are real numbers, then*

$$\int_{\frac{\pi}{2}}^{\pi} \left(\sum_{k=1}^{n} b_k \sin 2^k t \right)^4 dt \leq \frac{3\pi}{4} \left(\sum_{k=1}^{n} b_k^2 \right)^2.$$

Proof. If $n \in \{1, 2\}$, then the result holds. Henceforth, we assume that $n \geq 3$.

For each $j \in \mathbb{N}$ and $x \in \mathbb{R}$, let $S_j(x) = \sin 2^j(x)$. Then, using multinomial expansion and Lemma 7.4.1, we get

$$\begin{aligned}
&\int_{\frac{\pi}{2}}^{\pi}\left(\sum_{k=1}^{n} b_k S_k(t)\right)^4 dt \\
&= \sum_{\substack{(k_1,\ldots,k_n)\in\mathbb{N}_0^n \\ k_1+\cdots+k_n=4}} \int_{\frac{\pi}{2}}^{\pi} \frac{4!}{\prod_{r=1}^{n} k_r!} \prod_{j=1}^{n}\left(b_j S_j(t)\right)^{k_j} dt \\
&= \sum_{j=1}^{n}\int_{\frac{\pi}{2}}^{\pi}\left(b_j S_j(t)\right)^4 dt + 6\sum_{j=1}^{n-1}\sum_{k=j+1}^{n}\int_{\frac{\pi}{2}}^{\pi}\left(b_j S_j(t)\right)^2\left(b_k S_k(t)\right)^2 dt \\
&\quad +12\sum_{j=1}^{n-2} b_j^2 b_{j+1} b_{j+2}\int_{\frac{\pi}{2}}^{\pi}(S_j(t))^2 S_{j+1}(t) S_{j+2}(t)\, dt \\
&= \frac{3\pi}{16}\sum_{j=1}^{n} b_j^4 + \frac{6\pi}{8}\sum_{j=1}^{n-1}\sum_{k=j+1}^{n} b_j^2 b_k^2 - \frac{12\pi}{16}\sum_{j=1}^{n-2} b_j^2 b_{j+1} b_{j+2} \\
&\le \frac{3\pi}{16}\sum_{j=1}^{n} b_j^4 + \frac{6\pi}{8}\sum_{j=1}^{n-1}\sum_{k=j+1}^{n} b_j^2 b_k^2 + \frac{9\pi}{16}\sum_{j=1}^{n} b_j^4 \\
&\le \frac{3\pi}{4}\left\{\sum_{j=1}^{n} b_j^4 + 2\sum_{j=1}^{n-1}\sum_{k=j+1}^{n} b_j^2 b_k^2\right\} \\
&= \frac{3\pi}{4}\left(\sum_{k=1}^{n} b_k^2\right)^2.
\end{aligned}$$

□

Lemma 7.4.3. *If $b_1, \ldots, b_n$ are real numbers, then*

$$\left(\int_{\frac{\pi}{2}}^{\pi}\left|\sum_{k=1}^{n} b_k \sin 2^k t\right| dt\right)^2 \ge \frac{\pi}{12}\int_{\frac{\pi}{2}}^{\pi}\left(\sum_{k=1}^{n} b_k \sin 2^k t\right)^2 dt.$$

Proof. Let $g_n(t) = \sum_{k=1}^{n} b_k \sin 2^k t$ for each $t \in [\frac{\pi}{2}, \pi]$. Then, using the observation $2 = \frac{2}{3} + \frac{4}{3}$, Hölder's inequality (with $p = \frac{3}{2}$ and $q = 3$) and Lemma 7.4.2, we get

$$\begin{aligned}
\int_{\frac{\pi}{2}}^{\pi}(g_n(t))^2\, dt &\le \left(\int_{\frac{\pi}{2}}^{\pi}|g_n(t)|\, dt\right)^{\frac{2}{3}}\left(\int_{\frac{\pi}{2}}^{\pi}(g_n(t))^4\, dt\right)^{\frac{1}{3}} \\
&\le \left(\int_{\frac{\pi}{2}}^{\pi}|g_n(t)|\, dt\right)^{\frac{2}{3}}\left(\frac{3\pi}{4}\left(\sum_{k=1}^{n} b_k^2\right)^2\right)^{\frac{1}{3}} \\
&= \left(\int_{\frac{\pi}{2}}^{\pi}|g_n(t)|\, dt\right)^{\frac{2}{3}}\left(\frac{3\pi}{4}\left\{\int_{\frac{\pi}{2}}^{\pi}\frac{4}{\pi}(g_n(t))^2\, dt\right\}^2\right)^{\frac{1}{3}},
\end{aligned}$$

and the result follows. □

Theorem 7.4.4. *Let* $\sum_{k=1}^{\infty} b_k$ *be an absolutely convergent series of real numbers. Then there exists* $h \in L^1([0,\pi])$ *such that* $h(t) = \frac{1}{t}\sum_{k=1}^{\infty} b_k \sin 2^k t$ *for* μ_1*-almost all* $t \in [0,\pi]$ *if and only if* $\sum_{k=1}^{\infty}\sqrt{\sum_{j=k}^{\infty} b_j^2}$ *converges. In this case,*

$$\int_0^{\pi} |\, h \,|\; d\mu_1 \le 2\pi \sum_{k=1}^{\infty}\sqrt{\sum_{j=k}^{\infty} b_j^2}.$$

Proof. ($\Longleftarrow$) For each $j \in \mathbb{N}$ and $x \in [0,\pi]$, let $S_j(t) = \sin 2^j t$. For each $N \in \mathbb{N}$ we have

$$\begin{aligned}
&\int_{\frac{\pi}{2^{N+1}}}^{\pi} \left| \sum_{j=1}^{\infty} \frac{b_j S_j(t)}{t} \right| dt \\
&\le \sum_{k=0}^{N} \int_{\frac{\pi}{2^{k+1}}}^{\frac{\pi}{2^k}} \left| \sum_{j=1}^{k} \frac{b_j S_j(t)}{t} \right| dt + \sum_{k=0}^{N} \int_{\frac{\pi}{2^{k+1}}}^{\frac{\pi}{2^k}} \left| \sum_{j=k+1}^{\infty} \frac{b_j S_j(t)}{t} \right| dt. \qquad (7.4.1)
\end{aligned}$$

The first term on the right-hand side of (7.4.1) is bounded above by $\pi\sum_{j=1}^{\infty} |\, b_j \,|$:

$$\sum_{k=0}^{N} \int_{\frac{\pi}{2^{k+1}}}^{\frac{\pi}{2^k}} \left| \sum_{j=1}^{k} \frac{b_j S_j(t)}{t} \right| dt \le \pi \sum_{k=1}^{\infty} \frac{1}{2^{k+1}} \sum_{j=1}^{k} 2^j \,|\, b_j \,| = \pi \sum_{j=1}^{\infty} |\, b_j \,|. \qquad (7.4.2)$$

We will next show that the second term on the right-hand side of (7.4.1) is bounded above by $\sum_{k=1}^{\infty}\sqrt{\sum_{j=k}^{\infty} b_j^2}$. Since Lebesgue's Dominated Convergence Theorem yields

$$\int_{\frac{\pi}{2}}^{\pi} \left| \sum_{j=k+1}^{\infty} b_j S_{j-k}(t) \right|^2 dt = \frac{\pi}{4} \sum_{j=k+1}^{\infty} b_j^2 \quad \text{for all } k \in \mathbb{N}_0, \qquad (7.4.3)$$

we infer from the Cauchy-Schwarz inequality that

$$\begin{aligned}
&\sum_{k=0}^{N} \int_{\frac{\pi}{2^{k+1}}}^{\frac{\pi}{2^k}} \left| \sum_{j=k+1}^{\infty} \frac{b_j S_j(t)}{t} \right| dt \\
&= \sum_{k=0}^{N} \int_{\frac{\pi}{2}}^{\pi} \left| \sum_{j=k+1}^{\infty} \frac{b_j S_{j-k}(t)}{t} \right| dt \\
&\le \sum_{k=0}^{N} \sqrt{\int_{\frac{\pi}{2}}^{\pi} \left(\sum_{j=k+1}^{\infty} b_j S_{j-k}(t) \right)^2 dt} \sqrt{\int_{\frac{\pi}{2}}^{\pi} \frac{1}{t^2}\, dt} \\
&\le \sum_{k=1}^{\infty} \sqrt{\sum_{j=k}^{\infty} b_j^2}. \qquad (7.4.4)
\end{aligned}$$

Since N is arbitrary, we deduce from (7.4.1), (7.4.2) and (7.4.4) that

$$\sup_{N\in\mathbb{N}}\int_{\frac{\pi}{2^{N+1}}}^{\pi}\left|\sum_{k=1}^{\infty}\frac{b_kS_k(t)}{t}\right|dt\le 2\pi\sum_{k=1}^{\infty}\sqrt{\sum_{j=k}^{\infty}b_j^2}<\infty;$$

hence the conclusion follows from Theorem 3.7.1.

($\Longrightarrow$) Conversely, suppose that there exists $h\in L^1([0,\pi])$ such that $h(t)=\sum_{k=1}^{\infty}\frac{b_kS_k(t)}{t}$ for μ_1-almost all $t\in[0,\pi]$. Since $\sum_{k=1}^{\infty}|\,b_k\,|$ is assumed to be convergent, it suffices to prove that

$$\sup_{N\in\mathbb{N}}\sum_{k=1}^{N}\sqrt{\sum_{j=k}^{\infty}b_j^2}\le 7\left\{\int_0^{\pi}|\,h\,|\;d\mu_1+\pi\sum_{k=1}^{\infty}|\,b_k\,|\right\}. \tag{7.4.5}$$

But (7.4.5) is a consequence of (7.4.3), Lemma 7.4.3 and (7.4.2):

$$\begin{aligned}
&\sum_{k=1}^{N}\sqrt{\sum_{j=k}^{\infty}b_j^2}\\
&\le\sqrt{\frac{48}{\pi^2}}\sum_{k=0}^{N}\int_{\frac{\pi}{2}}^{\pi}\left|\sum_{j=k+1}^{\infty}b_jS_{j-k}(t)\right|dt\\
&=\sqrt{\frac{48}{\pi^2}}\sum_{k=0}^{N}\int_{\frac{\pi}{2^{k+1}}}^{\frac{\pi}{2^k}}2^k\left|\sum_{j=k+1}^{\infty}b_jS_j(t)\right|dt\\
&\le 7\sum_{k=0}^{N}\int_{\frac{\pi}{2^{k+1}}}^{\frac{\pi}{2^k}}\left|\sum_{j=k+1}^{\infty}\frac{b_jS_j(t)}{t}\right|dt\\
&\le 7\left\{\sum_{k=0}^{N}\int_{\frac{\pi}{2^{k+1}}}^{\frac{\pi}{2^k}}\left|\sum_{j=1}^{\infty}\frac{b_jS_j(t)}{t}\right|dt+\sum_{k=0}^{N}\int_{\frac{\pi}{2^{k+1}}}^{\frac{\pi}{2^k}}\left|\sum_{j=1}^{k}\frac{b_jS_j(t)}{t}\right|dt\right\}\\
&\le 7\left\{\int_0^{\pi}|\,h\,|\;d\mu_1+\pi\sum_{k=1}^{\infty}|\,b_k\,|\right\}.
\end{aligned}$$

□

The following example follows from Theorem 7.4.4.

Example 7.4.5. $\lim_{\delta\to 0^+}\int_{\delta}^{\pi}\frac{1}{t}\left|\sum_{k=1}^{\infty}\frac{\sin 2^k t}{k^{\frac{3}{2}}}\right|\;dt=\infty.$

We are now really to state and prove an integrability theorem for single cosine series.

Theorem 7.4.6. *Let $(a_k)_{k=0}^{\infty}$ be a sequence of real numbers such that $a_k - a_{k+1} = 0$ for all $k \in \mathbb{N}\backslash\{2^r : r \in \mathbb{N}\}$. If $\lim_{n\to\infty} a_n = 0$ and $\sum_{k=0}^{\infty} |a_k - a_{k+1}|$ converges, then $\frac{a_0}{2} + \sum_{k=1}^{\infty} a_k \cos kt$ is a Fourier series if and only if*

$$\sum_{k=1}^{\infty} \sqrt{\sum_{j=k}^{\infty} (a_{2^j} - a_{2^j+1})^2} < \infty.$$

Proof. Since Theorem 7.1.2(iii), (7.1.4) and our hypotheses yield

$$\sup_{t\in(0,\pi)} \left| \sum_{k=0}^{\infty} \lambda_k a_k \cos kt - \sum_{k=0}^{\infty} (a_{2^k} - a_{2^k+1}) \frac{\sin 2^k t}{2\tan\frac{t}{2}} \right| \leq \sum_{k=0}^{\infty} |\, a_k - a_{k+1}\,| < \infty,$$

the result follows from Theorem 7.4.4. □

Let f be given as in Theorem 7.3.7. The following example shows that f need not belong to $L^1(\mathbb{T})$.

Example 7.4.7. Let

$$a_0 = a_1 = \sum_{j=2}^{\infty} \frac{\chi_{\{2^r : r\in\mathbb{N}\}}(j)}{(\ln j)^{\frac{3}{2}}} \quad \text{and } a_k = \sum_{j=k}^{\infty} \frac{\chi_{\{2^r : r\in\mathbb{N}\}}(j)}{(\ln j)^{\frac{3}{2}}} \text{ for } k \in \mathbb{N}\backslash\{1\}.$$

Then $\frac{a_0}{2} + \sum_{k=1}^{\infty} a_k \cos kt$ is not the Fourier series of a function in $L^1(\mathbb{T})$.

Proof. This follows from Theorem 7.4.6. □

Our next aim is to prove an analogous version of Theorem 7.4.6 for single sine series. We need some lemmas.

Lemma 7.4.8. *The following statements are true.*

(i) *If $j, k, \ell \in \mathbb{N}$ with $j \leq k < \ell$ or $j < k \leq \ell$, then*

$$\int_{\frac{\pi}{2}}^{\pi} \cos 2^j t \cos^2 2^k t \cos 2^\ell t \; dt = \int_{\frac{\pi}{2}}^{\pi} \cos 2^j t \cos 2^k t \cos^2 2^\ell t \; dt = 0.$$

(ii) *If $j, k, \ell \in \mathbb{N}$ with $\ell > k+1 > j+1$ or $\ell > k > j+1$, then*

$$\int_{\frac{\pi}{2}}^{\pi} \cos^2 2^j t \cos 2^k t \cos 2^\ell t \; dt = 0.$$

(iii) *If $j \in \mathbb{N}$, then*

$$\int_{\frac{\pi}{2}}^{\pi} \cos^2 2^j t \cos 2^{j+1} t \cos 2^{j+2} t \; dt = \frac{\pi}{16}.$$

(iv) *If $j, k, \ell, p \in \mathbb{N}$ with $j < k < \ell < p$, then*

$$\int_{\frac{\pi}{2}}^{\pi} \cos 2^j t \cos 2^k t \cos 2^\ell t \cos 2^p t \, dt = 0.$$

Proof. Exercise. □

Lemma 7.4.9. *If $a_1, \ldots, a_n$ are real numbers, then*

$$\int_{\frac{\pi}{2}}^{\pi} \left(\sum_{k=1}^{n} a_k (1 - \cos 2^k t) \right)^2 dt = \frac{\pi}{2} \left(\sum_{k=1}^{n} a_k \right)^2 + \frac{\pi}{4} \sum_{k=1}^{n} a_k^2.$$

Proof. Exercise. □

Lemma 7.4.10. *If $a_1, \ldots, a_n$ are real numbers, then*

$$\int_{\frac{\pi}{2}}^{\pi} \left(\sum_{k=1}^{n} a_k (1 - \cos 2^k t) \right)^4 dt \le 8\pi \left(\left(\sum_{k=1}^{n} a_k \right)^2 + \sum_{k=1}^{n} a_k^2 \right)^2.$$

Proof. If $n \in \{1, 2\}$, then the result holds. Henceforth, we assume that $n \ge 3$. Since Lemma 7.4.8 holds, we can follow the proof of Lemma 7.4.2 to conclude that

$$\int_{\frac{\pi}{2}}^{\pi} \left(\sum_{k=1}^{n} a_k \cos 2^k t \right)^4 dt \le \frac{3\pi}{4} \left(\sum_{k=1}^{n} a_k^2 \right)^2;$$

therefore Minkowski inequality yields the desired result:

$$\begin{aligned}
&\left(\int_{\frac{\pi}{2}}^{\pi} \left(\sum_{k=1}^{n} a_k (1 - \cos 2^k t) \right)^4 dt \right)^{\frac{1}{4}} \\
&\le \left(\int_{\frac{\pi}{2}}^{\pi} \left(\sum_{k=1}^{n} a_k \right)^4 dt \right)^{\frac{1}{4}} + \left(\int_{\frac{\pi}{2}}^{\pi} \left(\sum_{k=1}^{n} a_k \cos 2^k t \right)^4 dt \right)^{\frac{1}{4}} \\
&\le \left(\frac{\pi}{2} \right)^{\frac{1}{4}} \left| \sum_{k=1}^{n} a_k \right| + \left(\frac{3\pi}{4} \right)^{\frac{1}{4}} \left(\sum_{k=1}^{n} a_k^2 \right)^{\frac{1}{2}} \\
&\le \left(\left(\frac{\pi}{2} \right)^{\frac{1}{4}} + \left(\frac{3\pi}{4} \right)^{\frac{1}{4}} \right) \left(\left(\sum_{k=1}^{n} a_k \right)^2 + \sum_{k=1}^{n} a_k^2 \right)^{\frac{1}{2}} \\
&\le 2^{\frac{3}{4}} \pi^{\frac{1}{4}} \left(\left(\sum_{k=1}^{n} a_k \right)^2 + \sum_{k=1}^{n} a_k^2 \right)^{\frac{1}{2}}.
\end{aligned}$$

□

The proof of the following lemma is similar to that of Lemma 7.4.3.

Lemma 7.4.11. *If $a_1, \ldots, a_n$ are real numbers, then*

$$\left(\int_{\frac{\pi}{2}}^{\pi} \left| \sum_{k=1}^{n} a_k(1-\cos 2^k t) \right| dt\right)^2 \geq \frac{\pi}{128} \int_{\frac{\pi}{2}}^{\pi} \left(\sum_{k=1}^{n} a_k(1-\cos 2^k t) \right)^2 dt.$$

Proof. For each $n \in \mathbb{N}$ and $t \in [\frac{\pi}{2}, \pi]$, let $f_n(t) = \sum_{k=1}^{n} a_k(1-\cos 2^k t)$. Then, using Hölder's inequality (with $p = \frac{3}{2}$ and $q = 3$), Lemmas 7.4.10 and 7.4.9, we obtain

$$\begin{aligned}
\int_{\frac{\pi}{2}}^{\pi} (f_n(t))^2\, dt &\leq \left(\int_{\frac{\pi}{2}}^{\pi} |f_n(t)|\, dt\right)^{\frac{2}{3}} \left(\int_{\frac{\pi}{2}}^{\pi} (f_n(t))^4\, dt\right)^{\frac{1}{3}} \\
&\leq \left(\int_{\frac{\pi}{2}}^{\pi} |f_n(t)|\, dt\right)^{\frac{2}{3}} \left(8\pi\left\{\left(\sum_{k=1}^{n} a_k\right)^2 + \sum_{k=1}^{n} a_k^2\right\}^2\right)^{\frac{1}{3}} \\
&\leq \left(\int_{\frac{\pi}{2}}^{\pi} |f_n(t)|\, dt\right)^{\frac{2}{3}} \left(8\pi\left\{\int_{\frac{\pi}{2}}^{\pi} \frac{4}{\pi}(f_n(t))^2\, dt\right\}^2\right)^{\frac{1}{3}},
\end{aligned}$$

and the result follows. □

The following theorem involves absolutely convergent cosine series.

Theorem 7.4.12. *Let $(a_k)_{k=1}^{\infty}$ be a sequence of real numbers such that $\sum_{k=1}^{\infty} |a_k|$ converges. Then there exists $\psi \in L^1([0,\pi])$ such that $\psi(t) = \frac{1}{t}\sum_{k=1}^{\infty} a_k(1-\cos 2^k t)$ for μ_1-almost all $t \in [0,\pi]$ if and only if*

$$\sum_{k=0}^{\infty} \sqrt{\left(\sum_{j=k+1}^{\infty} a_j\right)^2 + \sum_{j=k+1}^{\infty} a_j^2} < \infty.$$

Proof. ($\Longleftarrow$) For each $j \in \mathbb{N}$ and $x \in [0,\pi]$, let $C_j(x) = 1 - \cos 2^j x$. For any $N \in \mathbb{N}$, we have

$$\begin{aligned}
&\int_{\frac{\pi}{2^{N+1}}}^{\pi} \left| \sum_{j=1}^{\infty} \frac{a_j C_j(t)}{t} \right| dt \\
&\leq \sum_{k=0}^{N} \int_{\frac{\pi}{2^{k+1}}}^{\frac{\pi}{2^k}} \left| \sum_{j=1}^{k} \frac{a_j C_j(t)}{t} \right| dt + \sum_{k=0}^{N} \int_{\frac{\pi}{2^{k+1}}}^{\frac{\pi}{2^k}} \left| \sum_{j=k+1}^{\infty} \frac{a_j C_j(t)}{t} \right| dt. \quad (7.4.6)
\end{aligned}$$

The first term on the right-hand side of (7.4.6) is bounded above by $\pi \sum_{j=1}^{\infty} |a_j|$:

$$\sum_{k=0}^{N} \int_{\frac{\pi}{2^{k+1}}}^{\frac{\pi}{2^k}} \left| \sum_{j=1}^{k} \frac{a_j C_j(t)}{t} \right| dt \leq \pi \sum_{k=1}^{\infty} \sum_{j=1}^{k} \left| \frac{a_j\, 2^j}{2^{k+1}} \right| = \pi \sum_{j=1}^{\infty} |a_j|. \quad (7.4.7)$$

The second term on the right-hand side of (7.4.6) is bounded above by $\frac{\pi}{2}\sum_{k=0}^{\infty}\sqrt{\left(\sum_{j=k+1}^{\infty}a_j\right)^2+\sum_{j=k+1}^{\infty}a_j^2}$:

$$\begin{aligned}\sum_{k=0}^{N}\int_{\frac{\pi}{2^{k+1}}}^{\frac{\pi}{2^k}}\left|\sum_{j=k+1}^{\infty}\frac{a_jC_j(t)}{t}\right|dt &= \sum_{k=0}^{N}\int_{\frac{\pi}{2}}^{\pi}\left|\sum_{j=k+1}^{\infty}\frac{a_jC_{j-k}(t)}{t}\right|dt\\ &\le \sum_{k=0}^{N}\sqrt{\int_{\frac{\pi}{2}}^{\pi}\left|\sum_{j=k+1}^{\infty}a_jC_{j-k}(t)\right|^2dt}\\ &\le \frac{\pi}{2}\sum_{k=0}^{\infty}\sqrt{\left(\sum_{j=k+1}^{\infty}a_j\right)^2+\sum_{j=k+1}^{\infty}a_j^2}.\end{aligned}$$

As N is arbitrary, we deduce from the above inequalities that

$$\sup_{N\in\mathbb{N}}\int_{\frac{\pi}{2^{N+1}}}^{\pi}\left|\sum_{k=1}^{\infty}\frac{a_kC_k(t)}{t}\right|dt \le 2\pi\sum_{k=0}^{\infty}\sqrt{\left(\sum_{j=k+1}^{\infty}a_j\right)^2+\sum_{j=k+1}^{\infty}a_j^2}; \quad (7.4.8)$$

hence the conclusion follows from Theorem 3.7.1.

($\Longrightarrow$) Conversely, suppose that there exists $\psi \in L^1([0,\pi])$ such that $\psi(t) = \sum_{k=1}^{\infty}\frac{a_kC_k(t)}{t}$ for μ_1-almost all $t \in [0,\pi]$. Since $\sum_{k=1}^{\infty}|a_k|$ is assumed to be convergent, it suffices to prove that

$$\sup_{N\in\mathbb{N}}\sum_{k=0}^{N}\sqrt{\left(\sum_{j=k+1}^{\infty}a_j\right)^2+\sum_{j=k+1}^{\infty}a_j^2} \le 8\left\{\int_0^{\pi}|\psi|\,d\mu_1+\pi\sum_{k=1}^{\infty}|a_k|\right\}. \quad (7.4.9)$$

But (7.4.9) follows from Lemmas 7.4.9 and 7.4.11, and (7.4.7):

$$\begin{aligned}\sum_{k=0}^{N}\sqrt{\left(\sum_{j=k+1}^{\infty}a_j\right)^2+\sum_{j=k+1}^{\infty}a_j^2} &\le \frac{\sqrt{512}}{\pi}\sum_{k=0}^{N}\int_{\frac{\pi}{2}}^{\pi}\left|\sum_{j=k+1}^{\infty}a_jC_{j-k}(t)\right|dt\\ &\le \frac{23}{\pi}\sum_{k=0}^{N}\int_{\frac{\pi}{2^{k+1}}}^{\frac{\pi}{2^k}}2^k\left|\sum_{j=k+1}^{\infty}a_jC_j(t)\right|dt\\ &\le 8\sum_{k=0}^{N}\int_{\frac{\pi}{2^{k+1}}}^{\frac{\pi}{2^k}}\left|\sum_{j=k+1}^{\infty}\frac{a_jC_j(t)}{t}\right|dt\\ &\le 8\left\{\int_0^{\pi}|\psi|\,d\mu_1+\pi\sum_{k=1}^{\infty}|a_k|\right\}.\end{aligned}$$

□

The following example follows from Theorem 7.4.12.

Example 7.4.13. $\lim_{\delta\to 0+}\int_\delta^\pi \frac{1}{t}\left|\sum_{k=1}^\infty \frac{(-1)^k}{k^{\frac{3}{2}}}(1-\cos 2^k t)\right| dt = \infty$.

We are now ready to state and prove an analogous version of Theorem 7.4.6 for single sine series.

Theorem 7.4.14. *Let $(b_k)_{k=1}^\infty$ be a sequence of real numbers such that $b_k - b_{k+1} = 0$ for all $k \in \mathbb{N}\backslash\{2^r : r \in \mathbb{N}\}$. If $\sum_{k=1}^\infty |b_k - b_{k+1}|$ converges and $\lim_{n\to\infty} b_n = 0$, then $\sum_{k=1}^\infty b_k \sin kt$ is a Fourier series if and only if*

$$\sum_{k=0}^\infty \sqrt{\sum_{j=k+1}^\infty (b_{2^j} - b_{2^j+1})^2 + \left(\sum_{j=k+1}^\infty (b_{2^j} - b_{2^j+1})\right)^2} < \infty.$$

Proof. Since Theorem 7.1.2(iii), (7.1.3) and our hypotheses yield

$$\sup_{t\in(0,\pi)}\left|\sum_{k=1}^\infty b_k \sin kt - \sum_{k=1}^\infty (b_{2^k} - b_{2^k+1})\frac{1-\cos 2^k t}{2\tan\frac{t}{2}}\right| \le \sum_{k=1}^\infty |\, b_k - b_{k+1}\,| < \infty,$$

the result follows from Theorem 7.4.12. □

Example 7.4.15. Let $b_1 = 0$, let

$$b_k = \sum_{j=k}^\infty (-1)^{\frac{\ln j}{\ln 2}} \frac{\chi_{\{2^r : r\in\mathbb{N}\}}(j)}{(\ln j)^{\frac{3}{2}}} \quad \text{for} \quad k \in \mathbb{N}\backslash\{1\},$$

and let $B_j = b_{2^j} - b_{2^j+1}$ for $j = 1, 2, \ldots$. Since $\sum_{k=1}^\infty |b_k - b_{k+1}|$ converges, $\lim_{n\to\infty} b_n = 0$, and

$$\sum_{k=0}^\infty \sqrt{\sum_{j=k+1}^\infty B_j^2 + \left(\sum_{j=k+1}^\infty B_j\right)^2} \ge \sum_{k=0}^\infty \sqrt{\sum_{j=k+1}^\infty \frac{1}{j^3}} = \infty,$$

we infer from Theorem 7.4.14 that $\sum_{k=1}^\infty b_k \sin kt$ is not the Fourier series of a function in $L^1(\mathbb{T})$.

7.5 Boas' results

Let $(b_k)_{k=1}^\infty$ be a sequence of real numbers such that $\lim_{n\to\infty} b_n = 0$ and $\sum_{k=1}^\infty |b_k - b_{k+1}|$ converges. According to Examples 7.2.4, 7.2.5 and 7.4.15, $\sum_{k=1}^\infty b_k \sin kt$ need not be the Fourier series of a function in $L^1(\mathbb{T})$. In 1951, Boas [7] proved that $\lim_{\delta\to 0+}\int_\delta^\pi \sum_{k=1}^\infty b_k \sin kt\, dt$ exists if and only if $\sum_{k=1}^\infty \frac{b_k}{k}$ converges. The next theorem generalizes this particular result.

Theorem 7.5.1. *Let $(c_k)_{k=1}^{\infty}$ be a sequence of real numbers such that $\lim_{n\to\infty} c_n = 0$ and $\sum_{k=1}^{\infty} |\, c_k - c_{k+1} \,|$ converges. If $(\Phi_n)_{n=1}^{\infty}$ is a sequence in $C[a,b]$ satisfying*

$$\left\{ \sup_{n\in\mathbb{N}} ||\Phi_n||_{C[a,b]} + \sup_{a<x<b} \sup_{n\in\mathbb{N}} \left| \frac{\Phi_n(x)-\Phi_n(a)}{n(x-a)} \right| + \sup_{x\in[a,b]} \sup_{n\in\mathbb{N}} \left| \sum_{k=1}^{n} (x-a)\Phi_k(x) \right| \right\} < \infty,$$

then the following assertions hold.

(i) *If $x \in (a,b]$, then $\sum_{k=1}^{\infty} \frac{c_k\Phi_k(x)}{k}$ converges.*

(ii) $\lim_{\delta\to 0+} \left\{ \sum_{k=1}^{\lfloor \frac{1}{\delta} \rfloor} \frac{c_k\Phi_k(a)}{k} - \sum_{k=1}^{\infty} \frac{c_k\Phi_k(a+\delta)}{k} \right\} = 0.$

Proof. Assertion (i) is a consequence of Theorem 7.1.2(iii). To prove assertion (ii) we may further assume that the sequence $(c_k)_{k=1}^{\infty}$ is decreasing and

$$\left\{ \sup_{n\in\mathbb{N}} ||\Phi_n||_{C[a,b]} + \sup_{a<x<b} \sup_{n\in\mathbb{N}} \left| \frac{\Phi_n(x)-\Phi_n(a)}{n(x-a)} \right| + \sup_{x\in[a,b]} \sup_{n\in\mathbb{N}} \left| \sum_{k=1}^{n} (x-a)\Phi_k(x) \right| \right\} \le \frac{1}{2}. \tag{7.5.1}$$

Since $\lim_{n\to\infty} c_n = 0$, it remains to prove that

$$\left| \sum_{k=1}^{\lfloor \frac{1}{\delta} \rfloor} \frac{c_k\Phi_k(a)}{k} - \sum_{k=1}^{\infty} \frac{c_k\Phi_k(a+\delta)}{k} \right| \le \frac{1}{\lfloor \frac{1}{\delta} \rfloor} \sum_{k=1}^{\lfloor \frac{1}{\delta} \rfloor} c_k + 2c_{\lfloor \frac{1}{\delta} \rfloor} \tag{7.5.2}$$

for all $\delta \in (0,1)$. But (7.5.2) is a consequence of (7.5.1) and Theorem 7.1.2(iii):

$$\begin{aligned}
&\left| \sum_{k=1}^{\lfloor \frac{1}{\delta} \rfloor} \frac{c_k\Phi_k(a)}{k} - \sum_{k=1}^{\infty} \frac{c_k\Phi_k(a+\delta)}{k} \right| \\
&\le \sum_{k=1}^{\lfloor \frac{1}{\delta} \rfloor} c_k \left| \frac{\Phi_k(a+\delta)-\Phi_k(a)}{k} \right| + \left| \sum_{k=\lfloor \frac{1}{\delta} \rfloor+1}^{\infty} \frac{c_k\Phi_k(a+\delta)}{k} \right| \\
&\le \frac{1}{\lfloor \frac{1}{\delta} \rfloor} \sum_{k=1}^{\lfloor \frac{1}{\delta} \rfloor} c_k + \frac{1}{\lfloor \frac{1}{\delta} \rfloor} c_{\lfloor \frac{1}{\delta} \rfloor} \sup_{n\ge \lfloor \frac{1}{\delta} \rfloor+1} \left| \sum_{k=\lfloor \frac{1}{\delta} \rfloor+1}^{n} \Phi_k(a+\delta) \right| \\
&\le \frac{1}{\lfloor \frac{1}{\delta} \rfloor} \sum_{k=1}^{\lfloor \frac{1}{\delta} \rfloor} c_k + 2c_{\lfloor \frac{1}{\delta} \rfloor}.
\end{aligned}$$

□

The following results of Boas are immediate consequences of Theorem 7.5.1.

Theorem 7.5.2. *Let $(b_k)_{k=1}^{\infty}$ be a sequence of real numbers such that $\lim_{n\to\infty} b_n = 0$ and $\sum_{k=1}^{\infty} |b_k - b_{k+1}|$ converges. Then $\sum_{k=1}^{\infty} \frac{b_k}{k}$ converges if and only if $\lim_{x\to 0^+} \sum_{k=1}^{\infty} \frac{b_k \cos kx}{k}$ exists.*

Proof. Applying Theorem 7.5.1 with $[a, b] = [0, \pi]$, $c_k = b_k$ and $\Phi_k(t) = \cos kt$, we get the result. □

Theorem 7.5.3. *Let $(b_k)_{k=1}^{\infty}$ be a sequence of real numbers such that $\lim_{n\to\infty} b_n = 0$ and $\sum_{k=1}^{\infty} | b_k - b_{k+1} |$ converges. Then $\sum_{k=1}^{\infty} \frac{b_k}{k}$ converges if and only if $\lim_{\delta\to 0^+} \int_{\delta}^{\pi} \sum_{k=1}^{\infty} b_k \sin kt \, dt$ exists.*

Proof. Since $\lim_{n\to\infty} b_n = 0$ and $\sum_{k=1}^{\infty} |b_k - b_{k+1}|$ converges, it follows from Theorem 7.1.2 that the series $\sum_{k=1}^{\infty} \frac{(-1)^{k-1} b_k}{k}$ converges. Hence, by Theorem 7.1.9, it is easy to check that $\lim_{\delta\to 0^+} \int_{\delta}^{\pi} \sum_{k=1}^{\infty} b_k \sin kt \, dt$ exists if and only if $\lim_{\delta\to 0^+} \sum_{k=1}^{\infty} \frac{b_k \cos k\delta}{k}$ exists. An application of Theorem 7.5.2 yields the desired result. □

Theorem 7.5.4. *Let $(a_k)_{k=0}^{\infty}$ be a sequence of real numbers such that $\lim_{n\to\infty} a_n = 0$ and $\sum_{k=0}^{\infty} | a_k - a_{k+1} |$ converges. Then*

$$\lim_{\delta\to 0^+} \int_{\delta}^{\pi} \left\{ \frac{a_0}{2} + \sum_{k=1}^{\infty} a_k \cos kt \right\} dt \quad \textit{exists.}$$

Proof. By Theorem 7.1.9, $\lim_{\delta\to 0^+} \int_{\delta}^{\pi} \sum_{k=1}^{\infty} a_k \cos kt \, dt$ exists if and only if $\lim_{\delta\to 0^+} \sum_{k=1}^{\infty} \frac{a_k \sin k\delta}{k}$ exists. Now we apply Theorem 7.5.1 with $[a, b] = [0, \pi]$, $c_k = a_k$ and $\Phi_k(t) = \sin kt$ to conclude that $\lim_{\delta\to 0^+} \sum_{k=1}^{\infty} \frac{a_k \sin k\delta}{k}$ exists. □

Theorem 7.5.5. *If $\sum_{k=1}^{\infty} b_k$ is an absolutely convergent series of real numbers, then $\lim_{\delta\to 0^+} \int_{\delta}^{\pi} \frac{1}{t} \sum_{k=1}^{\infty} b_k \sin kt \, dt$ exists.*

Proof. This is an immediate consequence of Theorem 7.5.4, since $\sum_{k=1}^{\infty} \left| \sum_{j=k}^{\infty} b_j - \sum_{j=k+1}^{\infty} b_j \right|$ converges, $\lim_{n\to\infty} \sum_{j=n}^{\infty} b_j = 0$, and

$$\sup_{t\in(0,\pi)} \left| \sum_{k=1}^{\infty} \frac{b_k \sin kt}{t} - \sum_{k=1}^{\infty} \left(\sum_{j=k}^{\infty} b_j \right) \cos kt \right| \le 2 \sum_{k=1}^{\infty} | b_k | < \infty.$$

□

The following example sharpens Example 7.4.5.

Example 7.5.6. There exists a function $h \in HK[0, \pi] \backslash L^1[0, \pi]$ such that $h(t) = \frac{1}{t} \sum_{k=1}^{\infty} k^{-\frac{3}{2}} \sin 2^k t$ for all $t \in (0, \pi]$.

Proof. This follows from Theorem 7.5.5 and Example 7.4.5. □

The following example is a consequence of Theorem 7.5.5 and Example 7.4.7.

Example 7.5.7. Let $(a_k)_{k=0}^{\infty}$ be given as in Example 7.4.7. Then there exists $f \in HK[0,\pi]\backslash L^1[0,\pi]$ such that $f(t) = \frac{a_0}{2} + \sum_{k=1}^{\infty} a_k \cos kt$ for all $t \in (0,\pi]$.

The following example is a consequence of Theorem 7.5.3 and Example 7.4.15.

Example 7.5.8. Let $(b_k)_{k=1}^{\infty}$ be given as in Example 7.4.15. Then there exists $g \in HK[0,\pi]\backslash L^1[0,\pi]$ such that $g(t) = \sum_{k=1}^{\infty} b_k \sin kt$ for all $t \in [0,\pi]$.

7.6 On a result of Hardy and Littlewood concerning Fourier series

Let $f \in L^1(\mathbb{T})$ and suppose that

$$f(t) \sim \frac{a_0}{2} + \sum_{k=1}^{\infty}(a_k \cos kt + b_k \sin kt).$$

According to Example 7.3.9, the series $\sum_{k=1}^{\infty} \frac{a_k}{k}$ need not be convergent. In 1924, Hardy and Littlewood proved the following result.

Lemma 7.6.1. *Let $f \in L^1(\mathbb{T})$ and suppose that*

$$f(t) \sim \frac{a_0}{2} + \sum_{k=1}^{\infty}(a_k \cos kt + b_k \sin kt).$$

Then the series $\sum_{k=1}^{\infty} \frac{a_k}{k}$ converges if and only if

$$\lim_{\varepsilon\to 0+} \int_{\varepsilon}^{\pi} \left\{ \frac{1}{x} \int_{-x}^{x} f \, d\mu_1 \right\} dx$$

exists. In this case,

$$\lim_{\varepsilon\to 0+} \frac{1}{\pi} \int_{\varepsilon}^{\pi} \left\{ \left(\cot \frac{x}{2} \right) \int_{-x}^{x} f \, d\mu_1 \right\} dx = 2a_0 \ln 2 + 2\sum_{k=1}^{\infty} \frac{a_k}{k}.$$

Proof. According to Theorem 7.2.3,

$$\lim_{n\to\infty}\max_{x\in[0,\pi]}\left|\sum_{k=1}^{n}\frac{2a_k}{k}\sin kx-\int_{-x}^{x}\left(f-\frac{a_0}{2}\right)d\mu_1\right|=0. \tag{7.6.1}$$

Since the function $x \mapsto \frac{1}{x}-\frac{1}{2}\cot\frac{x}{2}$ is bounded on $(0,\pi)$, we deduce from (7.6.1) that

$$\lim_{\delta\to 0+}\int_{\delta}^{\pi}\left\{\frac{1}{x}\left(\int_{-x}^{x}\left(f-\frac{a_0}{2}\right)d\mu_1\right)-\sum_{k=1}^{\infty}\frac{a_k\sin kx}{k\tan\frac{x}{2}}\right\}dx \text{ exists.} \tag{7.6.2}$$

We will next prove that

$$\lim_{\delta\to 0+}\left\{\int_{\delta}^{\pi}\sum_{k=1}^{\infty}\frac{a_k\sin kx}{k\tan\frac{x}{2}}\,dx-\sum_{k=1}^{\infty}\frac{a_k}{k^2}\int_{\delta}^{\pi}\frac{1-\cos kx}{1-\cos x}\,dx\right\}=0. \tag{7.6.3}$$

For each $\delta\in(0,1)$ we use (7.6.1) and integration by parts to conclude that

$$\begin{aligned}
&\int_{\delta}^{\pi}\sum_{k=1}^{\infty}\frac{a_k\sin kx}{k\tan\frac{x}{2}}\,dx\\
&=\sum_{k=1}^{\infty}\int_{\delta}^{\pi}\frac{a_k\sin kx}{k\tan\frac{x}{2}}\,dx\\
&=-\sum_{k=1}^{\infty}\frac{2a_k}{k^2\tan\frac{\delta}{2}}(1-\cos k\delta)+\sum_{k=1}^{\infty}\frac{a_k}{k^2}\int_{\delta}^{\pi}\frac{1-\cos kx}{1-\cos x}\,dx\\
&=-\frac{2}{\tan\frac{\delta}{2}}\sum_{k=1}^{\infty}\int_{0}^{\delta}\frac{a_k\sin kx}{k}\,dx+\sum_{k=1}^{\infty}\frac{a_k}{k^2}\int_{\delta}^{\pi}\frac{1-\cos kx}{1-\cos x}\,dx.
\end{aligned} \tag{7.6.4}$$

Since (7.6.1) implies

$$\lim_{\delta\to 0+}\frac{1}{\tan\frac{\delta}{2}}\sum_{k=1}^{\infty}\int_{0}^{\delta}\frac{a_k}{k}\sin kx\,dx=\lim_{\delta\to 0+}\frac{1}{\tan\frac{\delta}{2}}\int_{0}^{\delta}\sum_{k=1}^{\infty}\frac{a_k}{k}\sin kx\,dx=0,$$

(7.6.3) follows from (7.6.4).

We will next prove that

$$\begin{aligned}
&\text{the series }\sum_{k=1}^{\infty}\frac{a_k}{k}\text{ converges}\\
\iff\ &\lim_{\delta\to 0+}\sum_{k=1}^{\infty}\frac{a_k}{k^2}\int_{\delta}^{\pi}\frac{1-\cos kx}{1-\cos x}\,dx\text{ exists.}
\end{aligned} \tag{7.6.5}$$

For each $\delta \in (0,1)$ we have

$$\begin{aligned}
&\left| \pi \sum_{k=1}^{\lfloor \frac{1}{\delta} \rfloor} \frac{a_k}{k} - \sum_{k=1}^{\infty} \frac{a_k}{k^2} \int_\delta^\pi \frac{1-\cos kx}{1-\cos x}\, dx \right| \\
&\le \left| \sum_{k=1}^{\lfloor \frac{1}{\delta} \rfloor} \frac{a_k}{k^2} \int_0^\delta \frac{1-\cos kx}{1-\cos x}\, dx \right| + \left| \sum_{k=\lfloor \frac{1}{\delta} \rfloor+1}^{\infty} \frac{a_k}{k^2} \int_\delta^\pi \frac{1-\cos kx}{1-\cos x}\, dx \right| \\
&\le \frac{\pi^2}{2\lfloor \frac{1}{\delta} \rfloor} \sum_{k=1}^{\lfloor \frac{1}{\delta} \rfloor} |a_k| + \left\{ \sup_{k \ge \lfloor \frac{1}{\delta} \rfloor} |a_k| \right\} \sum_{k=\lfloor \frac{1}{\delta} \rfloor+1}^{\infty} \frac{\pi^2}{k^2 \delta} \\
&\le \frac{\pi^2}{2\lfloor \frac{1}{\delta} \rfloor} \sum_{k=1}^{\lfloor \frac{1}{\delta} \rfloor} |a_k| + \left\{ \sup_{k \ge \lfloor \frac{1}{\delta} \rfloor} |a_k| \right\} \frac{2\pi^2}{\delta(\lfloor \frac{1}{\delta} \rfloor + 1)} \\
&\le \frac{\pi^2}{2\lfloor \frac{1}{\delta} \rfloor} \sum_{k=1}^{\lfloor \frac{1}{\delta} \rfloor} |a_k| + 2\pi^2 \left\{ \sup_{k \ge \lfloor \frac{1}{\delta} \rfloor} |a_k| \right\}. \qquad (7.6.6)
\end{aligned}$$

Since $\delta \in (0,1)$ is arbitrary, (7.6.5) follows from (7.6.6) and the Riemann Lebesgue Lemma.

It is now clear that the first part of the lemma follows from (7.6.5), (7.6.3) and (7.6.2). To prove the second part of the lemma, we suppose that $\sum_{k=1}^{\infty} \frac{a_k}{k}$ converges. In this case, we infer from (7.6.1), (7.6.3), Lemma 7.1.6 (with x replaced by $\pi - x$), Lebesgue's Dominated Convergence Theorem and (7.6.6) that

$$\begin{aligned}
&\lim_{\delta \to 0+} \int_\delta^\pi \left\{ \frac{1}{\tan \frac{x}{2}} \int_{-x}^{x} f\, d\mu_1 \right\} dx \\
&= a_0 \int_0^\pi x \cot \frac{x}{2}\, dx + 2 \lim_{\delta \to 0+} \int_\delta^\pi \sum_{k=1}^{\infty} \frac{a_k \sin kx}{k \tan \frac{x}{2}}\, dx \\
&= a_0 \int_0^\pi x \cot \frac{x}{2}\, dx + 2 \lim_{\delta \to 0+} \sum_{k=1}^{\infty} \int_\delta^\pi \frac{a_k}{k^2} \frac{1-\cos kx}{1-\cos x}\, dx \\
&= a_0 \int_0^\pi \sum_{k=1}^{\infty} \frac{2(-1)^{k-1}}{k} \sin kx \cot \frac{x}{2}\, dx + 2 \lim_{\delta \to 0+} \sum_{k=1}^{\infty} \int_\delta^\pi \frac{a_k}{k^2} \frac{1-\cos kx}{1-\cos x}\, dx \\
&= 2a_0 \sum_{k=1}^{\infty} \int_0^\pi \frac{(-1)^{k-1}}{k} \sin kx \cot \frac{x}{2}\, dx + 2 \lim_{\delta \to 0+} \sum_{k=1}^{\infty} \int_\delta^\pi \frac{a_k}{k^2} \frac{1-\cos kx}{1-\cos x}\, dx \\
&= 2a_0 \pi \ln 2 + 2\pi \sum_{k=1}^{\infty} \frac{a_k}{k}.
\end{aligned}$$

□

7.7 Notes and Remarks

There are many excellent monographs on trigonometric series; consult, for instance, Boas [8], Thomson [153], and Zygmund [165]. The formulation and proof of Theorem 7.1.2 is based on [103, Theorem 2.3]. Theorem 7.2.3 is due to W.H. Young [157, 158]. A more general version of Example 7.2.5 can be found in Perron [142].

Theorem 7.3.1 is also a special case of [103, Theorem 2.3]. W.H. Young [158, 161] proved Theorem 7.3.4. Theorem 7.3.7 and Corollary 7.3.8 are also due to W.H. Young [161]. For other results concerning Lebesgue integrability of trigonometric series; see, for example, Fridli [43], Leindler [112], and Móricz [128].

Proofs of Lemmas 7.4.2 and 7.4.3 depend on that of [47, Theorem 3.7.4]. A more general version of Theorem 7.4.6 can be obtained; see [8, Theorem 5.2.7]. Example 7.4.5 is taken from [8, p.19], proofs of Lemmas 7.4.10 and 7.4.11 depend on that of [47, Theorem 3.7.4], and the rest of Section 7.4 is based on the paper [107].

Theorem 7.5.1 is due to Lee [101, Theorem 2.2]. Theorems 7.5.2, 7.5.3, 7.5.4, and 7.5.5 are due to Boas [7]. For other related results involving improper Riemann integrals; consult, for example, Edmonds [32–35], Heywood [60], and Boas [8].

The proof of Lemma 7.6.1 is similar to that of [108, Theorem 1.2]. For other related results, see Hardy and Rogosinski [52, 53].

Chapter 8

Some applications of the Henstock-Kurzweil integral to double trigonometric series

In Chapter 7 we apply the Henstock-Kurzweil integral to study certain classes of single trigonometric series. In this chapter we use the double Henstock-Kurzweil integral to study double trigonometric series. We begin with the study of a sufficiently well-behaved conditionally convergent double series.

8.1 Regularly convergent double series

Definition 8.1.1. Let $(u_{k_1,k_2})_{(k_1,k_2)\in\mathbb{N}_0^2}$ be a double sequence of real numbers. The double series $\sum_{(k_1,k_2)\in\mathbb{N}_0^2} u_{k_1,k_2}$ *converges in Pringsheim's sense* to a real number s if for each $\varepsilon > 0$ there exists $N(\varepsilon) \in \mathbb{N}_0$ such that

$$\left|\sum_{k_1=0}^{n_1}\sum_{k_2=0}^{n_2} u_{k_1,k_2} - s\right| < \varepsilon$$

for all $(n_1, n_2) \in \mathbb{N}_0^2$ satisfying $\min\{n_1, n_2\} \geq N(\varepsilon)$.

Example 8.1.2. Let

$$u_{k_1,k_2} = \begin{cases} (-1)^{k_1} k_2 & \text{if } (k_1,k_2) \in \{0,1\}\times\mathbb{N}_0, \\ 0 & \text{otherwise.} \end{cases}$$

Then $\sum_{(k_1,k_2)\in\mathbb{N}_0^2} u_{k_1,k_2}$ converges in Pringsheim's sense to 0.

Proof. Let $\varepsilon > 0$ be given. If $(n_1, n_2) \in \mathbb{N}^2$ with $\min\{n_1, n_2\} \geq 2$, then

$$\left|\sum_{k_1=0}^{n_1}\sum_{k_2=0}^{n_2} u_{k_1,k_2}\right| = \left|\sum_{k_1=0}^{1}\sum_{k_2=0}^{n_2} (-1)^{k_1} k_2\right| = 0 < \varepsilon.$$

Since $\varepsilon > 0$ is arbitrary, the given double series converges in Pringsheim's sense to 0. □

Theorem 8.1.3. *The double series $\sum_{(k_1,k_2)\in\mathbb{N}_0^2} u_{k_1,k_2}$ of real numbers converges in Pringsheim's sense if and only if for each $\varepsilon > 0$ there exists $N(\varepsilon) \in \mathbb{N}_0$ such that*

$$\left|\sum_{k_1=0}^{q_1}\sum_{k_2=0}^{q_2} u_{k_1,k_2} - \sum_{k_1=0}^{p_1}\sum_{k_2=0}^{p_2} u_{k_1,k_2}\right| < \varepsilon \tag{8.1.1}$$

for all $(p_1,p_2),(q_1,q_2) \in \mathbb{N}_0^2$ satisfying $q_i \geq p_i$ $(i = 1,2)$ and $\min\{p_1,p_2\} \geq N(\varepsilon)$.

Proof. $(\Longrightarrow)$ This is obvious.

$(\Longleftarrow)$ We infer from (8.1.1) that $\left(\sum_{k_1=0}^{n}\sum_{k_2=0}^{n} u_{k_1,k_2}\right)_{n=0}^{\infty}$ is a Cauchy sequence of real numbers; hence the completeness of $\mathbb{R}$ implies that $\left(\sum_{k_1=0}^{n}\sum_{k_2=0}^{n} u_{k_1,k_2}\right)_{n=0}^{\infty}$ converges to some $L \in \mathbb{R}$. Using (8.1.1) again, we conclude that the double series $\sum_{(k_1,k_2)\in\mathbb{N}_0^2} u_{k_1,k_2}$ converges in Pringsheim's sense to L. □

We observe that if a double series $\sum_{(k_1,k_2)\in\mathbb{N}_0^2} u_{k_1,k_2}$ of real numbers converges in Pringsheim's sense, then the double sequence $(u_{n_1,n_2})_{(n_1,n_2)\in\mathbb{N}_0^2}$ may not be bounded (cf. Example 8.1.2). In order to overcome this deficiency, we use the following definition due to Hardy [50].

Definition 8.1.4. A double series $\sum_{(k_1,k_2)\in\mathbb{N}_0^2} u_{k_1,k_2}$ of real numbers *converges regularly* if for each $\varepsilon > 0$ there exists $N(\varepsilon) \in \mathbb{N}_0$ such that

$$\left|\sum_{k_1=p_1}^{q_1}\sum_{k_2=p_2}^{q_2} u_{k_1,k_2}\right| < \varepsilon$$

for all $(p_1,p_2),(q_1,q_2) \in \mathbb{N}_0^2$ with $q_i \geq p_i$ $(i = 1,2)$ and $\max\{p_1,p_2\} \geq N(\varepsilon)$.

Example 8.1.5. The double series $\sum_{(k_1,k_2)\in\mathbb{N}_0^2} \frac{(-1)^{k_1+k_2}}{(k_1+k_2+1)^2}$ converges regularly.

Proof. For each $(p_1,p_2),(q_1,q_2) \in \mathbb{N}_0^2$ satisfying $q_i \geq p_i$ $(i = 1,2)$, we have

$$\begin{aligned}\left|\sum_{k_1=p_1}^{q_1}\sum_{k_2=p_2}^{q_2} \frac{(-1)^{k_1+k_2}}{(k_1+k_2+1)^2}\right| &\leq \sum_{k_1=p_1}^{q_1}\left|\sum_{k_2=p_2}^{q_2} \frac{(-1)^{k_2}}{(k_1+k_2+1)^2}\right| \\ &\leq \sum_{k_1=p_1}^{\infty} \frac{4}{(k_1+p_2+1)(k_1+p_2+2)} \\ &= \frac{4}{p_1+p_2+1}.\end{aligned}$$

It is now clear that the given double series converges regularly. □

The following theorem follows from Definition 8.1.4.

Theorem 8.1.6. *Every absolutely convergent double series of real numbers is regularly convergent.*

Remark 8.1.7. Since the double series $\sum_{(k_1,k_2)\in\mathbb{N}_0^2} \frac{1}{(k_1+k_2+1)^2}$ is not convergent in Pringsheim's sense, Example 8.1.5 shows that the converse of Theorem 8.1.6 need not be true.

Theorem 8.1.8. *If the double series $\sum_{(k_1,k_2)\in\mathbb{N}_0^2} u_{k_1,k_2}$ of real numbers converges regularly, then it converges in Pringsheim's sense.*

Proof. This is an easy consequence of Definition 8.1.4 and Theorem 8.1.3. □

Example 8.1.2 shows that the converse of Theorem 8.1.8 is not true in general. The next theorem gives a necessary and sufficient condition for a double series of real numbers to be regularly convergent.

Theorem 8.1.9. *The double series $\sum_{(k_1,k_2)\in\mathbb{N}_0^2} u_{k_1,k_2}$ of real numbers is regularly convergent if and only if*

(i) $\sum_{(k_1,k_2)\in\mathbb{N}_0^2} u_{k_1,k_2}$ *converges in Pringsheim's sense, and*
(ii) *all the single series* $\sum_{k_1=0}^{\infty} u_{k_1,k_2}$ $(k_2 = 0,1,\dots)$ *and* $\sum_{k_2=0}^{\infty} u_{k_1,k_2}$ $(k_1 = 0,1,\dots)$ *converge.*

Proof. $(\Longrightarrow)$ Suppose that the double series $\sum_{(k_1,k_2)\in\mathbb{N}_0^2} u_{k_1,k_2}$ is regularly convergent. An application of Theorem 8.1.8 gives conclusion (i). Conclusion (ii) follows from Definition 8.1.4.

$(\Longleftarrow)$ Let $\varepsilon > 0$ be given. According to (i), there exists an integer $N_0 \geq 2$ such that

$$\left|\sum_{k_1=t_1}^{w_1}\sum_{k_2=t_2}^{w_2} u_{k_1,k_2}\right| < \frac{\varepsilon}{2}$$

for every integers w_1, w_2, t_1, t_2 satisfying $w_i \geq t_i \geq N_0$ $(i = 1,2)$. Using (ii), there exists an integer $N \geq N_0$ such that

$$s_i \geq r_i \geq N \ (i=1,2)$$
$$\Longrightarrow \max_{k_1=0,\dots,N_0-1}\left|\sum_{k_2=r_2}^{s_2} u_{k_1,k_2}\right| + \max_{k_2=0,\dots,N_0-1}\left|\sum_{k_1=r_1}^{s_1} u_{k_1,k_2}\right| < \frac{\varepsilon}{2N_0}.$$

To this end, we let $(p_1, p_2), (q_1, q_2) \in \mathbb{N}_0^2$ such that $q_i \geq p_i$ $(i = 1, 2)$ and $\max\{p_1, p_2\} \geq N$. There are five cases to consider.

Case 1: $p_1 \leq q_1 \leq N_0$.

In this case, we have $p_2 \geq N \geq N_0$ and so

$$\left| \sum_{k_1=p_1}^{q_1} \sum_{k_2=p_2}^{q_2} u_{k_1,k_2} \right| < \varepsilon. \tag{8.1.2}$$

Case 2: $p_2 \leq q_2 \leq N_0$.

Following the proof of case 1, we have (8.1.2) too.

Case 3: $p_1 \leq N_0 < q_1$.

In this case, $p_2 \geq N \geq N_0$ and so

$$\left| \sum_{k_1=p_1}^{q_1} \sum_{k_2=p_2}^{q_2} u_{k_1,k_2} \right| \leq \left| \sum_{k_1=p_1}^{N_0} \sum_{k_2=p_2}^{q_2} u_{k_1,k_2} \right| + \left| \sum_{k_1=N_0+1}^{q_1} \sum_{k_2=p_2}^{q_2} u_{k_1,k_2} \right| < \varepsilon.$$

Case 4: $p_2 \leq N_0 < q_2$.

Arguing as in the proof of case 3, we have (8.1.2) too.

Case 5: $q_i \geq p_i \geq N_0$ $(i = 1, 2)$.

In this case, we have (8.1.2) too.

Combining the above cases, we conclude that (8.1.2) holds. □

The following corollary shows that Fubini's Theorem holds for regularly convergent double series.

Corollary 8.1.10. *If the double series* $\sum_{(k_1,k_2)\in\mathbb{N}_0^2} u_{k_1,k_2}$ *of real numbers converges regularly, then the iterated series* $\sum_{k_1=0}^{\infty}\left\{\sum_{k_2=0}^{\infty} u_{k_1,k_2}\right\}$, $\sum_{k_2=0}^{\infty}\left\{\sum_{k_1=0}^{\infty} u_{k_1,k_2}\right\}$ *converge and*

$$\sum_{(k_1,k_2)\in\mathbb{N}_0^2} u_{k_1,k_2} = \sum_{k_1=0}^{\infty}\left\{\sum_{k_2=0}^{\infty} u_{k_1,k_2}\right\} = \sum_{k_2=0}^{\infty}\left\{\sum_{k_1=0}^{\infty} u_{k_1,k_2}\right\}.$$

Proof. According to Theorem 8.1.9, the single series

$$\sum_{k_1=0}^{\infty} u_{k_1,k_2} \ (k_2 = 0, 1, \dots) \text{ and } \sum_{k_2=0}^{\infty} u_{k_1,k_2} \ (k_1 = 0, 1, \dots)$$

converge. To prove that the iterated series $\sum_{k_1=0}^{\infty}\left\{\sum_{k_2=0}^{\infty} u_{k_1,k_2}\right\}$ converges, it suffices to observe that the regular convergence of

$\sum_{(k_1,k_2)\in\mathbb{N}_0^2} u_{k_1,k_2}$ implies

$$\sum_{k_1=p}^{q}\sum_{k_2=0}^{\infty} u_{k_1,k_2} \to 0 \text{ as } p,q\to\infty.$$

A similar reasoning shows that the iterated series $\sum_{k_2=0}^{\infty}\left\{\sum_{k_1=0}^{\infty} u_{k_1,k_2}\right\}$ converges.

Now we prove that

$$\left|\sum_{k_1=0}^{n_1}\sum_{k_2=0}^{n_2} u_{k_1,k_2} - \sum_{k_1=0}^{\infty}\left\{\sum_{k_2=0}^{\infty} u_{k_1,k_2}\right\}\right| \to 0 \text{ as } \min\{n_1,n_2\}\to\infty.$$

But this is easy, since the regular convergence of $\sum_{(k_1,k_2)\in\mathbb{N}_0^2} u_{k_1,k_2}$ implies

$$\begin{aligned}
&\left|\sum_{k_1=0}^{n_1}\sum_{k_2=0}^{n_2} u_{k_1,k_2} - \sum_{k_1=0}^{\infty}\left\{\sum_{k_2=0}^{\infty} u_{k_1,k_2}\right\}\right| \\
&\le \left|\sum_{k_1=0}^{n_1}\sum_{k_2=0}^{n_2} u_{k_1,k_2} - \sum_{k_1=0}^{n_1}\left\{\sum_{k_2=0}^{\infty} u_{k_1,k_2}\right\}\right| \\
&+ \left|\sum_{k_1=0}^{n_1}\left\{\sum_{k_2=0}^{\infty} u_{k_1,k_2}\right\} - \sum_{k_1=0}^{\infty}\left\{\sum_{k_2=0}^{\infty} u_{k_1,k_2}\right\}\right| \\
&\le \left|\sum_{k_1=0}^{n_1}\sum_{k_2=n_2+1}^{\infty} u_{k_1,k_2}\right| + \left|\sum_{k_1=n_1+1}^{\infty}\sum_{k_2=0}^{\infty} u_{k_1,k_2}\right| \\
&\to 0 \text{ as } \min\{n_1,n_2\}\to\infty.
\end{aligned}$$

Since a similar reasoning shows that

$$\sum_{k_1=0}^{\infty}\left\{\sum_{k_2=0}^{\infty} u_{k_1,k_2}\right\} = \sum_{k_2=0}^{\infty}\left\{\sum_{k_1=0}^{\infty} u_{k_1,k_2}\right\},$$

the corollary follows. □

Our next aim is to establish a useful generalized Dirichlet test for regularly convergent double series of real numbers. For any double sequence $(v_{k_1,k_2})_{(k_1,k_2)\in\mathbb{N}_0^2}$ of real numbers, we write

$$\Delta_{\emptyset}(v_{k_1,k_2}) = v_{k_1,k_2},$$

$$\Delta_{\{1\}}(v_{k_1,k_2}) = v_{k_1,k_2} - v_{k_1+1,k_2}, \quad \Delta_{\{2\}}(v_{k_1,k_2}) = v_{k_1,k_2} - v_{k_1,k_2+1},$$

and

$$\begin{aligned}
\Delta_{\{1,2\}}(v_{k_1,k_2}) &= \Delta_{\{1\}}(\Delta_{\{2\}}(v_{k_1,k_2})) \\
&= v_{k_1,k_2} - v_{k_1,k_2+1} + v_{k_1+1,k_2+1} - v_{k_1+1,k_2}.
\end{aligned}$$

The following observation is a consequence of the single summation by parts (Theorem 7.1.1).

Observation 8.1.11. Let $(u_{k_1,k_2})_{(k_1,k_2)\in\mathbb{N}_0^2}$ and $(v_{k_1,k_2})_{(k_1,k_2)\in\mathbb{N}_0^2}$ be two double sequences of real numbers. Then

$$\begin{aligned}
&\sum_{k_1=0}^{n_1}\sum_{k_2=0}^{n_2} u_{k_1,k_2}v_{k_1,k_2}\\
&=\sum_{k_1=0}^{n_1}\left\{v_{k_1,n_2}\sum_{k_2=0}^{n_2}u_{k_1,k_2}+\sum_{k_2=0}^{n_2-1}\Delta_{\{2\}}(v_{k_1,k_2})\sum_{j_2=0}^{k_2}u_{k_1,j_2}\right\}\\
&=\sum_{k_1=0}^{n_1}v_{k_1,n_2}\sum_{k_2=0}^{n_2}u_{k_1,k_2}+\sum_{k_2=0}^{n_2-1}\sum_{k_1=0}^{n_1}\Delta_{\{2\}}(v_{k_1,k_2})\sum_{j_2=0}^{k_2}u_{k_1,j_2}\\
&=\left\{v_{n_1,n_2}\sum_{k_1=0}^{n_1}\sum_{k_2=0}^{n_2}u_{k_1,k_2}+\sum_{k_1=0}^{n_1-1}\Delta_{\{1\}}(v_{k_1,n_2})\sum_{j_1=0}^{k_1}\sum_{k_2=0}^{n_2}u_{j_1,k_2}\right\}\\
&\quad+\sum_{k_2=0}^{n_2-1}\left\{\Delta_{\{2\}}(v_{n_1,k_2})\sum_{k_1=0}^{n_1}\sum_{j_2=0}^{k_2}u_{k_1,j_2}+\sum_{k_1=0}^{n_1-1}\Delta_{\{1,2\}}(v_{k_1,k_2})\sum_{j_1=0}^{k_1}\sum_{j_2=0}^{k_2}u_{j_1,j_2}\right\}\\
&=\sum_{\Gamma\subseteq\{1,2\}}\ \sum_{\substack{(k_1,k_2)\in\mathbb{N}_0^2\\ 0\le k_i\le n_i-1\,\forall i\in\Gamma\\ k_\ell=n_\ell\,\forall \ell\in\{1,2\}\setminus\Gamma}}\left\{\Delta_\Gamma(v_{k_1,k_2})\sum_{j_1=0}^{k_1}\sum_{j_2=0}^{k_2}u_{j_1,j_2}\right\}.
\end{aligned}$$

The following theorem is an immediate consequence of Observation 8.1.11.

Theorem 8.1.12. *Let $(v_{k_1,k_2})_{(k_1,k_2)\in\mathbb{N}_0^2}$ be a double sequence of real numbers such that $\lim_{\max\{n_1,n_2\}\to\infty} v_{n_1,n_2}=0$ and $\sum_{(k_1,k_2)\in\mathbb{N}_0^2}\left|\Delta_{\{1,2\}}(v_{k_1,k_2})\right|$ converges. If $(u_{k_1,k_2})_{(k_1,k_2)\in\mathbb{N}_0^2}$ is a double sequence of real numbers and $(n_1,n_2)\in\mathbb{N}_0^2$, then*

$$\begin{aligned}
&\sum_{k_1=0}^{n_1}\sum_{k_2=0}^{n_2} u_{k_1,k_2}v_{k_1,k_2}\\
&=\sum_{\Gamma\subseteq\{1,2\}}\ \sum_{\substack{(k_1,k_2)\in\mathbb{N}_0^2\\ 0\le k_i\le n_{i-1}\,\forall i\in\Gamma\\ k_\ell\ge n_\ell\,\forall \ell\in\{1,2\}\setminus\Gamma}}\left\{\Delta_{\{1,2\}}(v_{k_1,k_2})\sum_{\substack{(j_1,j_2)\in\mathbb{N}_0^2\\ 0\le j_i\le k_i\,\forall i\in\Gamma\\ 0\le j_\ell\le n_\ell\,\forall \ell\in\{1,2\}\setminus\Gamma}}u_{j_1,j_2}\right\}.
\end{aligned}\tag{8.1.3}$$

The next theorem, which is an easy consequence of Theorem 8.1.12, is a proper generalization of the classical Dirichlet test [122, Theorem 3].

Theorem 8.1.13. *Let* $(u_{k_1,k_2})_{(k_1,k_2)\in\mathbb{N}_0^2}$ *and* $(v_{k_1,k_2})_{(k_1,k_2)\in\mathbb{N}_0^2}$ *be two double sequences of real numbers such that* $\lim_{\max\{n_1,n_2\}\to\infty} v_{n_1,n_2} = 0$ *and*

$$\sum_{(k_1,k_2)\in\mathbb{N}_0^2} \left|\Delta_{\{1,2\}}(v_{k_1,k_2})\right| \max_{\substack{r_1=0,\dots,k_1\\ r_2=0,\dots,k_2}} \left|\sum_{j_1=0}^{r_1}\sum_{j_2=0}^{r_2} u_{j_1,j_2}\right| < \infty. \qquad (8.1.4)$$

Then the following assertions hold.

(i) *The double series* $\sum_{(k_1,k_2)\in\mathbb{N}_0^2} \Delta_{\{1,2\}}(v_{k_1,k_2}) \sum_{j_1=0}^{k_1}\sum_{j_2=0}^{k_2} u_{j_1,j_2}$ *converges absolutely.*

(ii) *If* $(n_1,n_2)\in\mathbb{N}_0^2$, *then*

$$\left|\sum_{k_1=0}^{n_1}\sum_{k_2=0}^{n_2} u_{k_1,k_2}v_{k_1,k_2} - \sum_{(k_1,k_2)\in\mathbb{N}_0^2} \Delta_{\{1,2\}}(v_{k_1,k_2}) \sum_{j_1=0}^{k_1}\sum_{j_2=0}^{k_2} u_{j_1,j_2}\right|$$

$$\leq \sum_{\emptyset\neq\Gamma\subset\{1,2\}} \sum_{\substack{(k_1,k_2)\in\mathbb{N}_0^2\\ k_i\geq 0\,\forall i\in\Gamma\\ k_\ell\geq n_\ell\,\forall \ell\in\{1,2\}\setminus\Gamma}} 3\left|\Delta_{\{1,2\}}(v_{k_1,k_2})\right| \max_{\substack{r_1=0,\dots,k_1\\ r_2=0,\dots,k_2}} \left|\sum_{j_1=0}^{r_1}\sum_{j_2=0}^{r_2} u_{j_1,j_2}\right|.$$

(iii) *For any* $(p_1,p_2),(q_1,q_2)\in\mathbb{N}_0^2$ *satisfying* $q_i\geq p_i$ $(i=1,2)$, *we have*

$$\left|\sum_{k_1=p_1}^{q_1}\sum_{k_2=p_2}^{q_2} u_{k_1,k_2}v_{k_1,k_2}\right|$$

$$\leq 16\sum_{k_1=p_1}^{\infty}\sum_{k_2=p_2}^{\infty} \left|\Delta_{\{1,2\}}(v_{k_1,k_2})\right| \max_{\substack{r_1=0,\dots,k_1\\ r_2=0,\dots,k_2}} \left|\sum_{j_1=0}^{r_1}\sum_{j_2=0}^{r_2} u_{j_1,j_2}\right|. \qquad (8.1.5)$$

(iv) *The double series* $\sum_{(k_1,k_2)\in\mathbb{N}_0^2} u_{k_1,k_2}v_{k_1,k_2}$ *converges regularly and*

$$\sum_{(k_1,k_2)\in\mathbb{N}_0^2} u_{k_1,k_2}v_{k_1,k_2} = \sum_{(k_1,k_2)\in\mathbb{N}_0^2} \Delta_{\{1,2\}}(v_{k_1,k_2}) \sum_{j_1=0}^{k_1}\sum_{j_2=0}^{k_2} u_{j_1,j_2}. \qquad (8.1.6)$$

Proof. Assertion (i) follows from (8.1.4). To prove assertion (ii), we let $(n_1,n_2)\in\mathbb{N}_0^2$ be arbitrary and consider two cases.

Case 1: Suppose that

$$\sup_{(n_1,n_2)\in\mathbb{N}_0^2} \max_{\substack{r_1=0,\dots,n_1\\ r_2=0,\dots,n_2}} \left|\sum_{j_1=0}^{r_1}\sum_{j_2=0}^{r_2} u_{j_1,j_2}\right| = 0.$$

In this case, it is clear that assertion (ii) holds.

Case 2: Suppose that

$$\sup_{(n_1,n_2)\in\mathbb{N}_0^2} \max_{\substack{r_1=0,\dots,n_1\\ r_2=0,\dots,n_2}} \left|\sum_{j_1=0}^{r_1}\sum_{j_2=0}^{r_2} u_{j_1,j_2}\right| > 0.$$

To prove (ii), we write

$$L_0 = \sum_{(k_1,k_2)\in\mathbb{N}_0^2} \Delta_{\{1,2\}}(v_{k_1,k_2}) \sum_{j_1=0}^{k_1}\sum_{j_2=0}^{k_2} u_{j_1,j_2}$$

and

$$L_{\Gamma,n_1,n_2} = \sum_{\substack{(k_1,k_2)\in\mathbb{N}_0^2\\ 0\le k_i\le n_i-1\,\forall\, i\in\Gamma\\ k_\ell\ge n_\ell\,\forall\,\ell\in\{1,2\}\setminus\Gamma}} \Delta_{\{1,2\}}(v_{k_1,k_2}) \sum_{\substack{(j_1,j_2)\in\mathbb{N}_0^2\\ 0\le j_i\le k_i\,\forall\, i\in\Gamma\\ 0\le j_\ell\le n_\ell\,\forall\,\ell\in\{1,2\}\setminus\Gamma}} u_{j_1,j_2}$$

for every $\Gamma \subseteq \{1,2\}$ and $(n_1,n_2)\in\mathbb{N}_0^2$. According to (8.1.3),

$$\begin{aligned}
&\left|\sum_{k_1=0}^{n_1}\sum_{k_2=0}^{n_2} u_{k_1,k_2}v_{k_1,k_2} - L_0\right|\\
&= \left|\sum_{\Gamma\subseteq\{1,2\}} L_{\Gamma,n_1,n_2} - L_0\right|\\
&\le \left|\sum_{k_1=0}^{n_1-1}\sum_{k_2=0}^{n_2-1} \Delta_{\{1,2\}}(v_{k_1,k_2}) \sum_{j_1=0}^{k_1}\sum_{j_2=0}^{k_2} u_{j_1,j_2} - L_0\right| + \sum_{\Gamma\subset\{1,2\}} |L_{\Gamma,n_1,n_2}|\\
&\le 3\sum_{\emptyset\neq\Gamma\subset\{1,2\}} \sum_{\substack{(k_1,k_2)\in\mathbb{N}_0^2\\ k_i\ge 0\,\forall\, i\in\Gamma\\ k_\ell\ge n_\ell\,\forall\,\ell\in\{1,2\}\setminus\Gamma}} |\Delta_{\{1,2\}}(v_{k_1,k_2})| \max_{\substack{r_1=0,\dots,k_1\\ r_2=0,\dots,k_2}} \left|\sum_{j_1=0}^{r_1}\sum_{j_2=0}^{r_2} u_{j_1,j_2}\right|.
\end{aligned}$$

Since (iii) follows from (8.1.3), and (iv) is an immediate consequence of (iii), (8.1.4) and (ii), the proof is complete. □

8.2 Double Fourier series

A function $f:\mathbb{R}^2 \longrightarrow \mathbb{R}$ is said to be in $L^1(\mathbb{T}^2)$ if f is 2π-periodic in each variable, and $f\in L^1([-\pi,\pi]^2)$. For any $f\in L^1(\mathbb{T}^2)$, we say that

$$\sum_{(k_1,k_2)\in\mathbb{N}_0^2} \lambda_{k_1}\lambda_{k_2}\Big\{a_{k_1,k_2}\cos k_1t_1\cos k_2t_2 + b_{k_1,k_2}\cos k_1t_1\sin k_2t_2$$
$$+c_{k_1,k_2}\sin k_1t_1\cos k_2t_2 + d_{k_1,k_2}\sin k_1t_1\sin k_2t_2\Big\}$$

is the double Fourier series of f, where $\lambda_0 = \frac{1}{2}$, $\lambda_k = 1$ $(k \in \mathbb{N})$, and

$$a_{k_1,k_2} = \frac{1}{\pi^2} \int_{[-\pi,\pi]^2} f(t_1,t_2) \cos k_1 t_1 \cos k_2 t_2 \; d\mu_2(t_1,t_2) \quad ((k_1,k_2) \in \mathbb{N}_0^2);$$

$$b_{k_1,k_2} = \frac{1}{\pi^2} \int_{[-\pi,\pi]^2} f(t_1,t_2) \cos k_1 t_1 \sin k_2 t_2 \; d\mu_2(t_1,t_2) \quad ((k_1,k_2) \in \mathbb{N}_0^2);$$

$$c_{k_1,k_2} = \frac{1}{\pi^2} \int_{[-\pi,\pi]^2} f(t_1,t_2) \sin k_1 t_1 \cos k_2 t_2 \; d\mu_2(t_1,t_2) \quad ((k_1,k_2) \in \mathbb{N}_0^2);$$

$$d_{k_1,k_2} = \frac{1}{\pi^2} \int_{[-\pi,\pi]^2} f(t_1,t_2) \sin k_1 t_1 \sin k_2 t_2 \; d\mu_2(t_1,t_2) \quad ((k_1,k_2) \in \mathbb{N}_0^2).$$

Here $a_{k_1,k_2}, b_{k_1,k_2}, c_{k_1,k_2}$ and d_{k_1,k_2} are known as Fourier coefficients of f, and we write

$$f(t_1,t_2) \sim \sum_{(k_1,k_2)\in\mathbb{N}_0^2} \lambda_{k_1}\lambda_{k_2} \Big\{ a_{k_1,k_2} \cos k_1 t_1 \cos k_2 t_2 + b_{k_1,k_2} \cos k_1 t_1 \sin k_2 t_2 + c_{k_1,k_2} \sin k_1 t_1 \cos k_2 t_2 + d_{k_1,k_2} \sin k_1 t_1 \sin k_2 t_2 \Big\}.$$

Moreover, for every $(n_1,n_2) \in \mathbb{N}_0^2$ and $(t_1,t_2) \in \mathbb{R}^2$, let

$$s_{n_1,n_2} f(t_1,t_2) := \sum_{k_1=0}^{n_1} \sum_{k_2=0}^{n_2} \lambda_{k_1}\lambda_{k_2} \Big\{ a_{k_1,k_2} \cos k_1 t_1 \cos k_2 t_2 + b_{k_1,k_2} \cos k_1 t_1 \sin k_2 t_2 + c_{k_1,k_2} \sin k_1 t_1 \cos k_2 t_2 + d_{k_1,k_2} \sin k_1 t_1 \sin k_2 t_2 \Big\}. \tag{8.2.1}$$

A two-dimensional analogue of Theorem 7.2.2 holds for double Fourier series.

Theorem 8.2.1 (Riemann Lebesgue Lemma). *Let $f \in L^1(\mathbb{T}^2)$ and assume that*

$$f(t_1,t_2) \sim \sum_{(k_1,k_2)\in\mathbb{N}_0^2} \lambda_{k_1}\lambda_{k_2} \Big\{ a_{k_1,k_2} \cos k_1 t_1 \cos k_2 t_2 + b_{k_1,k_2} \cos k_1 t_1 \sin k_2 t_2 + c_{k_1,k_2} \sin k_1 t_1 \cos k_2 t_2 + d_{k_1,k_2} \sin k_1 t_1 \sin k_2 t_2 \Big\}.$$

Then $\lim_{\max\{n_1,n_2\}\to\infty} a_{n_1,n_2} = \lim_{\max\{n_1,n_2\}\to\infty} b_{n_1,n_2} = \lim_{\max\{n_1,n_2\}\to\infty} c_{n_1,n_2} = \lim_{\max\{n_1,n_2\}\to\infty} d_{n_1,n_2} = 0.$

Proof. Exercise. □

We need the following lemmas in order to prove Theorem 8.2.4 below.

Lemma 8.2.2. *If $f \in L^1([-\pi,\pi]^2)$, then for each $\varepsilon > 0$ there exists a positive integer N such that*

$$\sup_{(x_1,x_2)\in[-\pi,\pi]^2} \left| \int_{[-\pi,\pi]^2} f(t_1,t_2) \prod_{i=1}^{2} \sum_{k_i=p_i}^{q_i} \frac{\sin k_i(t_i - x_i)}{k_i}\, d\mu_2(t_1,t_2) \right| < \varepsilon$$

for all positive integers p_1, q_1, p_2, q_2 satisfying $q_i \geq p_i$ $(i = 1, 2)$ and $\max\{p_1, p_2\} \geq N$.

Proof. Let $\varepsilon > 0$ be given. Since $f \in L^1([-\pi,\pi]^2)$ there exists $\eta \in (0, \frac{\pi}{8})$ such that

$$\sup_{(x_1,x_2)\in[-\pi,\pi]^2} \|f\chi_{U(x_1)\times U(x_2)}\|_{L^1([-\pi,\pi]^2} < \frac{\varepsilon}{2(6\pi)^2},$$

where $U(\theta)$ denotes the set $[-\pi,-\pi+\eta) \cup ([\theta-\eta,\theta+\eta] \cap [-\pi,\pi]) \cup (\pi-\eta,\pi]$. According to Lemma 7.1.5 and our choice of η, we select a positive integer N so that

$$\sup_{\theta\in[\eta,2\pi-\eta]} \left| \sum_{k=p}^{q} \frac{\sin k\theta}{k} \right| < \frac{\varepsilon}{2(1 + 6\pi\|f\|_{L^1([-\pi,\pi]^2)})}$$

whenever p, q are positive integers with $q \geq p \geq N$. Then, for every positive integers p_1, p_2, q_1, q_2 satisfying $q_i \geq p_i$ $(i = 1, 2)$ and $\max_{i=1,2} p_i \geq N$, we apply Lemma 7.1.7 to get

$$\begin{aligned}
&\sup_{(x_1,x_2)\in[-\pi,\pi]^2} \left| \sum_{k_1=p_1}^{q_1} \sum_{k_2=p_2}^{q_2} \int_{[-\pi,\pi]^2} f(t_1,t_2) \prod_{i=1}^{2} \frac{\sin k_i(t_i - x_i)}{k_i}\, d\mu_2(t_1,t_2) \right| \\
&\leq \sup_{(x_1,x_2)\in[-\pi,\pi]^2} \Bigg\{ (6\pi)^2 \|f\chi_{U(x_1)\times U(x_2)}\|_{L^1([-\pi,\pi]^2)} \\
&\qquad + 6\pi\|f - f\chi_{U(x_1)\times U(x_2)}\|_{L^1([-\pi,\pi]^2)} \min_{i=1,2} \sup_{\theta\in[\eta,2\pi-\eta]} \left| \sum_{k_i=p_i}^{q_i} \frac{\sin k_i\theta}{k_i} \right| \Bigg\} \\
&< \varepsilon.
\end{aligned}$$

□

Lemma 8.2.3. *Let $f \in L^1(\mathbb{T}^2)$ and assume that*

$$\begin{aligned}
f(t_1,t_2) \sim \sum_{(k_1,k_2)\in\mathbb{N}_0^2} \lambda_{k_1}\lambda_{k_2} \Big\{ &a_{k_1,k_2} \cos k_1t_1 \cos k_2t_2 + b_{k_1,k_2} \cos k_1t_1 \sin k_2t_2 \\
&+ c_{k_1,k_2} \sin k_1t_1 \cos k_2t_2 + d_{k_1,k_2} \sin k_1t_1 \sin k_2t_2 \Big\}.
\end{aligned}$$

Then the following double series

$$\sum_{(k_1,k_2)\in\mathbb{N}^2} \frac{a_{k_1,k_2}}{k_1k_2} \sin k_1x_1 \sin k_2x_2, \quad \sum_{(k_1,k_2)\in\mathbb{N}^2} \frac{b_{k_1,k_2}}{k_1k_2} \sin k_1x_1 \cos k_2x_2,$$

$$\sum_{(k_1,k_2)\in\mathbb{N}^2} \frac{c_{k_1,k_2}}{k_1k_2} \cos k_1x_1 \sin k_2x_2 \text{ and } \sum_{(k_1,k_2)\in\mathbb{N}^2} \frac{d_{k_1,k_2}}{k_1k_2} \cos k_1x_1 \cos k_2x_2$$

converge regularly for all $(x_1, x_2) \in [-\pi, \pi]^2$. *Moreover, the convergence is uniform on* $[-\pi, \pi]^2$.

Proof. Since $2\sin\alpha\cos\beta = \sin(\alpha+\beta) + \sin(\alpha-\beta)$ for every $\alpha, \beta \in \mathbb{R}$, the lemma follows from Lemma 8.2.2. □

The following theorem refines a result of W.H. Young concerning double Fourier series.

Theorem 8.2.4. *Let* $f \in L^1(\mathbb{T}^2)$ *and assume that*

$$f(t_1,t_2) \sim \sum_{(k_1,k_2)\in\mathbb{N}^2} \Big\{ a_{k_1,k_2} \cos k_1t_1 \cos k_2t_2 + b_{k_1,k_2} \cos k_1t_1 \sin k_2t_2 + c_{k_1,k_2} \sin k_1t_1 \cos k_2t_2 + d_{k_1,k_2} \sin k_1t_1 \sin k_2t_2 \Big\}. \quad (8.2.2)$$

Then the following assertions hold.

(i) $\lim_{\min\{n_1,n_2\}\to\infty} \|s_{n_1,n_2}f - f\|_{HK([-\pi,\pi]^2)} = 0$.

(ii) (Parseval's formula). *If* $g \in BV_{HK}([-\pi,\pi]^2)$, *then*

$$\int_{[-\pi,\pi]^2} fg \, d\mu_2 = \sum_{(k_1,k_2)\in\mathbb{N}^2} \int_{[-\pi,\pi]^2} g\Delta_{\{1,2\}}(s_{k_1-1,k_2-1}f) \, d\mu_2; \quad (8.2.3)$$

the double series on the right being regularly convergent.

Proof. In view of Fubini's Theorem, we may assume that both functions $t_2 \mapsto \int_{-\pi}^{\pi} f(t_1,t_2)\, d\mu_1(t_1)$ and $t_1 \mapsto \int_{-\pi}^{\pi} f(t_1,t_2)\, d\mu_1(t_2)$ belong to $L^1[-\pi,\pi]$. Then (8.2.2) implies that

$$\left\| \int_{-\pi}^{\pi} f(t_1,\cdot)\, d\mu_1(t_1) \right\|_{L^1[-\pi,\pi]} + \left\| \int_{-\pi}^{\pi} f(\cdot,t_2)\, d\mu_1(t_2) \right\|_{L^1[-\pi,\pi]} = 0. \quad (8.2.4)$$

Since $f \in L^1(\mathbb{T}^2)$ and (8.2.4) holds, we define the function $F : \mathbb{R}^2 \longrightarrow \mathbb{R}$ by setting

$$F(x_1,x_2) = \int_{[-\pi,x_1]\times[-\pi,x_2]} f\, d\mu_2 \quad ((x_1,x_2) \in [-\pi,\pi]^2),$$

and $F(x_1+2\pi, x_2) = F(x_1, x_2+2\pi) = F(x_1, x_2)$ whenever $(x_1, x_2) \in \mathbb{R}^2$. To determine the double Fourier series of F, we write

$$S_{\Gamma,W}(x_1,x_2) = \sum_{\substack{(k_1,k_2)\in\mathbb{N}_0^2 \\ k_i\geq 1\,\forall\, i\in\Gamma\cup W \\ k_j=0\,\forall\, j\in\{1,2\}\setminus(\Gamma\cup W)}} \frac{A_{k_1,k_2}}{2^{2-\mathrm{card}(W\cup\Gamma)}}\Big\{\prod_{i\in\Gamma}\sin k_i x_i\Big\}\Big\{\prod_{j\in W}\cos k_j x_j\Big\}$$

$$(\Gamma \subseteq \{1,2\}, W \subseteq \{1,2\}\setminus\Gamma, \text{and } (x_1,x_2)\in[-\pi,\pi]^2)$$

and let $\sum_{\Gamma\subseteq\{1,2\}}\sum_{W\subseteq\{1,2\}\setminus\Gamma} S_{\Gamma,W}(x_1,x_2)$ be the double Fourier series of F. In view of Lemma 8.2.3, it suffices to prove if $\Gamma \subseteq \{1,2\}$ and $W \subseteq \{1,2\}\setminus\Gamma$, then

$$S_{\Gamma,W}(x_1,x_2) = \sum_{(k_1,k_2)\in\mathbb{N}^2} c_{k_1,k_2,\Gamma}\prod_{i\in\Gamma}\frac{\sin k_i x_i}{k_i}\prod_{j\in W}\Big(-\frac{\cos k_j x_j}{k_j}\Big)\prod_{\ell\in\{1,2\}\setminus(\Gamma\cup W)}\frac{\cos k_\ell\pi}{k_\ell}, \tag{8.2.5}$$

where

$$c_{k_1,k_2,\Gamma} = \frac{1}{\pi^2}\int_{[-\pi,\pi]^2} f(t_1,t_2)\Big\{\prod_{i\in\Gamma}\cos k_i t_i\Big\}\Big\{\prod_{j\in\{1,2\}\setminus\Gamma}\sin k_j t_j\Big\}\,d\mu_2(t_1,t_2).$$

We consider two cases.

Case 1: $W \neq \{1,2\}\setminus\Gamma$.

Let

$$\psi_r(\alpha) = \begin{cases} \sin\alpha & \text{if } r\in\Gamma \text{ and } \alpha\in\mathbb{R}, \\ -\cos\alpha & \text{if } r\in\{1,2\}\setminus\Gamma \text{ and } \alpha\in\mathbb{R}, \end{cases}$$

and write $(\Gamma\cup W)' = \{1,2\}\setminus(\Gamma\cup W)$. Then, for each $(k_1,k_2)\in\mathbb{N}_0^2$ satisfying $\{i\in\{1,2\} : k_i\neq 0\} = \Gamma\cup W$, we have

$$\pi^2 A_{k_1,k_2} \prod_{j\in W}(-1)$$

$$= \int_{[-\pi,\pi]^2} F(t_1,t_2) \prod_{i\in\Gamma} \sin k_i t_i \prod_{j\in W} (-\cos k_j t_j)\, d\mu_2(t_1,t_2)$$

$$= \int_{[-\pi,\pi]^2} f(t_1,t_2) \prod_{i\in\Gamma\cup W} \frac{-\psi_i'(k_i\pi)+\psi_i'(k_i t_i)}{k_i} \prod_{\ell\in(\Gamma\cup W)'} (\pi - t_\ell)\, d\mu_2(t_1,t_2)$$

$$= \int_{[-\pi,\pi]^2} f(t_1,t_2) \prod_{i\in\Gamma\cup W} \frac{\psi_i'(k_i t_i)}{k_i} \prod_{\ell\in(\Gamma\cup W)'} (-t_\ell)\, d\mu_2(t_1,t_2) \quad \text{(by (8.2.2))}$$

$$= \int_{[-\pi,\pi]^2} f(t_1,t_2) \prod_{i\in\Gamma\cup W} \frac{\psi_i'(k_i t_i)}{k_i} \prod_{\ell\in(\Gamma\cup W)'} \sum_{r_\ell=1}^{\infty} \frac{2(-1)^{r_\ell}\sin r_\ell t_\ell}{r_\ell}\, d\mu_2(t_1,t_2)$$

$$= \pi^2 \sum_{\substack{(r_1,r_2)\in\mathbb{N}^2 \\ r_i=k_i\,\forall\, i\in\Gamma\cup W \\ r_j\geq 1\,\forall\, j\in(\Gamma\cup W)'}} \frac{c_{r_1,r_2,\Gamma}}{r_1 r_2} \prod_{\ell\in(\Gamma\cup W)'} 2(-1)^{r_\ell},$$

where the last equality follows from Lebesgue's Dominated Convergence Theorem.

Case 2: $W = \{1,2\}\backslash\Gamma$.

In this case, we can follow the proof of case 1 to conclude that

$$A_{k_1,k_2} \prod_{j\in\{1,2\}\backslash\Gamma} (-1) = \frac{c_{k_1,k_2,\Gamma}}{k_1 k_2} \quad ((k_1,k_2)\in\mathbb{N}^2).$$

Finally, we deduce from the above cases, Lemma 8.2.3 and Corollary 8.1.10 that (8.2.5) holds:

$$\sum_{\substack{(k_1,k_2)\in\mathbb{N}_0^2 \\ k_i\geq 1\,\forall\, i\in\Gamma\cup W \\ k_j=0\,\forall\, j\in(\Gamma\cup W)'}} \frac{A_{k_1,k_2}}{2^{2-\mathrm{card}(W\cup\Gamma)}} \prod_{i\in\Gamma} \sin k_i x_i \prod_{j\in W} \cos k_j x_j$$

$$= \sum_{\substack{(k_1,k_2)\in\mathbb{N}_0^2 \\ k_i\geq 1\,\forall\, i\in\Gamma\cup W \\ k_j=0\,\forall\, j\in(\Gamma\cup W)'}} \sum_{\substack{(r_1,r_2)\in\mathbb{N}^2 \\ r_i=k_i\,\forall\, i\in\Gamma\cup W \\ r_j\geq 1\,\forall\, j\in(\Gamma\cup W)'}} \frac{c_{r_1,r_2,\Gamma}}{r_1 r_2} \prod_{\ell\in(\Gamma\cup W)'} (-1)^{r_\ell} \prod_{i\in\Gamma\cup W} \psi_i(k_i x_i)$$

$$= \sum_{(k_1,k_2)\in\mathbb{N}^2} \frac{c_{k_1,k_2,\Gamma}}{k_1 k_2} \prod_{\ell\in(\Gamma\cup W)'} (-1)^{k_\ell} \prod_{i\in\Gamma\cup W} \psi_i(k_i x_i)$$

$$= \sum_{(k_1,k_2)\in\mathbb{N}^2} c_{k_1,k_2,\Gamma} \left\{\prod_{i\in\Gamma} \frac{\sin k_i x_i}{k_i}\right\} \left\{\prod_{j\in W} \left(-\frac{\cos k_j x_j}{k_j}\right) \prod_{\ell\in(\Gamma\cup W)'} \frac{\cos k_\ell \pi}{k_\ell}\right\}.$$

Since (ii) is an immediate consequence of (i), Theorem 6.5.9 and Lemma 8.2.3, the theorem is proved. □

The following corollary is a special case of Lemma 8.2.3.

Corollary 8.2.5. *Let f be given as in Theorem 8.2.4. Then the double series*

$$\sum_{(k_1,k_2)\in\mathbb{N}^2} \frac{d_{k_1,k_2}}{k_1 k_2}$$

is regularly convergent.

The following example follows from Corollary 8.2.5.

Example 8.2.6. The double sine series

$$\sum_{(k_1,k_2)\in\mathbb{N}^2} \frac{\sin k_1 t_1 \sin k_2 t_2}{(\ln(k_1 + k_2 + 2))^2}$$

is not the double Fourier series of a function in $L^1(\mathbb{T}^2)$.

8.3 Some examples of double Fourier series

An analogue of Theorem 8.1.3 holds for Lebesgue integrable functions.

Theorem 8.3.1. *Let $(h_{n_1,n_2})_{(n_1,n_2)\in\mathbb{N}_0^2}$ be a double sequence in $L^1([a_1,b_1]\times[a_2,b_2])$. The following conditions are equivalent.*

(i) *There exists $h \in L^1([a_1,b_1]\times[a_2,b_2])$ such that*

$$\lim_{\min\{n_1,n_2\}\to\infty} \left\| \sum_{k_1=0}^{n_1}\sum_{k_2=0}^{n_2} h_{k_1,k_2} - h \right\|_{L^1([a_1,b_1]\times[a_2,b_2])} = 0.$$

(ii) *For each $\varepsilon > 0$ there exists $N(\varepsilon) \in \mathbb{N}_0$ such that*

$$\left\| \sum_{k_1=0}^{q_1}\sum_{k_2=0}^{q_2} h_{k_1,k_2} - \sum_{k_1=0}^{p_1}\sum_{k_2=0}^{p_2} h_{k_1,k_2} \right\|_{L^1([a_1,b_1]\times[a_2,b_2])} < \varepsilon$$

for all $(p_1,p_2),(q_1,q_2) \in \mathbb{N}_0^2$ satisfying $q_i \geq p_i$ $(i = 1,2)$ and $\min\{p_1,p_2\} \geq N(\varepsilon)$.

The following theorem is a consequence of Theorem 8.3.1.

Theorem 8.3.2. *Let $(h_{n_1,n_2})_{(n_1,n_2)\in\mathbb{N}_0^2}$ be a double sequence in $L^1([a_1,b_1]\times[a_2,b_2])$ and suppose that $\sum_{(k_1,k_2)\in\mathbb{N}_0^2} h_{k_1,k_2}$ converges regularly in the L^1-norm; that is, for each $\varepsilon > 0$ there exists $N \in \mathbb{N}_0$ such that*

$$\left\| \sum_{k_1=p_1}^{q_1} \sum_{k_2=p_2}^{q_2} h_{k_1,k_2} \right\|_{L^1([a_1,b_1]\times[a_2,b_2])} < \varepsilon$$

for all $(p_1,p_2),(q_1,q_2) \in \mathbb{N}_0^2$ satisfying $q_i \geq p_i$ $(i = 1,2)$ and $\max\{p_1,p_2\} \geq N(\varepsilon)$. Then there exists $h \in L^1([a_1,b_1]\times[a_2,b_2])$ such that

$$\lim_{\min\{n_1,n_2\}\to\infty} \left\| \sum_{k_1=0}^{n_1} \sum_{k_2=0}^{n_2} h_{k_1,k_2} - h \right\|_{L^1([a_1,b_1]\times[a_2,b_2])} = 0.$$

We are now ready to state and prove another convergence theorem for Lebesgue integrals.

Theorem 8.3.3. *Let $(c_{k_1,k_2})_{(k_1,k_2)\in\mathbb{N}_0^2}$ be a double sequence of real numbers such that $\lim_{\max\{n_1,n_2\}\to\infty} c_{n_1,n_2} = 0$. If $(f_{k_1,k_2})_{(k_1,k_2)\in\mathbb{N}_0^2}$ is a sequence in $L^1([a_1,b_1]\times[a_2,b_2])$ and*

$$\sum_{(k_1,k_2)\in\mathbb{N}_0^2} \left|\Delta_{\{1,2\}}(c_{k_1,k_2})\right| \max_{\substack{r_1=0,\dots,k_1\\ r_2=0,\dots,k_2}} \left\| \sum_{j_1=0}^{r_1} \sum_{j_2=0}^{r_2} f_{j_1,j_2} \right\|_{L^1([a_1,b_1]\times[a_2,b_2])} < \infty,$$

then there exists $f \in L^1([a_1,b_1]\times[a_2,b_2])$ such that

$$\lim_{\min\{n_1,n_2\}\to\infty} \left\| \sum_{k_1=0}^{n_1} \sum_{k_2=0}^{n_2} c_{k_1,k_2} f_{k_1,k_2} - f \right\|_{L^1([a_1,b_1]\times[a_2,b_2])} = 0.$$

Proof. Following the proof of Theorem 8.1.13, we conclude that $\sum_{(k_1,k_2)\in\mathbb{N}_0^2} c_{k_1,k_2} f_{k_1,k_2}$ converges regularly in the L^1-norm. An application of Theorem 8.3.2 yields the result. □

The following two-dimensional analogue of Lemma 7.3.2 is a consequence of Theorems 8.1.12 and 8.1.13.

Lemma 8.3.4. *Let $(u_{k_1,k_2})_{(k_1,k_2)\in\mathbb{N}_0^2}$ be a double sequence of real numbers such that $\inf_{(k_1,k_2)\in\mathbb{N}_0^2} \Delta_{\{1,2\}}(u_{k_1,k_2}) \geq 0$ and $\lim_{\max\{n_1,n_2\}\to\infty} u_{n_1,n_2} = 0$. If $(v_{k_1,k_2})_{(k_1,k_2)\in\mathbb{N}_0^2}$ is a double sequence of non-negative numbers, then the double series $\sum_{(k_1,k_2)\in\mathbb{N}_0^2} u_{k_1,k_2} v_{k_1,k_2}$ converges if and only if $\sum_{(k_1,k_2)\in\mathbb{N}_0^2} \Delta_{\{1,2\}}(u_{k_1,k_2}) \sum_{j_1=0}^{k_1} \sum_{j_2=0}^{k_2} v_{j_1,j_2}$ converges.*

The following theorem is a two-dimensional analogue of Theorem 7.3.4.

Theorem 8.3.5. *Let $(b_{k_1,k_2})_{(k_1,k_2)\in\mathbb{N}^2}$ be a double sequence of real numbers such that $\inf_{(n_1,n_2)\in\mathbb{N}^2} \Delta_{\{1,2\}}(b_{n_1,n_2}) \geq 0$ and $\lim_{\max\{n_1,n_2\}\to\infty} b_{n_1,n_2} = 0$. Then $\sum_{(k_1,k_2)\in\mathbb{N}^2} b_{k_1,k_2} \sin k_1 t_1 \sin k_2 t_2$ is a double Fourier series if and only if $\sum_{(k_1,k_2)\in\mathbb{N}^2} \frac{b_{k_1,k_2}}{k_1 k_2}$ converges.*

Proof. ($\Longrightarrow$) This follows from Corollary 8.2.5.

($\Longleftarrow$) Suppose that $\sum_{(k_1,k_2)\in\mathbb{N}^2} \frac{b_{k_1,k_2}}{k_1 k_2}$ converges. In this case, we deduce from Lemmas 8.3.4 and 7.3.3 that

$$\sum_{(k_1,k_2)\in\mathbb{N}^2} |\Delta_{\{1,2\}}(b_{k_1,k_2})| \prod_{i=1}^{2} \max_{r_i=1,\ldots,k_i} \int_0^\pi \left| \sum_{j_i=1}^{r_i} \sin j_i t_i \right| dt_i$$

converges and hence the result follows from Theorem 8.3.3. □

The following theorem is a two-dimensional analogue of Corollary 7.3.8.

Theorem 8.3.6. *Let $(a_{k_1,k_2})_{(k_1,k_2)\in\mathbb{N}_0^2}$ be a double sequence of real numbers such that $\inf_{(n_1,n_2)\in\mathbb{N}_0^2} \Delta_{\{1,2\}}(\Delta_{\{1,2\}}(a_{n_1,n_2})) \geq 0$ and $\lim_{\max\{n_1,n_2\}\to\infty} a_{n_1,n_2} = 0$. Then there exists $f \in L^1(\mathbb{T}^2)$ such that $f(t_1,t_2) \sim \sum_{(k_1,k_2)\in\mathbb{N}_0^2} \lambda_{k_1}\lambda_{k_2} a_{k_1,k_2} \cos k_1 t_1 \cos k_2 t_2$.*

Proof. For each $(t_1,t_2) \in (0,\pi]^2$, it follows from part (iv) of Theorem 8.1.13 that $\sum_{(k_1,k_2)\in\mathbb{N}_0^2} \lambda_{k_1}\lambda_{k_2} a_{k_1,k_2} \cos k_1 t_1 \cos k_2 t_2$ converges regularly and

$$\sum_{(k_1,k_2)\in\mathbb{N}_0^2} \lambda_{k_1}\lambda_{k_2} a_{k_1,k_2} \cos k_1 t_1 \cos k_2 t_2$$
$$= \frac{1}{16} \sum_{(k_1,k_2)\in\mathbb{N}_0^2} \Delta_{\{1,2\}}(\Delta_{\{1,2\}}(a_{k_1,k_2})) \prod_{i=1}^{2} \frac{1-\cos(k_i+1)t_i}{1-\cos t_i}. \quad (8.3.1)$$

Let

$$f(t_1,t_2) = \begin{cases} \sum_{(k_1,k_2)\in\mathbb{N}_0^2} \lambda_{k_1}\lambda_{k_2} a_{k_1,k_2} \cos k_1 t_1 \cos k_2 t_2 & \text{if } (t_1,t_2) \in (0,\pi]^2, \\ 0 & \text{if } (t_1,t_2) \in [0,\pi]^2 \backslash (0,\pi]^2, \\ f(t_1,t_2+2\pi) = f(t_1+2\pi,t_2) & \text{if } (t_1,t_2) \in \mathbb{R}^2. \end{cases}$$

According to the definition of f and (8.3.1), f is 2π-periodic in each variable and $f(t_1,t_2) \geq 0$ for all $(t_1,t_2) \in [-\pi,\pi]^2$.

It remains to prove that $f \in L^1([-\pi,\pi]^2)$ and

$$f(t_1,t_2) \sim \sum_{(k_1,k_2)\in\mathbb{N}_0^2} \lambda_{k_1}\lambda_{k_2}a_{k_1,k_2}\cos k_1t_1\cos k_2t_2.$$

Since

$$\inf_{(n_1,n_2)\in\mathbb{N}_0^2} \Delta_{\{1,2\}}(\Delta_{\{1,2\}}(a_{n_1,n_2})) \geq 0 \text{ and } \lim_{\max\{n_1,n_2\}\to\infty} a_{n_1,n_2} = 0,$$

Lemma 8.3.4 implies that the double series

$$\sum_{(k_1,k_2)\in\mathbb{N}_0^2} (k_1+1)(k_2+1)\Delta_{\{1,2\}}(\Delta_{\{1,2\}}(a_{k_1,k_2})) < \infty;$$

the Monotone Convergence Theorem yields $f \in L^1([-\pi,\pi]^2)$ and

$$\begin{aligned}
&\int_{[-\pi,\pi]^2} f(t_1,t_2)\,d\mu_2(t_1,t_2)\\
&= \frac{1}{4}\sum_{(k_1,k_2)\in\mathbb{N}_0^2} \Delta_{\{1,2\}}(\Delta_{\{1,2\}}(a_{k_1,k_2}))\prod_{i=1}^{2}\int_0^\pi\left(\frac{1-\cos(k_i+1)t_i}{1-\cos t_i}\right)^2 dt_i\\
&= \frac{1}{4}\sum_{(k_1,k_2)\in\mathbb{N}_0^2} (k_1+1)(k_2+1)\Delta_{\{1,2\}}(\Delta_{\{1,2\}}(a_{k_1,k_2})).
\end{aligned}$$

Finally, for any $(p_1,p_2) \in \mathbb{N}_0^2$, it follows from Lebesgue's Dominated Convergence Theorem and part (iv) of Theorem 8.1.13 that

$$\begin{aligned}
&\int_{[-\pi,\pi]^2} f(t_1,t_2)\cos p_1t_1\cos p_2t_2\,d\mu_2(t_1,t_2)\\
&= \sum_{(k_1,k_2)\in\mathbb{N}_0^2} \Delta_{\{1,2\}}(\Delta_{\{1,2\}}(a_{k_1,k_2}))\prod_{i=1}^{2}\int_0^\pi\left(\frac{1-\cos(k_i+1)t_i}{2(1-\cos t_i)}\right)^2\cos p_it_i\,dt_i\\
&= 4\sum_{(k_1,k_2)\in\mathbb{N}_0^2} \Delta_{\{1,2\}}(\Delta_{\{1,2\}}(a_{k_1,k_2}))\prod_{i=1}^{2}\int_{[0,\pi]}\sum_{j_i=0}^{k_i} D_{j_i}(t_i)\cos p_it_i\,dt_i\\
&= 4\lambda_{p_1}\lambda_{p_2}\sum_{k_1=p_1}^{\infty}\sum_{k_2=p_2}^{\infty} \Delta_{\{1,2\}}(\Delta_{\{1,2\}}(a_{k_1,k_2}))\frac{\pi^2}{4}(k_1-p_1+1)(k_2-p_2+1)\\
&= \pi^2\lambda_{p_1}\lambda_{p_2}\sum_{k_1=p_1}^{\infty}\sum_{k_2=p_2}^{\infty} \Delta_{\{1,2\}}(a_{k_1,k_2})\\
&= \pi^2\lambda_{p_1}\lambda_{p_2}a_{p_1,p_2}.
\end{aligned}$$

This completes the proof of the theorem. □

8.4 A Lebesgue integrability theorem for double cosine series

The main aim of this section is to establish two-dimensional analogues of Theorems 7.4.4 and 7.4.6. We begin with two lemmas.

Lemma 8.4.1. *Let* $(N_1, N_2) \in \mathbb{N}^2$ *and assume that* $\{c_{k_1,k_2} : (k_1,k_2) \in \prod_{i=1}^{2}\{1,\ldots,N_i\}\} \subset \mathbb{R}$. *Then*

$$\int_{[\frac{\pi}{2},\pi]^2} \left(\sum_{k_1=1}^{N_1} \sum_{k_2=1}^{N_2} c_{k_1,k_2} \sin 2^{k_1} t_1 \sin 2^{k_2} t_2 \right)^4 d(t_1,t_2) \leq \left(\frac{3\pi}{4} \sum_{k_1=1}^{N_1} \sum_{k_2=1}^{N_2} c_{k_1,k_2}^2 \right)^2 .$$

Proof. For each $j \in \mathbb{N}$ and $x \in \mathbb{R}$, let$S_j(x) = \sin(2^j x)$. By Lemma 7.4.2,

$$\begin{aligned} &\int_{[\frac{\pi}{2},\pi]^2} \left(\sum_{k_1=1}^{N_1} \sum_{k_2=1}^{N_2} c_{k_1,k_2} S_{k_1}(t_1) S_{k_2}(t_2) \right)^4 d(t_1,t_2) \\ &= \int_{\frac{\pi}{2}}^{\pi} \left\{ \int_{\frac{\pi}{2}}^{\pi} \left\{ \sum_{k_1=1}^{N_1} \left(\sum_{k_2=1}^{N_2} c_{k_1,k_2} S_{k_2}(t_2) \right) S_{k_1}(t_1) \right\}^4 dt_1 \right\} dt_2 \\ &\leq \frac{3\pi}{4} \int_{\frac{\pi}{2}}^{\pi} \left\{ \sum_{k_1=1}^{N_1} \left(\sum_{k_2=1}^{N_2} c_{k_1,k_2} S_{k_2}(t_2) \right)^2 \right\}^2 dt_2. \end{aligned} \tag{8.4.1}$$

Now we obtain an upper bound for the right-hand side of (8.4.1):

$$\begin{aligned} &\int_{\frac{\pi}{2}}^{\pi} \left\{ \sum_{k_1=1}^{N_1} \left(\sum_{k_2=1}^{N_2} c_{k_1,k_2} S_{k_2}(t_2) \right)^2 \right\}^2 dt_2 \\ &= \int_{\frac{\pi}{2}}^{\pi} \sum_{k_1=1}^{N_1} \left(\sum_{k_2=1}^{N_2} c_{k_1,k_2} S_{k_2}(t_2) \right)^4 dt_2 \\ &+2 \sum_{j_1=1}^{N_1-1} \sum_{k_1=j_1+1}^{N_1} \int_{\frac{\pi}{2}}^{\pi} \left(\sum_{k_2=1}^{N_2} c_{j_1,k_2} S_{k_2}(t_2) \right)^2 \left(\sum_{k_2=1}^{N_2} c_{k_1,k_2} S_{k_2}(t_2) \right)^2 dt_2 \\ &\leq \int_{\frac{\pi}{2}}^{\pi} \sum_{k_1=1}^{N_1} \left(\sum_{k_2=1}^{N_2} c_{k_1,k_2} S_{k_2}(t_2) \right)^4 dt_2 \\ &+2 \sum_{j_1=1}^{N_1-1} \sum_{k_1=j_1+1}^{N_1} \prod_{r_1 \in \{j_1,k_1\}} \sqrt{\int_{\frac{\pi}{2}}^{\pi} \left(\sum_{k_2=1}^{N_2} c_{r_1,k_2} S_{k_2}(t_2) \right)^4 dt_2}. \end{aligned} \tag{8.4.2}$$

Combining (8.4.1), (8.4.2) and Lemma 7.4.2, we get

$$\int_{[\frac{\pi}{2},\pi]^2}\left(\sum_{k_1=1}^{N_1}\sum_{k_2=1}^{N_2}c_{k_1,k_2}S_{k_1}(t_1)S_{k_2}(t_2)\right)^4 d(t_1,t_2)$$

$$\leq \left(\frac{3\pi}{4}\right)^2\left\{\sum_{k_1=1}^{N_1}\left(\sum_{k_2=1}^{N_2}c_{k_1,k_2}^2\right)^2+2\sum_{j_1=1}^{N_1-1}\sum_{k_1=j_1+1}^{N_1}\left(\sum_{k_2=1}^{N_2}\sum_{k_2=1}^{N_2}c_{j_1,k_2}^2c_{k_1,k_2}^2\right)\right\}$$

$$=\left(\frac{3\pi}{4}\right)^2\left(\sum_{k_1=1}^{N_1}\sum_{k_2=1}^{N_2}c_{k_1,k_2}^2\right)^2.$$

□

The following lemma is a two-dimensional version of Lemma 7.4.3.

Lemma 8.4.2. *Let* $(N_1, N_2) \in \mathbb{N}^2$ *and suppose that* $\{c_{k_1,k_2} : (k_1,k_2) \in \prod_{i=1}^{2}\{1,\ldots,N_i\}\} \subset \mathbb{R}$. *Then*

$$\int_{[\frac{\pi}{2},\pi]^2}\left|\sum_{k_1=1}^{N_1}\sum_{k_2=1}^{N_2}c_{k_1,k_2}\sin 2^{k_1}t_1\sin 2^{k_2}t_2\right| d(t_1,t_2)$$

$$\geq \frac{\pi}{12}\sqrt{\int_{[\frac{\pi}{2},\pi]^2}\left|\sum_{k_1=1}^{N_1}\sum_{k_2=1}^{N_2}c_{k_1,k_2}\sin 2^{k_1}t_1\sin 2^{k_2}t_2\right|^2 d(t_1,t_2)}.$$

Proof. Let

$$g_{N_1,N_2}(t_1,t_2)=\sum_{k_1=1}^{N_1}\sum_{k_2=1}^{N_2}c_{k_1,k_2}\sin 2^{k_1}t_1\sin 2^{k_2}t_2 \quad \left((t_1,t_2)\in[\frac{\pi}{2},\pi]^2\right).$$

Then Hölder's inequality (with $p = \frac{3}{2}$ and $q = 3$) and Lemma 8.4.1 give

$$\begin{aligned}\|g_{N_1,N_2}\|^3_{L^2([\frac{\pi}{2},\pi]^2)} &\leq \|g_{N_1,N_2}\|_{L^1([\frac{\pi}{2},\pi]^2)}\|g_{N_1,N_2}\|^2_{L^4([\frac{\pi}{2},\pi]^2)}\\ &\leq \|g_{N_1,N_2}\|_{L^1([\frac{\pi}{2},\pi]^2)}\left(\frac{3\pi}{4}\sum_{k_1=1}^{N_1}\sum_{k_2=1}^{N_2}c_{k_1,k_2}^2\right)\\ &\leq \frac{12}{\pi}\|g_{N_1,N_2}\|_{L^1([\frac{\pi}{2},\pi]^2)}\|g_{N_1,N_2}\|^2_{L^2([\frac{\pi}{2},\pi]^2)}.\end{aligned}$$

□

We are now ready to state and prove a two-dimensional analogue of Theorem 7.4.4.

Theorem 8.4.3. *Let* $(b_{k_1,k_2})_{(k_1,k_2)\in\mathbb{N}^2}$ *be a double sequence of real numbers such that*

$$\sum_{(k_1,k_2)\in\mathbb{N}^2}\left\{|b_{k_1,k_2}|+\sqrt{\sum_{r_1=k_1}^{\infty}b_{r_1,k_2}^2}+\sqrt{\sum_{r_2=k_2}^{\infty}b_{k_1,r_2}^2}\right\}<\infty. \quad (8.4.3)$$

Then there exists $h \in L^1([0,\pi]^2)$ *such that*

$$h(t_1, t_2) = \sum_{(k_1,k_2)\in\mathbb{N}^2} b_{k_1,k_2} \frac{\sin 2^{k_1} t_1 \sin 2^{k_2} t_2}{t_1 t_2}$$

μ_2*-almost everywhere on* $[0,\pi]^2$ *if and only if*

$$\sum_{(k_1,k_2)\in\mathbb{N}^2} \sqrt{\sum_{r_1=k_1}^{\infty} \sum_{r_2=k_2}^{\infty} b_{r_1,r_2}^2} < \infty.$$

Proof. We write $S_j(x) = \sin 2^j(x)$ $(j \in \mathbb{N};\ x \in \mathbb{R})$. Clearly, it suffices to prove that

$$\sup_{(N_1,N_2)\in\mathbb{N}_0^2} \int_{[\frac{\pi}{2^{N_1+1}},\pi]\times[\frac{\pi}{2^{N_2+1}},\pi]} \left| \sum_{(k_1,k_2)\in\mathbb{N}^2} b_{k_1,k_2} \prod_{i=1}^{2} \frac{S_{k_i}(t_i)}{t_i} \right| d(t_1,t_2) < \infty$$

$$\iff \sum_{(k_1,k_2)\in\mathbb{N}^2} \sqrt{\sum_{r_1=k_1}^{\infty} \sum_{r_2=k_2}^{\infty} b_{r_1,r_2}^2} < \infty. \tag{8.4.4}$$

Let $(N_1, N_2) \in \mathbb{N}_0^2$ be given. For each $(k_1, k_2) \in \mathbb{N}_0^2$ we write

$$E_{k_1,k_2} = [\frac{\pi}{2^{k_1+1}}, \frac{\pi}{2^{k_1}}] \times [\frac{\pi}{2^{k_2+1}}, \frac{\pi}{2^{k_2}}]$$

to get

$$\begin{aligned}
&\int_{[\frac{\pi}{2^{N_1+1}},\pi]\times[\frac{\pi}{2^{N_2+1}},\pi]} \left| \sum_{(k_1,k_2)\in\mathbb{N}^2} b_{k_1,k_2} \prod_{i=1}^{2} \frac{S_{k_i}(t_i)}{t_i} \right| d(t_1,t_2) \\
&\leq \sum_{k_1=0}^{N_1} \sum_{k_2=0}^{N_2} \int_{E_{k_1,k_2}} \left| \sum_{j_1=1}^{k_1} \sum_{j_2=1}^{k_2} b_{j_1,j_2} \prod_{i=1}^{2} \frac{S_{j_i}(t_i)}{t_i} \right| d(t_1,t_2) \\
&+ \sum_{k_1=0}^{N_1} \sum_{k_2=0}^{N_2} \int_{E_{k_1,k_2}} \left| \sum_{j_1=k_1+1}^{\infty} \sum_{j_2=1}^{k_2} b_{j_1,j_2} \prod_{i=1}^{2} \frac{S_{j_i}(t_i)}{t_i} \right| d(t_1,t_2) \\
&+ \sum_{k_1=0}^{N_1} \sum_{k_2=0}^{N_2} \int_{E_{k_1,k_2}} \left| \sum_{j_1=1}^{k_1} \sum_{j_2=k_2+1}^{\infty} b_{j_1,j_2} \prod_{i=1}^{2} \frac{S_{j_i}(t_i)}{t_i} \right| d(t_1,t_2) \\
&+ \sum_{k_1=0}^{N_1} \sum_{k_2=0}^{N_2} \int_{E_{k_1,k_2}} \left| \sum_{j_1=k_1+1}^{\infty} \sum_{j_2=k_2+1}^{\infty} b_{j_1,j_2} \prod_{i=1}^{2} \frac{S_{j_i}(t_i)}{t_i} \right| d(t_1,t_2) \\
&= S_{(N_1,N_2),\emptyset} + S_{(N_1,N_2),\{1\}} + S_{(N_1,N_2),\{2\}} + S_{(N_1,N_2),\{1,2\}},
\end{aligned}$$

say.

The sum $S_{(N_1,N_2),\emptyset}$ is bounded above by $\pi^2 \sum_{(k_1,k_2)\in\mathbb{N}^2} |b_{k_1,k_2}|$:

$$
\begin{aligned}
&S_{(N_1,N_2),\emptyset} \\
&= \sum_{k_1=0}^{N_1} \sum_{k_2=0}^{N_2} \int_{\prod_{i=1}^{2}[\frac{\pi}{2^{k_i+1}}, \frac{\pi}{2^{k_i}}]} \left| \sum_{j_1=1}^{k_1} \sum_{j_2=1}^{k_2} b_{j_1,j_2} \frac{S_{j_1}(t_1) S_{j_2}(t_2)}{t_1 t_2} \right| d(t_1,t_2) \\
&\leq \sum_{k_1=0}^{N_1} \sum_{k_2=0}^{N_2} \frac{\pi^2}{2^{k_1+1} 2^{k_2+1}} \sum_{j_1=1}^{k_1} \sum_{j_2=1}^{k_2} |b_{j_1,j_2} 2^{j_1+j_2}| \\
&\leq \sum_{(k_1,k_2)\in\mathbb{N}^2} \frac{\pi^2}{2^{k_1+1} 2^{k_2+1}} \sum_{j_1=1}^{k_1} \sum_{j_2=1}^{k_2} |b_{j_1,j_2} 2^{j_1+j_2}| \\
&= \pi^2 \sum_{(k_1,k_2)\in\mathbb{N}^2} |\, b_{k_1,k_2} \,| .
\end{aligned}
\tag{8.4.5}
$$

Now we use Cauchy-Schwarz inequality to prove that

$$
S_{(N_1,N_2),\{1\}} \leq \pi \sum_{(k_1,k_2)\in\mathbb{N}^2} \sqrt{\sum_{r_1=k_1}^{\infty} b_{r_1,k_2}^2}. \tag{8.4.6}
$$

For each $(k_1,k_2) \in \mathbb{N}_0^2$ satisfying $0 \leq k_i \leq N_i$ $(i = 1,2)$, we have

$$
\begin{aligned}
&\int_{E_{k_1,k_2}} \left| \sum_{j_1=k_1+1}^{\infty} \sum_{j_2=1}^{k_2} b_{j_1,j_2} \prod_{i=1}^{2} \frac{S_{j_i}(t_i)}{t_i} \right| d(t_1,t_2) \\
&\leq \sum_{j_2=1}^{k_2} \frac{\pi 2^{j_2}}{2^{k_2+1}} \int_{\frac{\pi}{2^{k_1+1}}}^{\frac{\pi}{2^{k_1}}} \left| \sum_{j_1=k_1+1}^{\infty} b_{j_1,j_2} \frac{S_{j_1}(t_1)}{t_1} \right| dt_1 \\
&\leq \sum_{j_2=1}^{k_2} \frac{\pi 2^{j_2}}{2^{k_2+1}} \int_{\frac{\pi}{2}}^{\pi} \left| \sum_{j_1=k_1+1}^{\infty} b_{j_1,j_2} \frac{S_{j_1-k_1}(t_1)}{t_1} \right| dt_1 \\
&\leq \sum_{j_2=1}^{k_2} \frac{\pi 2^{j_2}}{2^{k_2+1}} \sqrt{\int_{\frac{\pi}{2}}^{\pi} \left| \sum_{j_1=k_1+1}^{\infty} b_{j_1,j_2} S_{j_1-k_1}(t_1) \right|^2 dt_1} \sqrt{\int_{\frac{\pi}{2}}^{\pi} \frac{1}{t^2}\, dt} \\
&\leq \sum_{j_2=1}^{k_2} \frac{\pi 2^{j_2}}{2^{k_2+1}} \sqrt{\sum_{j_1=k_1+1}^{\infty} b_{j_1,j_2}^2}
\end{aligned}
$$

and hence (8.4.6) holds:

$$\begin{aligned}
&S_{(N_1,N_2),\{1\}} \\
&:= \sum_{k_1=0}^{N_1}\sum_{k_2=0}^{N_2}\int_{E_{k_1,k_2}}\left|\sum_{j_1=k_1+1}^{\infty}\sum_{j_2=1}^{k_2}\frac{b_{j_1,j_2}S_{j_1}(t_1)S_{j_2}(t_2)}{t_1t_2}\right| d(t_1,t_2) \\
&\leq \sum_{(k_1,k_2)\in\mathbb{N}_0^2}\sum_{j_2=1}^{k_2}\frac{\pi 2^{j_2}}{2^{k_2+1}}\sqrt{\sum_{j_1=k_1+1}^{\infty} b_{j_1,j_2}^2} \\
&= \pi\sum_{(k_1,k_2)\in\mathbb{N}^2}\left(\frac{1}{2^{k_2}}-\frac{1}{2^{k_2+1}}\right)\sum_{j_2=1}^{k_2}2^{j_2}\sqrt{\sum_{j_1=k_1}^{\infty} b_{j_1,j_2}^2} \\
&= \pi\sum_{(k_1,k_2)\in\mathbb{N}^2}\sqrt{\sum_{r_1=k_1}^{\infty} b_{r_1,k_2}^2}.
\end{aligned}$$

Similarly, we have

$$S_{(N_1,N_2),\{2\}} \leq \pi\sum_{(k_1,k_2)\in\mathbb{N}^2}\sqrt{\sum_{r_2=k_2}^{\infty} b_{k_1,r_2}^2}. \tag{8.4.7}$$

As $(N_1, N_2) \in \mathbb{N}_0^2$ is arbitrary, we infer from (8.4.3), (8.4.5), (8.4.6) and (8.4.7) that

$$\sup_{(N_1,N_2)\in\mathbb{N}_0^2}\left\{S_{(N_1,N_2),\emptyset}+S_{(N_1,N_2),\{1\}}+S_{(N_1,N_2),\{2\}}\right\}<\infty. \tag{8.4.8}$$

Finally, for any $(N_1, N_2) \in \mathbb{N}_0^2$ we apply Cauchy-Schwartz inequality, Lemma 8.4.2, and and Lebesgue's Dominated Convergence Theorem to obtain

$$\begin{aligned}
&\sum_{k_1=0}^{N_1}\sum_{k_2=0}^{N_2}\int_{E_{k_1,k_2}}\left|\sum_{j_1=k_1+1}^{\infty}\sum_{j_2=k_2+1}^{\infty} b_{j_1,j_2}\prod_{i=1}^{2}\frac{S_{j_i}(t_i)}{t_i}\right| d(t_1,t_2) \\
&= \sum_{k_1=0}^{N_1}\sum_{k_2=0}^{N_2}\int_{[\frac{\pi}{2},\pi]^2}\left|\sum_{j_1=k_1+1}^{\infty}\sum_{j_2=k_2+1}^{\infty} b_{j_1,j_2}\prod_{i=1}^{2}\frac{S_{j_i-k_i}(t_i)}{t_i}\right| d(t_1,t_2) \\
&\leq \sum_{k_1=0}^{N_1}\sum_{k_2=0}^{N_2}\sqrt{\int_{[\frac{\pi}{2},\pi]^2}\left|\sum_{j_1=k_1+1}^{\infty}\sum_{j_2=k_2+1}^{\infty} b_{j_1,j_2}\prod_{i=1}^{2}S_{j_i-k_i}(t_i)\right|^2 d(t_1,t_2)} \\
&\leq \frac{12}{\pi}\sum_{k_1=0}^{N_1}\sum_{k_2=0}^{N_2}\int_{E_{k_1,k_2}}\left|\sum_{j_1=k_1+1}^{\infty}\sum_{j_2=k_2+1}^{\infty} b_{j_1,j_2}\prod_{i=1}^{2}\frac{S_{j_i}(t_i)}{t_i}\right| d(t_1,t_2).
\end{aligned}$$

Since $(N_1, N_2) \in \mathbb{N}_0^2$ is arbitrary, we conclude that

$$\sup_{(N_1,N_2)\in\mathbb{N}_0^2} S_{(N_1,N_2),\{1,2\}} < \infty \iff \sum_{(k_1,k_2)\in\mathbb{N}^2} \sqrt{\sum_{r_1=k_1}^{\infty}\sum_{r_2=k_2}^{\infty} b_{r_1,r_2}^2} < \infty; \quad (8.4.9)$$

hence (8.4.4) follows from (8.4.8) and (8.4.9). □

The following theorem is a two-dimensional analogue of Theorem 7.4.6.

Theorem 8.4.4. *Let $(a_{k_1,k_2})_{(k_1,k_2)\in\mathbb{N}_0^2}$ be a double sequence of real numbers such that $\lim_{\max\{n_1,n_2\}\to\infty} a_{n_1,n_2} = 0$, $\Delta_{\{1,2\}}(a_{k_1,k_2}) = 0$ $((k_1,k_2) \in \mathbb{N}_0^2\backslash\{2^r : r \in \mathbb{N}\}^2)$ and*

$$\sum_{(k_1,k_2)\in\mathbb{N}^2} \left\{\left|\Delta_{\{1,2\}}(a_{2^{k_1},2^{k_2}})\right| + \sqrt{\sum_{r_1=k_1}^{\infty}\left(\Delta_{\{1,2\}}(a_{2^{r_1},2^{k_2}})\right)^2}\right.$$
$$\left. + \sqrt{\sum_{r_2=k_2}^{\infty}\left(\Delta_{\{1,2\}}(a_{2^{k_1},2^{r_2}})\right)^2}\right\} < \infty. \quad (8.4.10)$$

Then $\sum_{(k_1,k_2)\in\mathbb{N}_0^2} \lambda_{k_1}\lambda_{k_2} a_{k_1,k_2} \cos k_1 t_1 \cos k_2 t_2$ is a double Fourier series if and only if

$$\sum_{(k_1,k_2)\in\mathbb{N}^2} \sqrt{\sum_{r_1=k_1}^{\infty}\sum_{r_2=k_2}^{\infty}\left(\Delta_{\{1,2\}}(a_{2^{r_1},2^{r_2}})\right)^2} < \infty.$$

Proof. Let $A_{k_1,k_2} := \Delta_{\{1,2\}}(a_{2^{k_1},2^{k_2}})$ $((k_1,k_2) \in \mathbb{N}^2)$. Then, for each $(t_1,t_2) \in (0,\pi]^2$, an application of Theorem 8.1.13 yields

$$\sum_{(k_1,k_2)\in\mathbb{N}_0^2} \lambda_{k_1}\lambda_{k_2} a_{k_1,k_2} \cos k_1 t_1 \cos k_2 t_2$$
$$= \frac{1}{4}\sum_{(k_1,k_2)\in\mathbb{N}^2} A_{k_1,k_2} \prod_{i=1}^{2}\left(\frac{\sin 2^{k_i}t_i \cos\frac{t_i}{2}}{\sin\frac{t_i}{2}} + \cos 2^{k_i}t_i\right)$$
$$= \sum_{\Gamma\subseteq\{1,2\}}\sum_{(k_1,k_2)\in\mathbb{N}^2} A_{k_1,k_2}\prod_{i\in\Gamma}\frac{\sin 2^{k_i}t_i\cos\frac{t_i}{2}}{2\sin\frac{t_i}{2}}\prod_{\ell\in\{1,2\}\backslash\Gamma}\frac{\cos 2^{k_\ell}t_\ell}{2}.$$

It is now clear that the theorem is a consequence of Theorem 8.4.3, since Theorem 7.4.4 implies

$$\lim_{\delta\to 0+}\int_{[\delta,\pi]^2}\left|\sum_{(k_1,k_2)\in\mathbb{N}^2} A_{k_1,k_2}\frac{\sin 2^{k_1}t_1\cos\frac{t_1}{2}}{2\sin\frac{t_1}{2}}\frac{\cos 2^{k_2}t_2}{2}\right| d(t_1,t_2)$$
$$\le \lim_{\delta\to 0+}\sum_{k_2=1}^{\infty}\int_{\delta}^{\pi}\left|\sum_{r_1=1}^{\infty} A_{r_1,k_2}\frac{\sin 2^{r_1}t_1}{\tan\frac{t_1}{2}}\right| dt_1$$
$$\le 4\pi\sum_{(k_1,k_2)\in\mathbb{N}^2}\sqrt{\sum_{r_1=k_1}^{\infty} A^2_{r_1,k_2}},$$

$$\lim_{\delta\to 0+}\int_{[\delta,\pi]^2}\left|\sum_{(k_1,k_2)\in\mathbb{N}^2} A_{k_1,k_2}\frac{\sin 2^{k_2}t_2\cos\frac{t_2}{2}}{2\sin\frac{t_2}{2}}\frac{\cos 2^{k_1}t_1}{2}\right| d(t_1,t_2)$$
$$\le \lim_{\delta\to 0+}\sum_{k_1=1}^{\infty}\int_{\delta}^{\pi}\left|\sum_{r_2=1}^{\infty} A_{k_1,r_2}\frac{\sin 2^{r_2}t_2}{\tan\frac{t_2}{2}}\right| dt_2$$
$$\le 4\pi\sum_{(k_1,k_2)\in\mathbb{N}^2}\sqrt{\sum_{r_2=k_2}^{\infty} A^2_{k_1,r_2}},$$

$$\lim_{\delta\to 0+}\int_{[\delta,\pi]^2}\left|\sum_{(k_1,k_2)\in\mathbb{N}^2} A_{k_1,k_2}\frac{\cos 2^{k_1}t_1\cos 2^{k_2}t_2}{4}\right| d(t_1,t_2)$$
$$\le \pi\sum_{(k_1,k_2)\in\mathbb{N}^2}\sqrt{\sum_{r_2=k_2}^{\infty} A^2_{k_1,r_2}},$$

and (8.4.10) holds. The proof is complete. □

Example 8.4.5. Let $(b_{j,k})_{(j,k)\in\mathbb{N}_0^2}$ be a double sequence of real numbers such that $b_{j,k} = 0 \quad ((j,k) \in \mathbb{N}_0^2\setminus\{2^r : r \in \mathbb{N}\}^2)$ and $b_{2^j,2^k} = \frac{1}{(j+k)^3} \quad ((j,k)\in\mathbb{N}^2)$. Then the double series

$$\sum_{(k_1,k_2)\in\mathbb{N}_0^2}\lambda_{k_1}\lambda_{k_2}\left(\sum_{r=k_1}^{\infty}\sum_{s=k_2}^{\infty} b_{r,s}\right)\cos k_1t_1\cos k_2t_2 \tag{8.4.11}$$

converges regularly for all $(t_1,t_2)\in(0,\pi]^2$. However, (8.4.11) is not the double Fourier series of a function in $L^1(\mathbb{T}^2)$.

Proof. This is a consequence of Theorems 8.1.13(iv) and 8.4.4. □

8.5 A Lebesgue integrability theorem for double sine series

The main aim of this section is to prove a two-dimensional analogue of Theorem 7.4.12. We need the following modification of Lemma 7.4.10.

Lemma 8.5.1. *Let* $(N_1, N_2) \in \mathbb{N}^2$ *and assume that* $\{c_{k_1,k_2} : (k_1,k_2) \in \prod_{i=1}^{2}\{1,\ldots,N_i\}\} \subset \mathbb{R}$. *Then*

$$\int_{[\frac{\pi}{2},\pi]^2}\left(\sum_{k_1=1}^{N_1}\sum_{k_2=1}^{N_2} c_{k_1,k_2}(1-\cos 2^{k_1}t_1)(1-\cos 2^{k_2}t_2)\right)^4 d(t_1,t_2)$$

$$\leq 64\pi^2\left\{\sum_{k_1=1}^{N_1}\sum_{k_2=1}^{N_2} c_{k_1,k_2}^2 + \sum_{k_1=1}^{N_1}\left(\sum_{k_2=1}^{N_2} c_{k_1,k_2}\right)^2 + \sum_{k_2=1}^{N_2}\left(\sum_{k_1=1}^{N_1} c_{k_1,k_2}\right)^2 + \left(\sum_{k_2=1}^{N_2}\sum_{k_1=1}^{N_1} c_{k_1,k_2}\right)^2\right\}^2.$$

Proof. We write $C_j(x) := 1 - \cos 2^j(x)$ $(j \in \mathbb{N};\ x \in \mathbb{R})$,

$$f_1(t_2) = \sum_{k_1=1}^{N_1}\left(\sum_{k_2=1}^{N_2} c_{k_1,k_2}C_{k_2}(t_2)\right)^2 \text{ and } f_2(t_2) = \left(\sum_{k_1=1}^{N_1}\sum_{k_2=1}^{N_2} c_{k_1,k_2}C_{k_2}(t_2)\right)^2$$

so that Lemma 7.4.10 and Cauchy-Schwarz inequality yield

$$\int_{[\frac{\pi}{2},\pi]^2}\left(\sum_{k_1=1}^{N_1}\sum_{k_2=1}^{N_2} c_{k_1,k_2}C_{k_1}(t_1)C_{k_2}(t_2)\right)^4 d(t_1,t_2)$$

$$= \int_{\frac{\pi}{2}}^{\pi}\left\{\int_{\frac{\pi}{2}}^{\pi}\left(\sum_{k_1=1}^{N_1}\left(\sum_{k_2=1}^{N_2} c_{k_1,k_2}C_{k_2}(t_2)\right)C_{k_1}(t_1)\right)^4 dt_1\right\} dt_2$$

$$\leq 8\pi\int_{\frac{\pi}{2}}^{\pi}\left(f_2(t_2)+f_1(t_2)\right)^2 dt_2$$

$$\leq 8\pi\left(\sqrt{\int_{\frac{\pi}{2}}^{\pi}(f_1(t_2))^2\,dt_2} + \sqrt{\int_{\frac{\pi}{2}}^{\pi}(f_2(t_2))^2\,dt_2}\right)^2.$$

Now we obtain an upper bound for the integral $\int_{\frac{\pi}{2}}^{\pi}(f_1(t_2))^2\,dt_2$:

$$\int_{\frac{\pi}{2}}^{\pi}(f_1(t_2))^2\,dt_2$$

$$= \int_{\frac{\pi}{2}}^{\pi}\sum_{k_1=1}^{N_1}\left(\sum_{k_2=1}^{N_2} c_{k_1,k_2}C_{k_2}(t_2)\right)^4 dt_2$$

$$+2\sum_{j_1=1}^{N_1-1}\sum_{k_1=j_1+1}^{N_1}\int_{\frac{\pi}{2}}^{\pi}\left(\left\{\sum_{k_2=1}^{N_2} c_{j_1,k_2}C_{k_2}(t_2)\right\}\left\{\sum_{k_2=1}^{N_2} c_{k_1,k_2}C_{k_2}(t_2)\right\}\right)^2 dt_2.$$

For each $\ell \in \{1, \ldots, N_1\}$, Lemma 7.4.10 yields

$$\int_{\frac{\pi}{2}}^{\pi} \left(\sum_{k_2=1}^{N_2} c_{\ell,k_2} C_{k_2}(t_2) \right)^4 dt_2 \leq 8\pi \left(\left(\sum_{k_2=1}^{N_2} c_{\ell,k_2} \right)^2 + \sum_{k_2=1}^{N_2} c_{\ell,k_2}^2 \right)^2;$$

hence, for each $j_1 \in \{1, \ldots, N_1\}$ and $k_1 \in \{j_1 + 1, \ldots, N_1\}$, it follows from the Cauchy Schwarz inequality that

$$\int_{\frac{\pi}{2}}^{\pi} \left(\sum_{k_2=1}^{N_2} c_{j_1,k_2} C_{k_2}(t_2) \right)^2 \left(\sum_{k_2=1}^{N_2} c_{k_1,k_2} C_{k_2}(t_2) \right)^2 dt_2$$

$$\leq \sqrt{\int_{\frac{\pi}{2}}^{\pi} \left(\sum_{k_2=1}^{N_2} c_{j_1,k_2} C_{k_2}(t_2) \right)^4 dt_2} \sqrt{\int_{\frac{\pi}{2}}^{\pi} \left(\sum_{k_2=1}^{N_2} c_{k_1,k_2} C_{k_2}(t_2) \right)^4 dt_2}$$

$$\leq 8\pi \left\{ \left(\sum_{k_2=1}^{N_2} c_{j_1,k_2} \right)^2 + \sum_{k_2=1}^{N_2} c_{j_1,k_2}^2 \right\} \left\{ \left(\sum_{k_2=1}^{N_2} c_{k_1,k_2} \right)^2 + \sum_{k_2=1}^{N_2} c_{k_1,k_2}^2 \right\}.$$

Combining the above inequalities, we get

$$\int_{\frac{\pi}{2}}^{\pi} (f_1(t_2))^2 \, dt_2 \leq 8\pi \left(\sum_{k_1=1}^{N_1} \left\{ \left(\sum_{k_2=1}^{N_2} c_{k_1,k_2} \right)^2 + \sum_{k_2=1}^{N_2} c_{k_1,k_2}^2 \right\} \right)^2.$$

Finally, since a similar argument gives

$$\int_{\frac{\pi}{2}}^{\pi} (f_2(t_2))^2 \, dt_2 \leq 8\pi \left\{ \sum_{k_2=1}^{N_2} \left(\sum_{k_1=1}^{N_1} c_{k_1,k_2} \right)^2 + \left(\sum_{k_2=1}^{N_2} \sum_{k_1=1}^{N_1} c_{k_1,k_2} \right)^2 \right\}^2,$$

the result follows. □

We need the following lemmas in order to establish a two-dimensional version of Lemma 7.4.11; see Lemma 8.5.3.

Lemma 8.5.2. *Let* $(N_1, N_2) \in \mathbb{N}^2$ *and suppose that* $\{c_{k_1,k_2} : (k_1, k_2) \in \prod_{i=1}^{2} \{1, \ldots, N_i\}\} \subset \mathbb{R}$. *Then*

$$\int_{[\frac{\pi}{2},\pi]^2} \left(\sum_{k_1=1}^{N_1} \sum_{k_2=1}^{N_2} c_{k_1,k_2} (1 - \cos 2^{k_1} t_1)(1 - \cos 2^{k_2} t_2) \right)^2 d(t_1, t_2)$$

$$= \frac{\pi^2}{4} \left(\sum_{k_1=1}^{N_1} \sum_{k_2=1}^{N_2} c_{k_1,k_2} \right)^2 + \frac{\pi^2}{8} \sum_{k_1=1}^{N_1} \left(\sum_{k_2=1}^{N_2} c_{k_1,k_2} \right)^2$$

$$+ \frac{\pi^2}{8} \sum_{k_2=1}^{N_2} \left(\sum_{k_1=1}^{N_1} c_{k_1,k_2} \right)^2 + \frac{\pi^2}{16} \sum_{k_1=1}^{N_1} \sum_{k_2=1}^{N_2} c_{k_1,k_2}^2.$$

Proof. Exercise. □

Lemma 8.5.3. *Let* $(N_1, N_2) \in \mathbb{N}^2$ *and assume that* $\{c_{k_1,k_2} : (k_1,k_2) \in \prod_{i=1}^{2}\{1,\dots,N_i\}\} \subset \mathbb{R}$. *Then*

$$\left(\int_{[\frac{\pi}{2},\pi]^2} \left|\sum_{k_1=1}^{N_1}\sum_{k_2=1}^{N_2} c_{k_1,k_2}(1-\cos 2^{k_1}t_1)(1-\cos 2^{k_2}t_2)\right| d(t_1,t_2)\right)^2$$

$$\geq \left(\frac{\pi}{128}\right)^2 \int_{[\frac{\pi}{2},\pi]^2} \left(\sum_{k_1=1}^{N_1}\sum_{k_2=1}^{N_2} c_{k_1,k_2}(1-\cos 2^{k_1}t_1)(1-\cos 2^{k_2}t_2)\right)^2 d(t_1,t_2).$$

Proof. We write

$$f_{N_1,N_2}(t_1,t_2) := \sum_{k_1=1}^{N_1}\sum_{k_2=1}^{N_2} c_{k_1,k_2}\prod_{i=1}^{2}(1-\cos 2^{k_i}t_i) \quad ((t_1,t_2)\in[\frac{\pi}{2},\pi]^2).$$

Then, using Hölder's inequality (with $p=\frac{3}{2}$ and $q=3$), Lemmas 8.5.1 and 8.5.2, we get

$$\begin{aligned}\|f_{N_1,N_2}\|^3_{L^2([\frac{\pi}{2},\pi]^2)} &\leq \|f_{N_1,N_2}\|^2_{L^1([\frac{\pi}{2},\pi]^2)}\|f_{N_1,N_2}\|_{L^4([\frac{\pi}{2},\pi]^2)}\\ &\leq \|f_{N_1,N_2}\|^2_{L^1([\frac{\pi}{2},\pi]^2)} \times 64\pi^2 \times \left(\frac{16}{\pi^2}\|f_{N_1,N_2}\|_{L^2([\frac{\pi}{2},\pi]^2)}\right)^2,\end{aligned}$$

and the result follows. □

The following theorem is a two-dimensional analogue of Theorem 7.4.12.

Theorem 8.5.4. *Let* $(a_{k_1,k_2})_{(k_1,k_2)\in\mathbb{N}_0^2}$ *be a double sequence of real numbers such that*

$$\sum_{(k_1,k_2)\in\mathbb{N}_0^2}\left\{|a_{k_1,k_2}| + \sqrt{\sum_{j_1=k_1+1}^{\infty} a^2_{j_1,k_2} + \left(\sum_{j_1=k_1+1}^{\infty} a_{j_1,k_2}\right)^2} + \sqrt{\sum_{j_2=k_2+1}^{\infty} a^2_{k_1,j_2} + \left(\sum_{j_2=k_2+1}^{\infty} a_{k_1,j_2}\right)^2}\right\} < \infty. \qquad (8.5.1)$$

Then there exists $h \in L^1([0,\pi]^2)$ *such that*

$$h(t_1,t_2) = \frac{1}{t_1t_2}\sum_{(k_1,k_2)\in\mathbb{N}_0^2} a_{k_1,k_2}(1-\cos 2^{k_1}t_1)(1-\cos 2^{k_2}t_2)$$

μ_2*-almost everywhere on* $[0,\pi]^2$ *if and only if*

$$\sum_{(k_1,k_2)\in\mathbb{N}_0^2}\left\{\sum_{j_1=k_1+1}^{\infty}\sum_{j_2=k_2+1}^{\infty} a^2_{j_1,j_2} + \sum_{j_1=k_1+1}^{\infty}\left(\sum_{r_2=k_2+1}^{\infty} a_{j_1,r_2}\right)^2 + \sum_{j_2=k_2+1}^{\infty}\left(\sum_{r_1=k_1+1}^{\infty} a_{r_1,j_2}\right)^2 + \left(\sum_{j_1=k_1+1}^{\infty}\sum_{j_2=k_2+1}^{\infty} a_{j_1,j_2}\right)^2\right\}^{\frac{1}{2}} < \infty.$$

Proof. For each $j \in \mathbb{N}$ and $x \in \mathbb{R}$, we set $C_j(x) := 1 - \cos 2^j(x)$, and let $(N_1, N_2) \in \mathbb{N}_0^2$ be arbitrary. For each $(k_1, k_2) \in \mathbb{N}_0^2$ we write $E_{k_1,k_2} = [\frac{\pi}{2^{k_1+1}}, \frac{\pi}{2^{k_1}}] \times [\frac{\pi}{2^{k_2+1}}, \frac{\pi}{2^{k_2}}]$. Then

$$\int_{[\frac{\pi}{2^{N_1+1}},\pi]\times[\frac{\pi}{2^{N_2+1}},\pi]} \left| \sum_{(k_1,k_2)\in\mathbb{N}_0^2} a_{k_1,k_2} \prod_{i=1}^{2} C_{k_i}(t_i) \right| d(t_1,t_2)$$
$$\le \sum_{k_1=0}^{N_1} \sum_{k_2=0}^{N_2} \int_{E_{k_1,k_2}} \left| \sum_{j_1=0}^{k_1} \sum_{j_2=0}^{k_2} a_{j_1,j_2} \prod_{i=1}^{2} \frac{C_{j_i}(t_i)}{t_1 t_2} \right| d(t_1,t_2)$$
$$+ \sum_{k_1=0}^{N_1} \sum_{k_2=0}^{N_2} \int_{E_{k_1,k_2}} \left| \sum_{j_1=k_1+1}^{\infty} \sum_{j_2=0}^{k_2} a_{j_1,j_2} \prod_{i=1}^{2} \frac{C_{j_i}(t_i)}{t_i} \right| d(t_1,t_2)$$
$$+ \sum_{k_1=0}^{N_1} \sum_{k_2=0}^{N_2} \int_{E_{k_1,k_2}} \left| \sum_{j_1=0}^{k_1} \sum_{j_2=k_2+1}^{\infty} a_{j_1,j_2} \prod_{i=1}^{2} \frac{C_{j_i}(t_i)}{t_i} \right| d(t_1,t_2)$$
$$+ \sum_{k_1=0}^{N_1} \sum_{k_2=0}^{N_2} \int_{E_{k_1,k_2}} \left| \sum_{j_1=k_1+1}^{\infty} \sum_{j_2=k_2+1}^{\infty} a_{j_1,j_2} \prod_{i=1}^{2} \frac{C_{j_i}(t_i)}{t_i} \right| d(t_1,t_2)$$
$$=: C_{(N_1,N_2),\emptyset} + C_{(N_1,N_2),\{1\}} + C_{(N_1,N_2),\{2\}} + C_{(N_1,N_2),\{1,2\}}.$$

The rest of the proof is similar to that of Theorem 8.4.3. The term $C_{(N_1,N_2),\emptyset}$ is bounded above by $\pi^2 \sum_{(k_1,k_2)\in\mathbb{N}_0^2} |a_{k_1,k_2}|$:

$$C_{(N_1,N_2)\emptyset} \le \sum_{k_1=0}^{N_1} \sum_{k_2=0}^{N_2} \sum_{j_1=0}^{k_1} \sum_{j_2=0}^{k_2} \frac{\pi^2 2^{j_1+j_2} |a_{j_1,j_2}|}{2^{k_1+k_2+2}} \le \pi^2 \sum_{(k_1,k_2)\in\mathbb{N}_0^2} |a_{k_1,k_2}|. \tag{8.5.2}$$

We will next prove that

$$C_{(N_1,N_2),\{1\}} \le \pi \sum_{k_1=0}^{\infty} \sum_{k_2=0}^{\infty} \sqrt{\sum_{r_1=k_1+1}^{\infty} a_{r_1,k_2}^2 + \left(\sum_{r_1=k_1+1}^{\infty} a_{r_1,k_2} \right)^2}. \tag{8.5.3}$$

To prove (8.5.3), we let $(k_1, k_2) \in \mathbb{N}_0^2$ and compute:

$$\int_{E_{k_1,k_2}} \left| \sum_{j_1=k_1+1}^{\infty} \sum_{j_2=0}^{k_2} \frac{a_{j_1,j_2} C_{j_1}(t_1) C_{j_2}(t_2)}{t_1 t_2} \right| d(t_1, t_2)$$
$$\leq \sum_{j_2=0}^{k_2} \frac{\pi 2^{j_2}}{2^{k_2+1}} \int_{[\frac{\pi}{2},\pi]} \left| \sum_{r_1=k_1+1}^{\infty} \frac{a_{r_1,j_2} C_{r_1-k_1}(t_1)}{t_1} \right| dt_1$$
$$\leq \sum_{j_2=0}^{k_2} \frac{\pi 2^{j_2}}{2^{k_2+1}} \sqrt{\int_{[\frac{\pi}{2},\pi]} \left| \sum_{r_1=k_1+1}^{\infty} a_{r_1,j_2} C_{r_1-k_1}(t_1) \right|^2 dt_1}$$
$$\leq \sum_{j_2=0}^{k_2} \frac{\pi 2^{j_2}}{2^{k_2+1}} \sqrt{\sum_{r_1=k_1+1}^{\infty} a_{r_1,j_2}^2 + \left(\sum_{r_1=k_1+1}^{\infty} a_{r_1,j_2} \right)^2}$$

and so

$$C_{(N_1,N_2),\{1\}} \leq \sum_{k_1=0}^{N_1} \sum_{k_2=0}^{N_2} \sum_{j_2=0}^{k_2} \frac{\pi 2^{j_2}}{2^{k_2+1}} \sqrt{\sum_{r_1=k_1+1}^{\infty} a_{r_1,j_2}^2 + \left(\sum_{r_1=k_1+1}^{\infty} a_{r_1,j_2} \right)^2}$$
$$\leq \pi \sum_{k_1=0}^{\infty} \sum_{k_2=0}^{\infty} \sqrt{\sum_{r_1=k_1+1}^{\infty} a_{r_1,k_2}^2 + \left(\sum_{r_1=k_1+1}^{\infty} a_{r_1,k_2} \right)^2}.$$

A similar argument shows that

$$C_{(N_1,N_2),\{2\}} \leq \pi \sum_{k_1=0}^{\infty} \sum_{k_2=0}^{\infty} \sqrt{\sum_{r_2=k_2+1}^{\infty} a_{k_1,r_2}^2 + \left(\sum_{r_2=k_2+1}^{\infty} a_{k_1,r_2} \right)^2}. \quad (8.5.4)$$

Since $(N_1, N_2) \in \mathbb{N}_0^2$ is arbitrary, it follows from (8.5.1), (8.5.2), (8.5.3) and (8.5.4) that

$$\sup_{(N_1,N_2)\in\mathbb{N}_0^2} \left\{ C_{(N_1,N_2),\emptyset} + C_{(N_1,N_2),\{1\}} + C_{(N_1,N_2),\{2\}} \right\} < \infty. \quad (8.5.5)$$

In view of (8.5.5), it remains to prove that

$$\sup_{(N_1,N_2)\in\mathbb{N}_0^2} C_{(N_1,N_2),\{1,2\}} < \infty$$
$$\iff \sum_{(k_1,k_2)\in\mathbb{N}_0^2} \left\{ \sum_{j_1=k_1+1}^{\infty} \sum_{j_2=k_2+1}^{\infty} a_{j_1,j_2}^2 + \sum_{j_1=k_1+1}^{\infty} \left(\sum_{r_2=k_2+1}^{\infty} a_{j_1,r_2} \right)^2 \right.$$
$$\left. + \sum_{j_2=k_2+1}^{\infty} \left(\sum_{r_1=k_1+1}^{\infty} a_{r_1,j_2} \right)^2 + \left(\sum_{j_1=k_1+1}^{\infty} \sum_{j_2=k_2+1}^{\infty} a_{j_1,j_2} \right)^2 \right\}^{\frac{1}{2}} < \infty. \quad (8.5.6)$$

For any $(N_1, N_2) \in \mathbb{N}_0^2$, we apply Cauchy Schwarz inequality, Lemmas 8.5.3 and 8.5.2 to obtain

$$\begin{aligned}
&\sum_{k_1=0}^{N_1}\sum_{k_2=0}^{N_2}\int_{E_{k_1,k_2}}\left|\sum_{j_1=k_1+1}^{\infty}\sum_{j_2=k_2+1}^{\infty} a_{j_1,j_2}\prod_{i=1}^{2}\frac{C_{j_i}(t_i)}{t_i}\right| d(t_1,t_2)\\
&= \sum_{k_1=0}^{N_1}\sum_{k_2=0}^{N_2}\int_{[\frac{\pi}{2},\pi]^2}\left|\sum_{j_1=k_1+1}^{\infty}\sum_{j_2=k_2+1}^{\infty} a_{j_1,j_2}\prod_{i=1}^{2}\frac{C_{j_i-k_i}(t_i)}{t_i}\right| d(t_1,t_2)\\
&\le \sum_{k_1=0}^{N_1}\sum_{k_2=0}^{N_2}\sqrt{\int_{[\frac{\pi}{2},\pi]^2}\left|\sum_{j_1=k_1+1}^{\infty}\sum_{j_2=k_2+1}^{\infty} a_{j_1,j_2}\prod_{i=1}^{2}C_{j_i-k_i}(t_i)\right|^2 d(t_1,t_2)}\\
&\le 128\pi\sum_{k_1=0}^{N_1}\sum_{k_2=0}^{N_2}\int_{[\frac{\pi}{2},\pi]^2}\left|\sum_{j_1=k_1+1}^{\infty}\sum_{j_2=k_2+1}^{\infty} a_{j_1,j_2}\prod_{i=1}^{2}C_{j_i-k_i}(t_i)\right| d(t_1,t_2)\\
&= 128\pi\sum_{k_1=0}^{N_1}\sum_{k_2=0}^{N_2}\int_{E_{k_1,k_2}}\left|\sum_{j_1=k_1+1}^{\infty}\sum_{j_2=k_2+1}^{\infty} a_{j_1,j_2}\prod_{i=1}^{2}\frac{C_{j_i}(t_i)}{t_i}\right| d(t_1,t_2).
\end{aligned} \tag{8.5.7}$$

It is now clear that (8.5.6) is a consequence of (8.5.7) and Lemma 8.5.2. □

The next theorem is a two-dimensional analogue of Theorem 7.4.14.

Theorem 8.5.5. *Let* $(b_{k_1,k_2})_{(k_1,k_2)\in\mathbb{N}^2}$ *be a double sequence of real numbers such that* $\lim_{\max\{k_1,k_2\}\to\infty} b_{k_1,k_2} = 0$,

$$\Delta_{\{1,2\}}(b_{k_1,k_2}) = 0 \qquad ((k_1,k_2) \in \mathbb{N}^2\backslash\{2^r : r \in \mathbb{N}\}^2)$$

and

$$\begin{aligned}
\sum_{(k_1,k_2)\in\mathbb{N}^2}&\left\{|B_{k_1,k_2}| + \sqrt{\sum_{p=k_1+1}^{\infty} B_{p,k_2}^2 + \left(\sum_{p=k_1+1}^{\infty} B_{p,k_2}\right)^2}\right.\\
&\left.+\sqrt{\sum_{q=k_2+1}^{\infty} B_{k_1,q}^2 + \left(\sum_{q=k_2+1}^{\infty} B_{k_1,q}\right)^2}\right\} < \infty,
\end{aligned}$$

where $B_{j,k} := \Delta_{\{1,2\}}(b_{2^j,2^k}) \quad (j,k \in \mathbb{N})$. *Then* $\sum_{(k_1,k_2)\in\mathbb{N}^2} b_{k_1,k_2}\sin k_1t_1 \sin k_2t_2$ *is a double Fourier series if and only if*

$$\begin{aligned}
\sum_{(k_1,k_2)\in\mathbb{N}^2}&\left\{\sum_{p=k_1+1}^{\infty}\sum_{q=k_2+1}^{\infty} B_{p,q}^2 + \sum_{p=k_1+1}^{\infty}\left(\sum_{q=k_2+1}^{\infty} B_{p,q}\right)^2\right.\\
&\left.+\sum_{q=k_2+1}^{\infty}\left(\sum_{p=k_1+1}^{\infty} B_{p,q}\right)^2 + \left(\sum_{p=k_1+1}^{\infty}\sum_{q=k_2+1}^{\infty} B_{p,q}\right)^2\right\}^{\frac{1}{2}} < \infty.
\end{aligned}$$

Proof. Let $C_j(x) := 1 - \cos 2^j x$ $(j \in \mathbb{N};\ x \in \mathbb{R})$ and $S_j(x) := \sin 2^j x$ $(j \in \mathbb{N};\ x \in \mathbb{R})$. Then, for each $(t_1, t_2) \in (0, \pi)^2$, we deduce from Theorem 8.1.13 and Lemma 7.1.4 that

$$\sum_{(k_1,k_2)\in\mathbb{N}^2} b_{k_1,k_2} \sin k_1 t_1 \sin k_2 t_2$$

$$= \frac{1}{4} \sum_{(k_1,k_2)\in\mathbb{N}^2} B_{k_1,k_2} \left(\frac{C_{k_1}(t_1)}{\tan \frac{t_1}{2}} + S_{k_1}(t_1) \right) \left(\frac{C_{k_2}(t_2)}{\tan \frac{t_2}{2}} + S_{k_2}(t_2) \right)$$

$$= \frac{1}{4} \sum_{(k_1,k_2)\in\mathbb{N}^2} \frac{B_{k_1,k_2} C_{k_1}(t_1) C_{k_2}(t_2)}{\tan \frac{t_1}{2} \tan \frac{t_2}{2}} + \frac{1}{4} \sum_{(k_1,k_2)\in\mathbb{N}^2} \frac{B_{k_1,k_2} C_{k_1}(t_1)}{\tan \frac{t_1}{2}} S_{k_2}(t_2)$$

$$+ \frac{1}{4} \sum_{(k_1,k_2)\in\mathbb{N}^2} \frac{B_{k_1,k_2} C_{k_2}(t_2)}{\tan \frac{t_2}{2}} S_{k_1}(t_1) + \frac{1}{4} \sum_{(k_1,k_2)\in\mathbb{N}^2} B_{k_1,k_2} S_{k_1}(t_1) S_{k_2}(t_2).$$

Since

$$\lim_{\delta\to 0+} \int_{[\delta,\pi]^2} \left| \sum_{(k_1,k_2)\in\mathbb{N}^2} B_{k_1,k_2} S_{k_1} S_{k_2} \right| d(t_1,t_2) \le \pi^2 \sum_{(k_1,k_2)\in\mathbb{N}^2} |B_{k_1,k_2}|,$$

(7.4.8) implies that

$$\lim_{\delta\to 0+} \int_{[\delta,\pi]^2} \left| \sum_{(k_1,k_2)\in\mathbb{N}^2} \frac{B_{k_1,k_2} C_{k_1}(t_1)}{2\tan \frac{t_1}{2}} S_{k_2}(t_2) \right| d(t_1,t_2)$$

$$\le \lim_{\delta\to 0+} \pi \sum_{k_2=1}^{\infty} \int_{\delta}^{\pi} \left| \sum_{k_1=1}^{\infty} \frac{B_{k_1,k_2} C_{k_1}(t_1)}{t_1} \right| dt_1$$

$$\le 2\pi^2 \sum_{k_2=1}^{\infty} \sqrt{\sum_{k_1=1}^{\infty} B_{k_1,k_2}^2 + \left(\sum_{k_1=1}^{\infty} B_{k_1,k_2} \right)^2}$$

and

$$\lim_{\delta\to 0+} \int_{[\delta,\pi]^2} \left| \sum_{(k_1,k_2)\in\mathbb{N}^2} \frac{B_{k_1,k_2} C_{k_2}(t_2)}{2\tan \frac{t_2}{2}} S_{k_1}(t_1) \right| d(t_1,t_2)$$

$$\le 2\pi^2 \sum_{k_1=1}^{\infty} \sqrt{\sum_{k_2=1}^{\infty} B_{k_1,k_2}^2 + \left(\sum_{k_2=1}^{\infty} B_{k_1,k_2} \right)^2},$$

an application of Theorem 8.5.4 completes the proof. □

Example 8.5.6. Let

$$a_{k_1,k_2} = \begin{cases} \frac{(-1)^{r+s}}{(r+s)^3} & \text{if } (k_1,k_2) \in \{(2^r, 2^s) : (r,s) \in \mathbb{N}^2\}, \\ 0 & \text{otherwise} \end{cases}$$

and let $b_{k_1,k_2} = \sum_{r=k_1}^{\infty} \sum_{s=k_2}^{\infty} a_{r,s}$ $((k_1,k_2) \in \mathbb{N}^2)$. Then the double series $\sum_{(k_1,k_2)\in\mathbb{N}^2} b_{k_1,k_2} \sin k_1 t_1 \sin k_2 t_2$ converges regularly for all $(t_1,t_2) \in [0,\pi]^2$. However, it is not the double Fourier series of a function in $L^1(\mathbb{T}^2)$.

Proof. This is an easy consequence of Theorem 8.5.5. □

8.6 A convergence theorem for Henstock-Kurzweil integrals

In this section, we establish a new convergence theorem for Henstock-Kurzweil integrals. Since the Riesz-Fischer Theorem does not hold for $HK([a_1,b_1] \times [a_2,b_2])$ (cf. Example 4.5.5), it is not surprising that the proof of the following result is more involved.

Theorem 8.6.1. *Let $(c_{k_1,k_2})_{(k_1,k_2)\in\mathbb{N}_0^2}$ be a double sequence of real numbers, let $(\phi_{1,j})_{j=0}^{\infty}$ (resp. $(\phi_{2,j})_{j=0}^{\infty}$) be a sequence in $L^1[a_1,b_1]$ (resp. $L^1[a_2,b_2]$), and let $(C_n)_{n=1}^{\infty}$ be a sequence of real numbers. If*

$$\max_{\Gamma\subseteq\{1,2\}} \sum_{(k_1,k_2)\in\mathbb{N}_0^2} |c_{k_1,k_2}| \prod_{j\in\{1,2\}\setminus\Gamma} ||\phi_{j,k_j}||_{HK[a_j,b_j]} \prod_{i\in\Gamma} ||\phi_{i,k_i}||_{L^1[a_i+\frac{b_i-a_i}{n+1},b_i]} \leq C_n \tag{8.6.1}$$

for all $n \in \mathbb{N}$, then there exists $\phi \in HK([a_1,b_1] \times [a_2,b_2])$ such that

$$\lim_{\min\{n_1,n_2\}\to\infty} \left\| \sum_{k_1=0}^{n_1} \sum_{k_2=0}^{n_2} c_{k_1,k_2}(\phi_{1,k_1} \otimes \phi_{2,k_2}) - \phi \right\|_{HK([a_1,b_1]\times[a_2,b_2])} = 0. \tag{8.6.2}$$

Proof. First we apply (8.6.1) with $\Gamma = \emptyset$ to conclude that the double series

$$\sum_{(k_1,k_2)\in\mathbb{N}_0^2} c_{k_1,k_2} \int_{[s,t]\times[u,v]} \phi_{1,k_1} \otimes \phi_{2,k_2} \, d\mu_2$$

converges absolutely for all $[s,t] \times [u,v] \in \mathcal{I}_2([a_1,b_1] \times [a_2,b_2])$. Therefore we can define an interval function $\Phi : \mathcal{I}_2([a_1,b_1] \times [a_2,b_2]) \longrightarrow \mathbb{R}$ by setting

$$\Phi([s,t] \times [u,v]) = \sum_{(k_1,k_2)\in\mathbb{N}_0^2} c_{k_1,k_2} \int_{[s,t]\times[u,v]} \phi_{1,k_1} \otimes \phi_{2,k_2} \, d\mu_2.$$

Using (8.6.1) with $\Gamma = \{1, 2\}$, it follows from the Riesz-Fischer Theorem that there exists a function $\phi : [a_1, b_1] \times [a_2, b_2] \longrightarrow \mathbb{R}$ with the following property:

$$(x_1, x_2) \in \prod_{i=1}^{2} (a_i, b_i)$$

$$\Longrightarrow \phi \in L^1(\prod_{i=1}^{2}[x_i, b_i]) \text{ and } \int_{\prod_{i=1}^{2}[x_i,b_i]} \phi \, d\mu_2 = \Phi(\prod_{i=1}^{2}[x_i, b_i]). \quad (8.6.3)$$

It remains to prove that $\phi \in HK([a_1, b_1] \times [a_2, b_2]))$ and Φ is the indefinite Henstock-Kurzweil integral of ϕ. In view of assertion (8.6.3), Theorems 5.2.1(iii) and 5.5.9, it suffices to prove that

$$V_{\mathcal{HK}}\Phi(Z_{\Gamma,n}) = 0 \qquad (n \in \mathbb{N};\ \Gamma \subset \{1, 2\}), \qquad (8.6.4)$$

where

$$([a_1, b_1] \times [a_2, b_2]) \backslash ((a_1, b_1] \times (a_2, b_2]) = \bigcup_{n\in\mathbb{N}} \bigcup_{\Gamma\subset\{1,2\}} Z_{\Gamma,n},$$

$$Z_{\Gamma,n} := \prod_{k=1}^{2} Z_{\Gamma,n,k} \text{ and } Z_{\Gamma,n,k} := \begin{cases} \{a_k\} & \text{if } k \in \{1, 2\}\backslash\Gamma, \\ [a_k + \frac{b_k - a_k}{n+1}, b_k] & \text{if } k \in \Gamma. \end{cases}$$

Proof of (8.6.4). Let $\varepsilon > 0$ be given. For each $n \in \mathbb{N}$ and $\Gamma \subset \{1, 2\}$, we use (8.6.1) to select a positive integer $K = K(\Gamma, n, \varepsilon)$ so that the following conditions are satisfied:

$$\max_{\ell=1,2} \sum_{\substack{(k_1,k_2)\in\mathbb{N}_0^2 \\ k_\ell>K}} |c_{k_1,k_2}| \prod_{j\in\{1,2\}\backslash\Gamma} ||\phi_{j,k_j}||_{HK[a_j,b_j]} \prod_{i\in\Gamma} ||\phi_{i,k_i}||_{L^1[a_i+\frac{b_i-a_i}{n+1},b_i]}$$

$$< \frac{\varepsilon}{4}; \qquad (8.6.5)$$

there exists $\eta(\Gamma, n) > 0$ so that

$$\max_{j\in\{1,2\}\backslash\Gamma} \max_{k_j=0,\ldots,K} \sup_{\substack{[u,v]\subseteq[a_j,b_j] \\ 0<v-u<\eta(\Gamma,n)}} \|\phi_{j,k_j}\|_{HK[u,v]}$$

$$< \frac{\varepsilon}{2}\Big(1 + \max_{\substack{k_1=0,\ldots,K \\ k_2=0,\ldots,K}} |c_{k_1,k_2}|\Big)^{-1} \Big(1 + \prod_{i\in\Gamma} ||\phi_{i,n}||_{L^1[a_i+\frac{b_i-a_i}{n+1},b_i]}\Big)^{-1}. \quad (8.6.6)$$

Define a gauge $\delta_{\Gamma,n}$ on $Z_{\Gamma,n}$ by setting $\delta_{\Gamma,n}(x_1, x_2) := \eta(\Gamma, n)$, and let $P_{\Gamma,n}$ be a $Z_{\Gamma,n}$-tagged $\delta_{\Gamma,n}$-fine Perron subpartition of $[a_1, b_1] \times [a_2, b_2]$. We claim that

$$\sum_{(t,I)\in P_{\Gamma,n}} |\Phi(I)| < \varepsilon. \qquad (8.6.7)$$

We consider three cases.

Case 1: $\Gamma = \emptyset$.

In this case, (8.6.7) follows from the obvious equality $\mathrm{card}(P_{\emptyset,n}) = 1$, (8.6.5) and (8.6.6).

Case 2: $\Gamma = \{1\}$.

In this case, our choice of $P_{\{1\},n}$ implies that $\mu_1(I_1 \cap U_1) = 0$ whenever $((x_1, x_2), I_1 \times I_2)$ and $((y_1, y_2), U_1 \times U_2)$ are two distinct elements of $P_{\{1\},n}$. Letting

$$S_{k_1,k_2}(P_{\{1\},n}) = |\, c_{k_1,k_2} \,| \sum_{(t,I)\in P_{\{1\},n}} \|\phi_{1,k_1}\|_{L^1[a_1+\frac{b_1-a_1}{n+1},b_1]} \|\phi_{2,k_2}\|_{HK[a_2,b_2]},$$

we infer that (8.6.5) and (8.6.6) that

$$\begin{aligned}
&\sum_{(t,I)\in P_{\{1\},n}} |\Phi(I)| \\
&\leq \sum_{(k_1,k_2)\in \mathbb{N}_0^2} \sum_{(t,I)\in P_{\{1\},n}} \left| \int_I \phi_{1,k_1} \otimes \phi_{2,k_2} \, d\mu_2 \right| \\
&\leq \sum_{k_1=0}^{K} \sum_{k_2=0}^{K} S_{k_1,k_2}(P_{\{1\},n}) + \sum_{k_1=0}^{K} \sum_{k_2=K+1}^{\infty} S_{k_1,k_2}(P_{\{1\},n}) \\
&\quad + \sum_{k_1=K+1}^{\infty} \sum_{k_2=0}^{K} S_{k_1,k_2}(P_{\{1\},n}) + \sum_{k_1=K+1}^{\infty} \sum_{k_2=K+1}^{\infty} S_{k_1,k_2}(P_{\{1\},n}) \\
&< \varepsilon.
\end{aligned}$$

Case 3: $\Gamma = \{2\}$.

In this case, we follow the proof of case 2 to obtain (8.6.7).

Combining the above cases, we conclude that (8.6.7) holds. Since $\varepsilon > 0$ is arbitrary, (8.6.4) is established. It is now clear that (8.6.1) implies (8.6.2). □

The following theorem is a partial converse of Theorem 8.6.1.

Theorem 8.6.2. *If the following conditions are satisfied:*

(i) $(c_{n_1,n_2})_{(n_1,n_2)\in\mathbb{N}_0^2}$ *is a double sequence of non-negative numbers;*
(ii) *for each* $i \in \{1, 2\}$ *the function* $h_i : [a_i, b_i] \longrightarrow \mathbb{R}$ *is positive and decreasing on* $(a_i, b_i]$*;*

(iii) *for each* $i \in \{1,2\}$ *both sequences* $(\phi_{i,n})_{n=0}^{\infty}$ *and* $(\phi_{i,n}h_i)_{n=0}^{\infty}$ *belong to* $L^1[a_i,b_i]$ *with*

$$\inf_{n\in\mathbb{N}_0} \min\left\{\int_{a_i}^{b_i} \phi_{i,n}h_i\,d\mu_1,\ \min_{x_i\in[a_i,b_i]}\int_{a_i}^{x_i}\phi_{i,n}\,d\mu_1\right\} \geq 0; \quad (8.6.8)$$

(iv) *there exists* $\phi \in HK([a_1,b_1]\times[a_2,b_2])$ *such that*

$$\lim_{\min\{n_1,n_2\}\to\infty}\left\|\sum_{k_1=0}^{n_1}\sum_{k_2=0}^{n_2} c_{k_1,k_2}(\phi_{1,k_1}\otimes\phi_{2,k_2}) - \phi\right\|_{HK([a_1,b_1]\times[a_2,b_2])} = 0;$$

(v) $\phi(h_1\otimes h_2) \in HK([a_1,b_1]\times[a_2,b_2])$,

then the double series

$$\sum_{(k_1,k_2)\in\mathbb{N}_0^2} c_{k_1,k_2}\prod_{i=1}^{2}\int_{a_i}^{b_i}\phi_{i,k_i}h_i\,d\mu_1 \quad \textit{converges.}$$

Proof. In view of (i), (iii) and (v), it suffices to prove that

$$\begin{aligned}&\sup_{(N_1,N_2)\in\mathbb{N}_0^2}\sum_{k_1=0}^{N_1}\sum_{k_2=0}^{N_2} c_{k_1,k_2}\prod_{i=1}^{2}\int_{a_i}^{b_i}\phi_{i,k_i}h_i\,d\mu_1\\ &\leq 4\left\|\phi(h_1\otimes h_2)\right\|_{HK([a_1,b_1]\times[a_2,b_2])}.\end{aligned} \quad (8.6.9)$$

To prove (8.6.9), we let $(N_1,N_2)\in\mathbb{N}_0^2$ be arbitrary and write

$$W(\Gamma,(k_1,k_2),(y_1,y_2)) = \prod_{j\in\Gamma}\frac{1}{h_j(y_j)}\int_{y_j}^{b_j}\phi_{j,k_j}h_j\,d\mu_1 \prod_{\ell\in\{1,2\}\setminus\Gamma}\int_{a_\ell}^{y_\ell}\phi_{\ell,k_\ell}\,d\mu_1$$

$$((k_1,k_2)\in\mathbb{N}_0^2;\ (y_1,y_2)\in(a_1,b_1]\times(a_2,b_2];\ \Gamma\subseteq\{1,2\}).$$

We will first prove that

$$\begin{aligned}&\lim_{\substack{(y_1,y_2)\to(a_1,a_2)\\(y_1,y_2)\in(a_1,b_1]\times(a_2,b_2]}}\sum_{\Gamma\subseteq\{1,2\}}\sum_{k_1=0}^{N_1}\sum_{k_2=0}^{N_2} c_{k_1,k_2}W(\Gamma,(k_1,k_2),(y_1,y_2))\prod_{i=1}^{2}h_i(y_i)\\ &=\sum_{k_1=0}^{N_1}\sum_{k_2=0}^{N_2} c_{k_1,k_2}\prod_{i=1}^{2}\int_{a_i}^{b_i}\phi_{i,k_i}h_i\,d\mu_1.\end{aligned} \quad (8.6.10)$$

To prove (8.6.10) we consider two cases.

Case 1: $\Gamma = \{1,2\}$.

In this case, it follows from (iii) that

$$\lim_{\substack{(y_1,y_2)\to(a_1,a_2)\\(y_1,y_2)\in(a_1,b_1]\times(a_2,b_2]}} \sum_{k_1=0}^{N_1}\sum_{k_2=0}^{N_2} c_{k_1,k_2} W(\{1,2\},(k_1,k_2),(y_1,y_2)) \prod_{i=1}^{2} h_i(y_i)$$
$$= \sum_{k_1=0}^{N_1}\sum_{k_2=0}^{N_2} c_{k_1,k_2} \prod_{i=1}^{2} \int_{a_i}^{b_i} \phi_{i,k_i} h_i \, d\mu_1.$$

Case 2: $\Gamma \subset \{1,2\}$.

In this case, we choose any $(k_1,k_2) \in \{0,\ldots,N_1\} \times \{0,\ldots,N_2\}$. Then (ii), Theorem 6.1.7 and (iii) yield

$$\lim_{\substack{(y_1,y_2)\to(a_1,a_2)\\(y_1,y_2)\in(a_1,b_1]\times(a_2,b_2]}} W(\Gamma,(k_1,k_2),(y_1,y_2)) \prod_{i=1}^{2} h_i(y_i)$$
$$= \lim_{\substack{(y_1,y_2)\to(a_1,a_2)\\(y_1,y_2)\in(a_1,b_1]\times(a_2,b_2]}} \left\{\prod_{j\in\Gamma} \int_{y_j}^{b_j} \phi_{j,k_j} h_j \, d\mu_1\right\}\left\{\prod_{\ell\in\{1,2\}\setminus\Gamma} \int_{\theta_\ell}^{y_\ell} \phi_{\ell,k_\ell} h_\ell \, d\mu_1\right\}$$
$$\text{(for some } \theta_\ell \in [a_\ell, y_\ell] \ (\ell \in \{1,2\}\setminus\Gamma))$$
$$= 0.$$

Now we have

$$\inf_{(k_1,k_2)\in\mathbb{N}_0^2} \sum_{\Gamma\subseteq\{1,2\}} W(\Gamma,(k_1,k_2),(y_1,y_2)) \geq 0 \tag{8.6.11}$$

because $(k_1,k_2) \in \mathbb{N}_0^2$, (ii), Theorem 6.1.7 and (iii) imply

$$\sum_{\Gamma\subseteq\{1,2\}} W(\Gamma,(k_1,k_2),(y_1,y_2))$$
$$= \sum_{\Gamma\subseteq\{1,2\}} \prod_{j\in\Gamma} \int_{y_j}^{v_j} \phi_{j,k_j} \, d\mu_1 \prod_{\ell\in\{1,2\}\setminus\Gamma} \int_{a_\ell}^{y_\ell} \phi_{\ell,k_\ell} \, d\mu_1$$
$$\text{(for some } (v_1,v_2) \in [y_1,b_1]\times[y_2,b_2])$$
$$= \prod_{j=1}^{2} \int_{a_j}^{v_j} \phi_{j,k_j} \, d\mu_1$$
$$\geq 0.$$

Finally, since (i), (ii), (8.6.10) and (8.6.11) hold, it remains to prove that

$$\sup_{(y_1,y_2)\in(a_1,b_1]\times(a_2,b_2]} \sum_{(k_1,k_2)\in\mathbb{N}_0^2} c_{k_1,k_2} \sum_{\Gamma\subseteq\{1,2\}} W(\Gamma,(k_1,k_2),(y_1,y_2)) \prod_{i=1}^{2} h_i(y_i)$$
$$\leq 4\left\|\phi(h_1\otimes h_2)\right\|_{HK([a_1,b_1]\times[a_2,b_2])}. \tag{8.6.12}$$

Let $(y_1, y_2) \in (a_1, b_1] \times (a_2, b_2]$ be arbitrary. We consider the following cases.

Case α: $\Gamma = \{1,2\}$.

In this case, (iv), (ii), Theorem 6.5.12 and (v) yield

$$\begin{aligned}
&\left|\sum_{(k_1,k_2)\in\mathbb{N}_0^2} c_{k_1,k_2}\, W(\{1,2\},(k_1,k_2),(y_1,y_2))\, h_1(y_1)h_2(y_2)\right| \\
&= \left|(HK)\int_{[y_1,b_1]\times[y_2,b_2]} \phi(h_1\otimes h_2)\right| \\
&\leq \left\|\phi(h_1\otimes h_2)\right\|_{HK([a_1,b_1]\times[a_2,b_2])}.
\end{aligned} \tag{8.6.13}$$

Case β: $\Gamma = \emptyset$.

We use (iv), (ii), Theorem 6.5.13 and (v) to obtain (8.6.13):

$$\begin{aligned}
&\left|\sum_{(k_1,k_2)\in\mathbb{N}_0^2} c_{k_1,k_2}\, W(\emptyset,(k_1,k_2),(y_1,y_2))\, h_1(y_1)h_2(y_2)\right| \\
&= \left|(HK)\int_{[a_1,y_1]\times[a_2,y_2]} \phi\right| h_1(y_1)h_2(y_2) \\
&= \left|(HK)\int_{[\xi_1,y_1]\times[\xi_2,y_2]} \phi(h_1\otimes h_2)\right| \qquad \text{(for some } (\xi_1,\xi_2)\in\prod_{i=1}^{2}[a_i,y_i]) \\
&\leq \left\|\phi(h_1\otimes h_2)\right\|_{HK([a_1,b_1]\times[a_2,b_2])}.
\end{aligned}$$

Case γ: $\Gamma \subset \{1,2\}$ is non-empty.

Without loss of generality, we may assume that $\Gamma = \{1\}$. Then we infer from (iv), (ii), Theorem 6.5.12 and (v) that

$$\left| \sum_{(k_1,k_2)\in\mathbb{N}_0^2} c_{k_1,k_2} W(\{1\},(k_1,k_2),(y_1,y_2))\, h_1(y_1)h_2(y_2) \right|$$
$$= h_2(y_2) \left| (HK)\int_{[y_1,b_1]\times[a_2,y_2]} \phi(t_1,t_2)h_1(t_1)\, d(t_1,t_2) \right|. \quad (8.6.14)$$

Finally, it follows from (8.6.14), Fubini's Theorem, (ii), Theorem 6.1.6 and (v) that

$$\left| \sum_{(k_1,k_2)\in\mathbb{N}_0^2} c_{k_1,k_2} W(\{1\},(k_1,k_2),(y_1,y_2))\, h_1(y_1)h_2(y_2) \right|$$
$$= \left| (HK)\int_{[y_1,b_1]\times[\xi_2,y_2]} \phi(h_1\otimes h_2) \right| \qquad (\text{for some } \xi_2\in[a_2,y_2])$$
$$\leq \left\| \phi(h_1\otimes h_2) \right\|_{HK([a_1,b_1]\times[a_2,b_2])};$$

that is (8.6.13) holds. Combining the above cases yields (8.6.12) to be proved. □

8.7 Applications to double Fourier series

The aim of this section is to generalize Theorem 7.5.5 concerning single sine series. The proof of the following lemma is left to the reader.

Lemma 8.7.1. *If $x \geq 0$ and $p \in [0,2)$, then*

$$0 \leq \int_0^x \frac{\sin\theta}{\theta^p}\, d\mu_1(\theta) \leq \int_0^\pi \frac{\sin\theta}{\theta^p}\, d\mu_1(\theta). \quad (8.7.1)$$

The following lemma is a simple consequence of Lemma 8.7.1.

Lemma 8.7.2. *If $k \in \mathbb{N}$, $x \geq 0$ and $p \in [0,2)$, then*

$$0 \leq \frac{1}{k^{p-1}} \int_0^x \frac{\sin kt}{t^p}\, d\mu_1(t) \leq \int_0^\pi \frac{\sin\theta}{\theta}\, d\theta.$$

The following theorem is a substantial generalization of Theorem 7.5.5.

Theorem 8.7.3. *Let $g \in L^1(\mathbb{T}^2)$ and assume that*

$$g(t_1,t_2) \sim \sum_{(k_1,k_2)\in\mathbb{N}^2} b_{k_1,k_2} \sin k_1t_1 \sin k_2t_2.$$

If the double series $\sum_{(k_1,k_2)\in\mathbb{N}^2} |b_{k_1,k_2}|\, k_1^{\alpha_1-1} k_2^{\alpha_2-1}$ *converges for some* $(\alpha_1,\alpha_2)\in[0,1]^2$*, then there exists* $g_{\alpha_1,\alpha_2}\in HK([0,\pi]^2)$ *such that*

$$\lim_{\min\{n_1,n_2\}\to\infty} \|s_{n_1,n_2}g_{\alpha_1,\alpha_2} - g_{\alpha_1,\alpha_2}\|_{HK([0,\pi]^2} = 0.$$

Proof. Since $p\in[0,1]$ and Lemma 8.7.2 imply

$$\sup_{k\in\mathbb{N}}\sup_{x\in[0,\pi]}\left\{\frac{1}{k^{p-1}}\left|\int_0^x \frac{\sin kt}{t^p}\,dt\right|\right\} + \sup_{k\in\mathbb{N}}\sup_{\delta\in(0,\pi]} \delta^p \int_\delta^\pi \left|\frac{\sin kt}{t^p}\right| dt < \infty,$$

a simple application of Theorem 8.6.1 yields the result. □

If we begin with a double sine series with non-negative coefficients, we obtain a sharper version of Theorem 8.7.3.

Theorem 8.7.4. *Let* $(\beta_1,\beta_2)\in(0,1)^2$*, let* $g\in L^1(\mathbb{T}^2)$ *and suppose that*

$$g(t_1,t_2)\sim\sum_{(k_1,k_2)\in\mathbb{N}^2} b_{k_1,k_2}\sin k_1t_1\sin k_2t_2.$$

If $b_{k_1,k_2}\geq 0$ *for all* $(k_1,k_2)\in\mathbb{N}^2$*, then the double series* $\sum_{(k_1,k_2)\in\mathbb{N}^2} b_{k_1,k_2}k_1^{\beta_1-1}k_2^{\beta_2-1}$ *converges if and only if for each* $(\alpha_1,\alpha_2)\in[0,\pi)^2$ *there exists* $g_{\alpha_1,\alpha_2,\beta_1,\beta_2}\in HK([\alpha_1,\pi]\times[\alpha_2,\pi])$ *such that*

$$g_{\alpha_1,\alpha_2,\beta_1,\beta_2}(t_1,t_2) = \frac{g(t_1,t_2)}{(t_1-\alpha_1)^{\beta_1}(t_2-\alpha_2)^{\beta_2}}$$

for all $(t_1,t_2)\in(\alpha_1,\pi]\times(\alpha_2,\pi]$.

Proof. ($\Longrightarrow$) The proof is similar to that of Theorem 8.7.3.

($\Longleftarrow$) For this part of the proof we assume that $(\alpha_1,\alpha_2)=(0,0)$ and $(\beta_1,\beta_2)\in(0,1]^2$. Since

$$\inf_{(k_1,k_2)\in\mathbb{N}^2} b_{k_1,k_2}\geq 0,\ \inf_{n\in\mathbb{N}}\frac{1}{n^{p-1}}\int_0^\pi \frac{\sin nu}{u^p}\,du \geq \int_0^{2\pi}\frac{\sin\theta}{\theta^p}\,d\theta > 0\quad (p\in(0,1])$$

and

$$\inf_{x\in[0,\pi]}\inf_{n\in\mathbb{N}}\int_0^x \sin nt\,dt\geq 0,$$

the conclusion follows from Theorems 8.2.4(i) and 8.6.2. □

We remark that Theorem 8.7.4 is not true if $\beta_1=\beta_2=1$. However, the proofs of Theorems 8.7.3 and 8.7.4 give the following result.

Theorem 8.7.5. *Let $g \in L^1(\mathbb{T}^2)$ and suppose that*

$$g(t_1, t_2) \sim \sum_{(k_1,k_2)\in\mathbb{N}^2} b_{k_1,k_2} \sin k_1 t_1 \sin k_2 t_2.$$

If $b_{k_1,k_2} \geq 0$ for all $(k_1, k_2) \in \mathbb{N}^2$, then the double series $\sum_{(k_1,k_2)\in\mathbb{N}^2} b_{k_1,k_2}$ converges if and only if there exists $g_{0,0} \in HK([0,\pi]^2)$ such that

$$g_{0,0}(t_1, t_2) = \frac{g(t_1, t_2)}{t_1 t_2}$$

for all $(t_1, t_2) \in (0, \pi]^2$.

Our next theorem is a two-dimensional analogue of Lemma 7.6.1.

Theorem 8.7.6. *Let $f \in L^1(\mathbb{T}^2)$ and suppose that*

$$f(t_1, t_2) \sim \sum_{(k_1,k_2)\in\mathbb{N}^2} \Big\{ a_{k_1,k_2} \cos k_1 t_1 \cos k_2 t_2 + b_{k_1,k_2} \cos k_1 t_1 \sin k_2 t_2$$
$$+ c_{k_1,k_2} \sin k_1 t_1 \cos k_2 t_2 + d_{k_1,k_2} \sin k_1 t_1 \sin k_2 t_2 \Big\}.$$

If $a_{k_1,k_2} \geq 0$ for all $(k_1, k_2) \in \mathbb{N}^2$, then the double series $\sum_{(k_1,k_2)\in\mathbb{N}^2} \frac{a_{k_1,k_2}}{k_1 k_2}$ converges if and only if there exists $f_{0,0} \in HK([0,\pi]^2)$ such that

$$f_{0,0}(x_1, x_2) = \frac{1}{x_1 x_2} \int_{[-x_1,x_1]\times[-x_2,x_2]} f \, d\mu_2$$

for all $(x_1, x_2) \in (0, \pi]^2$.

Proof. According to Theorem 8.2.4,

$$\int_{\prod_{i=1}^2 [-x_i,x_i]} f \, d\mu_2 \sim \sum_{(k_1,k_2)\in\mathbb{N}^2} 2a_{k_1,k_2} \prod_{i=1}^{2} \frac{\sin k_i x_i}{k_i}.$$

An appeal to Theorem 8.7.5 completes the proof. □

The following example shows that Theorem 8.6.1 is beyond the realm of Lebesgue integration.

Example 8.7.7. The function $g_2 : [0,\pi]^2 \longrightarrow \mathbb{R}$ is defined by

$$g_2(t_1, t_2) = \begin{cases} \displaystyle\sum_{k_1=0}^{\infty} \sum_{k_2=1}^{\infty} \frac{1}{(k_1 + k_2)^3} \frac{\sin(2^{k_1} t_1) \sin(2^{k_2} t_2)}{\tan \frac{t_1}{2} \tan \frac{t_2}{2}} & \text{if } (t_1, t_2) \in (0,\pi)^2, \\ 0 & \text{if } (t_1, t_2) \in [0,\pi]^2 \backslash (0,\pi)^2. \end{cases}$$

Then $g_2 \in HK([0,\pi]^2) \backslash L^1([0,\pi]^2)$ and

$$(HK) \int_{[0,\pi]^2} g_2 = \frac{\pi^4}{6}.$$

Proof. By Theorems 8.7.3 and 6.5.9, $g_2 \in HK([0,\pi]^2)$ and

$$\begin{aligned}(HK)\int_{[0,\pi]^2} g_2 &= \sum_{k_1=0}^{\infty}\sum_{k_2=1}^{\infty}\frac{1}{(k_1+k_2)^3}\int_{[0,\pi]^2}\prod_{i=1}^{2}\frac{\sin(2^{k_i}t_i)}{\tan\frac{t_i}{2}}\,d(t_1,t_2)\\ &= \pi^2\sum_{k_1=0}^{\infty}\sum_{k_2=1}^{\infty}\frac{1}{(k_1+k_2)^3}\\ &- \pi^2\sum_{r=1}^{\infty}\sum_{\substack{(k_1,k_2)\in\mathbb{N}_0^2\\ k_2\geq 1\\ k_1+k_2=r}}\frac{1}{(k_1+k_2)^3}\\ &= \pi^2\sum_{r=1}^{\infty}\frac{r}{r^3}\\ &= \frac{\pi^4}{6}.\end{aligned}$$

It remains to prove that $g_2 \notin L^1([0,\pi]^2)$. But this is an easy consequence of Theorem 8.4.3, since the double series $\sum_{(j,k)\in\mathbb{N}^2}\frac{1}{(j+k)^6}$ converges,

$$\sup_{k\in\mathbb{N}}\sum_{j=1}^{\infty}\sqrt{\sum_{p=j}^{\infty}\frac{1}{(p+k)^6}}+\sup_{j\in\mathbb{N}}\sum_{k=1}^{\infty}\sqrt{\sum_{q=k}^{\infty}\frac{1}{(j+q)^6}}<\infty$$

and

$$\sum_{(j,k)\in\mathbb{N}^2}\sqrt{\sum_{p=j}^{\infty}\sum_{q=k}^{\infty}\frac{1}{(p+q)^6}}\geq\sum_{(j,k)\in\mathbb{N}^2}\frac{1}{5(j+k)^2}=\infty.$$

□

Exercise 8.7.8. Let $(a_{k_1,k_2})_{(k_1,k_2)\in\mathbb{N}_0^2}$ be a double sequence of real numbers such that $\lim_{\max\{n_1,n_2\}\to\infty} a_{n_1,n_2} = 0$ and $\sum_{(k_1,k_2)\in\mathbb{N}_0^2}|\Delta_{1,2}(a_{k_1,k_2})|$ converges. Prove that

(i) the double series $\sum_{(k_1,k_2)\in\mathbb{N}_0^2}\lambda_{k_1}\lambda_{k_2}a_{k_1,k_2}\cos k_1t_1\cos k_2t_2$ converges regularly for all $(t_1,t_2)\in(0,\pi]^2$;

(ii) there exists $f \in HK([0,\pi]^2)$ such that

$$f(t_1,t_2)=\sum_{(k_1,k_2)\in\mathbb{N}_0^2}\lambda_{k_1}\lambda_{k_2}a_{k_1,k_2}\cos k_1t_1\cos k_2t_2$$

for all $(t_1,t_2)\in(0,\pi]^2$;

(iii)

$$\left|(HK)\int_{[0,\pi]^2}\left\{\sum_{k_1=0}^{n_1}\sum_{k_2=0}^{n_2}a_{k_1,k_2}\prod_{i=1}^{2}\lambda_i\cos k_it_i-f(t_1,t_2)\right\}d(t_1,t_2)\right|\to 0$$

as $\min\{n_1,n_2\}\to\infty$.

8.8 Another convergence theorem for Henstock-Kurzweil integrals

In this section we give another convergence theorem for Henstock-Kurzweil integrals whose proof depends on Theorem 8.3.3. In fact, the proof of this result is essentially based on that of Exercise 8.7.8.

Theorem 8.8.1. *Let* $(\phi_{1,n})_{n=0}^{\infty}$ *(resp.* $(\phi_{2,n})_{n=0}^{\infty}$*) be a sequence in* $L^1[a_1,b_1]$ *(resp.* $L^1[a_2,b_2]$*) and suppose that the set*

$$\Xi := \left\{r \in \{1,2\} : \sup_{n\in\mathbb{N}_0} \left\|\sum_{k=0}^{n} \phi_{r,k}\right\|_{HK[a_r,b_r]} < \infty\right\}$$

is non-empty. If $(c_{k_1,k_2})_{(k_1,k_2)\in\mathbb{N}_0^2}$ *is a double sequence of real numbers such that*

$$\sum_{(k_1,k_2)\in\mathbb{N}_0^2} \left\{ |\Delta_{\Gamma\cup\Xi}(c_{k_1,k_2})| \prod_{\ell\in\{1,2\}\setminus(\Gamma\cup\Xi)} \|\phi_{\ell,k_\ell}\|_{HK[a_\ell,b_\ell]} \right.$$
$$\left. \prod_{j\in\Gamma} \max_{q_j=0,\ldots,k_j} \left\|\sum_{r_j=0}^{q_j} \phi_{j,r_j}\right\|_{L^1[a_j+\frac{b_j-a_j}{n+1},b_j]} \right\} \tag{8.8.1}$$

converges for every $\Gamma \subseteq \{1,2\}$ *and* $n \in \mathbb{N}$*, then the following assertions hold.*

(i) *The double series*

$$\sum_{(k_1,k_2)\in\mathbb{N}_0^2} c_{k_1,k_2} \int_I \prod_{i=1}^{2} \phi_{i,k_i}(t_i)\, d\mu_2(t_1,t_2)$$

converges regularly for all $I \in \mathcal{I}_2([a_1,b_1]\times[a_2,b_2])$.

(ii) *There exists* $f \in HK([a_1,b_1]\times[a_2,b_2])$ *such that*

$$\lim_{\min\{n_1,n_2\}\to\infty} \left\| \sum_{k_1=0}^{n_1}\sum_{k_2=0}^{n_2} c_{k_1,k_2} \bigotimes_{i=1}^{2} \phi_{i,k_i} - f \right\|_{HK([a_1,b_1]\times[a_2,b_2])} = 0. \tag{8.8.2}$$

Proof. We may assume that

$$\Xi = \left\{r \in \{1,2\} : 0 < \sup_{n\in\mathbb{N}_0} \left\|\sum_{k=0}^{n} \phi_{r,k}\right\|_{HK[a_r,b_r]} \le 1\right\}. \tag{8.8.3}$$

(i) Clearly, it suffices to prove that

$$\left\| \sum_{k_1=p_1}^{q_1}\sum_{k_2=p_2}^{q_2} c_{k_1,k_2} \bigotimes_{i=1}^{2} \phi_{i,k_i} \right\|_{HK([a_1,b_1]\times[a_2,b_2])}$$
$$\to 0 \text{ as } \max\{|||(p_1,p_2)|||, |||(q_1,q_2)|||\} \to \infty. \tag{8.8.4}$$

Let $\kappa_0 > 0$ be given. Using (8.8.1) with $\Gamma = \emptyset$, we select $N \in \mathbb{N}$ so that

$$\max_{i=1,2} \sum_{\substack{(k_1,k_2)\in\mathbb{N}_0^2 \\ k_i \geq N}} |\Delta_\Xi(c_{k_1,k_2})| \prod_{\ell\in\{1,2\}\setminus\Xi} ||\phi_{\ell,k_\ell}||_{HK[a_\ell,b_\ell]}$$
$$< \frac{\kappa_0}{16}\left(\max_{r\in\Xi}\sup_{n\in\mathbb{N}_0}\left|\left|\sum_{k=0}^{n}\phi_{r,k}\right|\right|_{HK[a_r,b_r]}\right)^{-1}. \qquad (8.8.5)$$

Hence, for any $(p_1,p_2),(q_1,q_2) \in \mathbb{N}_0^2$ satisfying $q_i \geq p_i$ $(i = 1,2)$ and $\max\{p_1,p_2\} \geq N$, we write $\Xi' = \{1,2\}\setminus\Xi$ so that

$$\left|\left|\sum_{k_1=p_1}^{q_1}\sum_{k_2=p_2}^{q_2} c_{k_1,k_2}\bigotimes_{i=1}^{2}\phi_{i,k_i}\right|\right|_{HK([a_1,b_1]\times[a_2,b_2])}$$
$$\leq \sum_{\substack{(r_1,r_2)\in\mathbb{N}_0^2 \\ r_j=0\,\forall\, j\in\Xi \\ p_\ell\leq r_\ell\leq q_\ell\,\forall\,\ell\in\,\Xi'}} \left|\left|\sum_{\substack{(k_1,k_2)\in\mathbb{N}_0^2 \\ p_j\leq k_j\leq q_j\,\forall\, j\in\Xi \\ k_\ell=r_\ell\,\forall\,\ell\in\Xi'}} c_{k_1,k_2}\bigotimes_{j\in\Xi}\phi_{j,k_j}\bigotimes_{\ell\in\Xi'}\phi_{\ell,k_\ell}\right|\right|_{HK(\prod_{i=1}^{2}[a_i,b_i])}$$

(by triangle inequality)

$$\leq 16\max_{r\in\Xi}\sup_{n\in\mathbb{N}_0}\left|\left|\sum_{k=0}^{n}\phi_{r,k}\right|\right|_{HK[a_r,b_r]}\sum_{k_1=p_1}^{\infty}\sum_{k_2=p_2}^{\infty}|\Delta_\Xi(c_{\mathbf{k}})|\prod_{\ell\in\Xi'}||\phi_{\ell,k_\ell}||_{HK[a_\ell,b_\ell]}$$

(by summation by parts and triangle inequality)

$$< \kappa_0 \qquad \text{(by (8.8.5))}.$$

Since $\kappa_0 > 0$ is arbitrary, (8.8.4) follows.

We are now ready to prove (ii). Applying (8.8.1) with $\Gamma = \{1,2\}$, we see that

$$\sum_{(k_1,k_2)\in\mathbb{N}_0^2}|\Delta_{\{1,2\}}(c_{k_1,k_2})|\prod_{j=1}^{2}\max_{q_j=0,\ldots,k_j}\left|\left|\sum_{r_j=0}^{q_j}\phi_{j,r_j}\right|\right|_{L^1[a_j+\frac{b_j-a_j}{n+1},b_j]} \qquad (8.8.6)$$

converges for every $n \in \mathbb{N}$. Hence, by Theorem 8.3.3, there exists a function $f : [a_1,b_1]\times[a_2,b_2] \longrightarrow \mathbb{R}$ such that

$$f \in L^1(I) \text{ and } \int_I f\, d\mu_2 = \sum_{(k_1,k_2)\in\mathbb{N}_0^2} c_{k_1,k_2}\int_I \phi_{1,k_1}\otimes\phi_{2,k_2}\, d\mu_2 \qquad (8.8.7)$$

whenever $I \in \mathcal{I}_2([a_1,b_1]\times[a_2,b_2])$ with $I \subset (a_1,b_1]\times(a_2,b_2]$.

Since (i) holds, we can define the interval function $G : \mathcal{I}_2([a_1,b_1]\times[a_2,b_2]) \longrightarrow \mathbb{R}$ by setting

$$G([u_1,v_1]\times[u_2,v_2]) = \sum_{(k_1,k_2)\in\mathbb{N}_0^2} c_{k_1,k_2}\prod_{i=1}^{2}\int_{u_i}^{v_i}\phi_{i,k_i}\,d\mu_1.$$

We will next show that $f \in HK([a_1,b_1]\times[a_2,b_2])$ and G is the indefinite Henstock-Kurzweil integral of f. In view of (8.8.7), Theorem 5.5.9 and (8.8.4), it remains to prove that

$$V_{\mathcal{HK}}G\Big(([a_1,b_1]\times[a_2,b_2])\backslash((a_1,b_1]\times(a_2,b_2])\Big) = 0.$$

Thanks to Theorem 5.2.1(iii), it suffices to prove that

$$V_{\mathcal{HK}}G(Z_{\Gamma,n}) = 0$$

whenever $n \in \mathbb{N}$ and $\Gamma \subset \{1,2\}$, where

$$([a_1,b_1]\times[a_2,b_2])\backslash((a_1,b_1]\times(a_2,b_2]) = \bigcup_{n\in\mathbb{N}}\bigcup_{\Gamma\subset\{1,2\}} Z_{\Gamma,n},$$

$$Z_{\Gamma,n} := \prod_{k=1}^{2} Z_{\Gamma,n,k} \text{ and } Z_{\Gamma,n,k} := \begin{cases} \{a_k\} & \text{if } k \in \{1,2\}\backslash\Gamma, \\ [a_k + \frac{b_k-a_k}{n+1}, b_k] & \text{if } k \in \Gamma. \end{cases}$$

For each $\varepsilon > 0$ we use (8.8.1) to pick a positive integer K so that

$$\sum_{\substack{(k_1,k_2)\in\mathbb{N}_0^2 \\ (k_1,k_2)\notin[0,K]^2}} \Bigg\{ |\Delta_{\Gamma\cup\Xi}(c_{k_1,k_2})| \prod_{\ell\in\{1,2\}\backslash(\Gamma\cup\Xi)} \|\phi_{\ell,k_\ell}\|_{HK[a_\ell,b_\ell]}$$

$$\prod_{j\in\Gamma}\max_{q_j=0,\ldots,k_j}\Bigg\|\sum_{r_j=0}^{q_j}\phi_{j,r_j}\Bigg\|_{L^1(Z_{\Gamma,n,j})}\Bigg\}$$

$$< \frac{\varepsilon}{2}\Bigg(1 + \max_{r\in\Xi}\sup_{n\in\mathbb{N}_0}\Bigg\|\sum_{k=0}^{n}\phi_{r,k}\Bigg\|_{HK[a_r,b_r]}\Bigg)^{-1}. \tag{8.8.8}$$

Employing the uniform continuity of indefinite Henstock-Kurzweil integral, we select an $\eta(\Gamma,n) > 0$ so that

$$\max_{i\in\{1,2\}\backslash\Gamma}\max_{k_i=0,\ldots,K}\sup_{\substack{[u,v]\subseteq[a_i,b_i] \\ 0<v-u<\eta(\Gamma,n)}} \|\phi_{i,k_i}\|_{HK[u,v]}$$

$$< \frac{\varepsilon}{2}\Bigg(1 + \sum_{k_1=0}^{N_1}\sum_{k_2=0}^{N_2}|\Delta_{\Gamma\cup\Xi}(c_{k_1,k_2})|\prod_{\ell\in\Gamma}\Bigg\|\sum_{j_\ell=0}^{k_\ell}\phi_{\ell,j_\ell}\Bigg\|_{L^1(Z_{\Gamma,n,\ell})}\Bigg)^{-1}. \tag{8.8.9}$$

Define a gauge $\delta_{\Gamma,n}$ on $Z_{\Gamma,n}$ by setting $\delta_{\Gamma,n}(x_1,x_2) := \eta(\Gamma,n)$, and let $P_{\Gamma,n}$ be a $Z_{\Gamma,n}$-tagged $\delta_{\Gamma,n}$-fine Perron subpartition of $[a_1,b_1]\times[a_2,b_2]$. Since all integrals are real-valued, it suffices to prove that

$$\left|\sum_{(t,I)\in P_{\Gamma,n}}\sum_{(k_1,k_2)\in\mathbb{N}_0^2} c_{k_1,k_2}\int_I \phi_{1,k_1}\otimes\phi_{2,k_2}\,d\mu_2\right| < \varepsilon. \tag{8.8.10}$$

Proof of (8.8.10). Write $T=\Gamma\cup\Xi$ and $\Phi_{i,k_i}=\sum_{j_i=0}^{k_i}\phi_{i,j_i}$ $(i=1,2)$. According to Theorems 7.1.2(iii) and 8.1.13(iv),

$$\begin{aligned}
&\left|\sum_{(t,I)\in P_{\Gamma,n}}\sum_{(k_1,k_2)\in\mathbb{N}_0^2} c_{k_1,k_2}\int_I \phi_{1,k_1}\otimes\phi_{2,k_2}\,d\mu_2\right|\\
&=\left|\sum_{(k_1,k_2)\in\mathbb{N}_0^2}\sum_{(t,I)\in P_{\Gamma,n}}\Delta_T(c_{k_1,k_2})\int_I\Big(\prod_{i\in T}\Phi_{i,k_i}\otimes\prod_{j\in\{1,2\}\setminus T}\phi_{j,k_j}\Big)\,d\mu_2\right|\\
&\le\sum_{k_1=0}^{K}\sum_{k_2=0}^{K}|\Delta_T(c_{k_1,k_2})|\left|\sum_{(t,I)\in P_{\Gamma,n}}\int_I\Big(\prod_{i\in T}\Phi_{i,k_i}\otimes\prod_{j\in\{1,2\}\setminus T}\phi_{j,k_j}\Big)\,d\mu_2\right|\\
&\quad+\sum_{\substack{(k_1,k_2)\in\mathbb{N}_0^2\\(k_1,k_2)\notin[0,K]^2}}|\Delta_T(c_{k_1,k_2})|\left|\sum_{(t,I)\in P_{\Gamma,n}}\int_I\Big(\prod_{i\in T}\Phi_{i,k_i}\otimes\prod_{j\in\{1,2\}\setminus T}\phi_{j,k_j}\Big)\,d\mu_2\right|\\
&=R_1+R_2,
\end{aligned}$$

say. To complete the proof of (8.8.10), we have to establish the following claims.

Claim 1: If $\Gamma=\emptyset$, then $\max\{R_1,R_2\}<\frac{\varepsilon}{2}$.

Since $\mathrm{card}(P_{\emptyset,n})=1$, (8.8.9) and (8.8.3) imply that $R_1<\frac{\varepsilon}{2}$. Likewise, we infer from (8.8.8) and (8.8.3) that $R_2<\frac{\varepsilon}{2}$. Thus, claim 1 holds.

Claim 2: If $\Gamma\subset\{1,2\}$ is non-empty, then $\max\{R_1,R_2\}<\frac{\varepsilon}{2}$.

In this case, our choice of $P_{\Gamma,n}$ implies that $\mu_{\mathrm{card}(\Gamma)}(\prod_{i\in\Gamma}I_i\cap\prod_{i\in\Gamma}U_i)=0$ whenever $((x_1,x_2),I_1\times I_2)$ and $((y_1,y_2),U_1\times U_2)$ are two distinct elements of $P_{\Gamma,n}$. Combining this with (8.8.9) and (8.8.3), we get

$$\begin{aligned}
R_1\le\sum_{k_1=0}^{K}\sum_{k_2=0}^{K}\Bigg\{&|\Delta_T(c_{k_1,k_2})|\prod_{i\in\{1,2\}\setminus T}\|\phi_{i,k_i}\|_{HK[a_i,b_i]}\\
&\prod_{j\in\Xi\setminus\Gamma}\|\Phi_{j,k_j}\|_{HK[a_j,b_j]}\prod_{\ell\in\Gamma}\|\Phi_{\ell,k_\ell}\|_{L^1(Z_{\Gamma,n,\ell})}\Bigg\}<\frac{\varepsilon}{2}.
\end{aligned}$$

A similar reasoning shows that $R_2 < \frac{\varepsilon}{2}$. The proof is complete. □

The following corollary, which improves parts (ii) and (iii) of Exercise 8.7.8, is an immediate consequence of Theorems 8.1.13 and 8.8.1.

Corollary 8.8.2. *Let $S \subseteq \{1,2\}$ be a non-empty set and let $(c_{k_1,k_2})_{(k_1,k_2)\in\mathbb{N}_0^2}$ be a double sequence of real numbers such that $\lim_{\max\{n_1,n_2\}\to\infty} c_{n_1,n_2} = 0$. If the double series*

$$\sum_{\substack{(k_1,k_2)\in\mathbb{N}_0^2 \\ k_i>0\,\forall\, i\in\{1,2\}\setminus(\Gamma\cup S)}} \left|\frac{\Delta_{\Gamma\cup S}(c_{k_1,k_2})}{\prod_{i\in\{1,2\}\setminus(\Gamma\cup S)} k_i}\right| \quad \textit{converges for every } \Gamma\subseteq\{1,2\},$$

then the following assertions hold.

(i) *The double series*

$$\sum_{(k_1,k_2)\in\mathbb{N}_0^2} c_{k_1,k_2}\Big\{\prod_{i\in S}\cos k_i x_i\Big\}\Big\{\prod_{j\in\{1,2\}\setminus S}\sin k_j x_j\Big\}$$

converges regularly for all $(x_1,x_2)\in(0,\pi]^2$.

(ii) *Let*

$$f_S(t_1,t_2) := \begin{cases} \sum_{(k_1,k_2)\in\mathbb{N}_0^2} c_{k_1,k_2}\{\prod_{i\in S}\cos k_i t_i\}\{\prod_{j\in\{1,2\}\setminus S}\sin k_j t_j\} \\ \qquad\qquad\qquad\qquad \textit{if } (t_1,t_2)\in(0,\pi]^2, \\ 0 \qquad\qquad\qquad\qquad \textit{otherwise,} \end{cases}$$

and let

$$f_{n_1,n_2,S}(t_1,t_2) = \sum_{k_1=0}^{n_1}\sum_{k_2=0}^{n_2} c_{k_1,k_2}\Big\{\prod_{i\in S}\cos k_i t_i\Big\}\Big\{\prod_{j\in\{1,2\}\setminus S}\sin k_j t_j\Big\}.$$

Then $f_S\in HK([-\pi,\pi]^2)$ and

$$\lim_{\min\{n_1,n_2\}\to\infty} \|f_{n_1,n_2,S} - f_S\|_{HK([-\pi,\pi]^2} = 0.$$

8.9 A two-dimensional analogue of Boas' theorem

The aim of this section is to prove a useful two-dimensional analogue of Theorem 7.5.2. We need the following lemmas.

Lemma 8.9.1. *Let $(c_{k_1,k_2})_{(k_1,k_2)\in\mathbb{N}^2}$ be a double sequence of real numbers such that*

$$\lim_{\max\{n_1,n_2\}\to\infty} c_{n_1,n_2} = 0 \tag{8.9.1}$$

and

$$\sum_{(k_1,k_2)\in\mathbb{N}^2} \left\{ \left| \frac{\Delta_{\{2\}}(c_{k_1,k_2})}{k_1} \right| + \left| \frac{\Delta_{\{1\}}(c_{k_1,k_2})}{k_2} \right| + \left| \Delta_{\{1,2\}}(c_{k_1,k_2}) \right| \right\} < \infty. \tag{8.9.2}$$

If $j \in \{1,2\}$ and $\alpha_j \in \mathbb{N}$, then the series

$$\sum_{\substack{(k_1,k_2)\in\mathbb{N}^2 \\ k_j=\alpha_j}} c_{k_1,k_2} \left(\prod_{\substack{i=1 \\ i\neq j}}^{2} k_i \right)^{-1}$$

is absolutely convergent.

Proof. Exercise. □

Lemma 8.9.2. *Let $(c_{k_1,k_2})_{(k_1,k_2)\in\mathbb{N}^2}$ be a double sequence of real numbers such that (8.9.1) and (8.9.2) hold. If $j \in \{1,2\}$, then*

$$\lim_{n_j\to\infty} \sum_{\substack{(k_1,k_2)\in\mathbb{N}^2 \\ k_j=n_j}} |c_{k_1,k_2}| \left(\prod_{\substack{i=1 \\ i\neq j}}^{2} k_i \right)^{-1} = 0.$$

Proof. Exercise. □

The following theorem, which is a generalization of [101, Theorem 4.3], is a two-dimensional analogue of Theorem 7.5.1.

Theorem 8.9.3. *Let $(c_{k_1,k_2})_{(k_1,k_2)\in\mathbb{N}^2}$ be a double sequence of real numbers such that*

$$\lim_{\max\{n_1,n_2\}\to\infty} c_{n_1,n_2} = 0 \tag{8.9.3}$$

and

$$\sum_{(k_1,k_2)\in\mathbb{N}^2} \left\{ \left| \frac{\Delta_{\{2\}}(c_{k_1,k_2})}{k_1} \right| + \left| \frac{\Delta_{\{1\}}(c_{k_1,k_2})}{k_2} \right| + \left| \Delta_{\{1,2\}}(c_{k_1,k_2}) \right| \right\} < \infty. \tag{8.9.4}$$

If $(\varphi_{1,n})_{n=1}^{\infty}$ (resp. $(\varphi_{2,n})_{n=1}^{\infty}$) is a sequence in $C[a_1,b_1]$ (resp. $C[a_2,b_2]$) such that

$$\max_{i=1,2}\left\{\sup_{n\in\mathbb{N}}||\varphi_{i,n}||_{C[a_i,b_i]}+\sup_{a_i<x_i<b_i}\sup_{n\in\mathbb{N}}\left|\frac{\varphi_{i,n}(x_i)-\varphi_{i,n}(a_i)}{n(x_i-a_i)}\right|\right.$$
$$\left.+\sup_{x_i\in[a_i,b_i]}\sup_{n\in\mathbb{N}}\left|\sum_{k=1}^{n}(x_i-a_i)\varphi_{i,k}(x_i)\right|\right\}<\infty,$$

then the following statements are true:

(i) *the double series*

$$\sum_{(k_1,k_2)\in\mathbb{N}^2}\frac{c_{k_1,k_2}}{k_1k_2}\varphi_{1,k_1}(x_1)\varphi_{2,k_2}(x_2)$$

converges regularly whenever $(x_1,x_2)\in([a_1,b_1]\times[a_2,b_2])\backslash\{(a_1,a_2)\}$;

(ii) *the function*

$$(x_1,x_2)\mapsto\sum_{(k_1,k_2)\in\mathbb{N}^2}\frac{c_{k_1,k_2}}{k_1k_2}\varphi_{1,k_1}(x_1)\varphi_{2,k_2}(x_2)$$

is continuous on $([a_1,b_1]\times[a_2,b_2])\backslash\{(a_1,a_2)\}$;

(iii) *if the limit*

$$\lim_{\substack{(x_1,x_2)\to(a_1,a_2)\\(x_1,x_2)\in(a_1,b_1)\times(a_2,b_2)}}\sum_{(k_1,k_2)\in\mathbb{N}^2}\frac{c_{k_1,k_2}}{k_1k_2}\varphi_{1,k_1}(x_1)\varphi_{2,k_2}(x_2)\quad\textit{exists},$$

then

$$\sum_{(k_1,k_2)\in\mathbb{N}^2}\frac{c_{k_1,k_2}}{k_1k_2}\varphi_{1,k_1}(a_1)\varphi_{2,k_2}(a_2)$$

converges regularly;

(iv) *if the double series*

$$\sum_{(k_1,k_2)\in\mathbb{N}^2}\frac{c_{k_1,k_2}}{k_1k_2}\varphi_{1,k_1}(a_1)\varphi_{2,k_2}(a_2)$$

is regularly convergent, the the function

$$(x_1,x_2)\mapsto\sum_{(k_1,k_2)\in\mathbb{N}^2}\frac{c_{k_1,k_2}}{k_1k_2}\varphi_{1,k_1}(x_1)\varphi_{2,k_2}(x_2)$$

is continuous on $[a_1,b_1]\times[a_2,b_2]$.

Proof. We may assume that

$$\max_{i=1,2}\left\{\sup_{n\in\mathbb{N}}||\varphi_{i,n}||_{C[a_i,b_i]}+\sup_{a_i<x_i<b_i}\sup_{n\in\mathbb{N}}\left|\frac{\varphi_{i,n}(x_i)-\varphi_{i,n}(a_i)}{n(x_i-a_i)}\right|\right.$$

$$\left.+\sup_{x_i\in[a_i,b_i]}\sup_{n\in\mathbb{N}}\left|\sum_{k=1}^{n}(x_i-a_i)\varphi_{i,k}(x_i)\right|\right\}\le\frac{1}{2}. \qquad (8.9.5)$$

(i) We consider three cases.

Case 1: $(x_1,x_2)\in(a_1,b_1]\times(a_2,b_2]$.

In this case, the result follows from Theorem 8.1.13 (iv).

Case 2: $(x_1,a_2)\in(a_1,b_1]\times[a_2,b_2]$.

Let $(p_1,p_2),(q_1,q_2)\in\mathbb{N}^2$ with $q_i>p_i$ $(i=1,2)$. According to the triangle inequality and (8.9.5),

$$\left|\sum_{k_1=p_1+1}^{q_1}\sum_{k_2=p_2+1}^{q_2}\frac{c_{k_1,k_2}}{k_1k_2}\varphi_{1,k_1}(x_1)\varphi_{2,k_2}(a_2)\right|$$

$$\le\sum_{k_2=p_2+1}^{q_2}\frac{1}{2k_2}\left|\sum_{k_1=p_1+1}^{q_1}\frac{c_{k_1,k_2}}{k_1}\varphi_{1,k_1}(x_1)\right|. \qquad (8.9.6)$$

Since Theorem 7.1.2(ii), triangle inequality and (8.9.5) imply that

$$\sup_{n\in\mathbb{N}}\left|\sum_{k_1=1}^{n}\frac{(x_1-a_1)\varphi_{1,k_1}(x_1)}{k_1}\right|\le 3\sum_{k_1=1}^{\infty}\left(\frac{1}{k_1}-\frac{1}{k_1+1}\right)\times\frac{1}{2}<2, \qquad (8.9.7)$$

we deduce from (8.9.6) and Theorem 7.1.2(ii) that

$$\left|\sum_{k_1=p_1+1}^{q_1}\sum_{k_2=p_2+1}^{q_2}\frac{c_{k_1,k_2}}{k_1k_2}\varphi_{1,k_1}(x_1)\varphi_{2,k_2}(a_2)\right|$$

$$\le 4\sum_{k_2=p_2+1}^{q_2}\sum_{k_2=p_2+1}^{q_2}\left|\frac{\Delta_{\{1\}}(c_{k_1,k_2})}{k_2(x_1-a_1)}\right|. \qquad (8.9.8)$$

Since (8.9.8) and (8.9.4) hold, it is now easy to show that the double series

$$\sum_{(k_1,k_2)\in\mathbb{N}^2}\frac{c_{k_1,k_2}}{k_1k_2}\varphi_{1,k_1}(x_1)\varphi_{2,k_2}(a_2)$$

converges regularly.

Case 3: $(a_1, x_2) \in [a_1, b_1] \times (a_2, b_2]$.

In this case, we modify the proof of case 2 to conclude that the double series

$$\sum_{(k_1,k_2)\in\mathbb{N}^2} \frac{c_{k_1,k_2}}{k_1 k_2} \varphi_{1,k_1}(a_1)\varphi_{2,k_2}(x_2)$$

converges regularly.

Combining the above cases yields (i) to be proved.

Since the proof of (ii) is similar to that of (i), it remains to prove (iii) and (iv). Due to assertion (i), we define a function $\varphi : \big([a_1, b_1] \times [a_2, b_2]\big)\backslash\{(a_1, a_2)\} \longrightarrow \mathbb{R}$ by setting

$$\varphi(x_1, x_2) = \sum_{(k_1,k_2)\in\mathbb{N}^2} c_{k_1,k_2} \prod_{i=1}^{2} \frac{\varphi_{i,k_i}(x_i)}{k_i}.$$

In view of Lemmas 8.9.1, 8.9.2, and Theorem 8.1.9, it suffices to prove that

$$\lim_{\substack{(\delta_1,\delta_2)\to(0,0)\\ (\delta_1,\delta_2)\in(0,1)^2}} \left| \sum_{k_1=1}^{\lfloor\frac{1}{\delta_1}\rfloor} \sum_{k_2=1}^{\lfloor\frac{1}{\delta_2}\rfloor} c_{k_1,k_2} \prod_{i=1}^{2} \frac{\varphi_{i,k_i}(a_i)}{k_i} - \varphi(a_1+\delta_1, a_2+\delta_2) \right| = 0, \tag{8.9.9}$$

$$\lim_{\delta_1\to 0+} \left| \sum_{k_1=1}^{\lfloor\frac{1}{\delta_1}\rfloor} \sum_{k_2=1}^{\infty} \frac{c_{k_1,k_2}}{k_1 k_2} \prod_{i=1}^{2} \varphi_{i,k_i}(a_i) - \varphi(a_1+\delta_1, a_2) \right| = 0 \tag{8.9.10}$$

and

$$\lim_{\delta_2\to 0+} \left| \sum_{k_1=1}^{\infty} \sum_{k_2=1}^{\lfloor\frac{1}{\delta_2}\rfloor} \frac{c_{k_1,k_2}}{k_1 k_2} \prod_{i=1}^{2} \varphi_{i,k_i}(a_i) - \varphi(a_1, a_2+\delta_2) \right| = 0. \tag{8.9.11}$$

Proof of (8.9.9). For each $(\delta_1, \delta_2) \in (0,1)^2$, we estimate the left-hand side of (8.9.9):

$$\left| \sum_{k_1=1}^{\lfloor \frac{1}{\delta_1} \rfloor} \sum_{k_2=1}^{\lfloor \frac{1}{\delta_2} \rfloor} \frac{c_{k_1,k_2}}{k_1 k_2} \prod_{i=1}^{2} \varphi_{i,k_i}(a_i) - \varphi(a_1+\delta_1, a_2+\delta_2) \right|$$

$$\leq \left| \sum_{k_1=1}^{\lfloor \frac{1}{\delta_1} \rfloor} \sum_{k_2=1}^{\lfloor \frac{1}{\delta_2} \rfloor} \frac{c_{k_1,k_2}}{k_1 k_2} \left(\prod_{i=1}^{2} \varphi_{i,k_i}(a_i) - \prod_{i=1}^{2} \varphi_{i,k_i}(a_i+\delta_i) \right) \right|$$

$$+ \left| \sum_{k_1=\lfloor \frac{1}{\delta_1} \rfloor+1}^{\infty} \sum_{k_2=1}^{\lfloor \frac{1}{\delta_2} \rfloor} \frac{c_{k_1,k_2}}{k_1 k_2} \prod_{i=1}^{2} \varphi_{i,k_i}(a_i+\delta_i) \right|$$

$$+ \left| \sum_{k_2=\lfloor \frac{1}{\delta_2} \rfloor+1}^{\infty} \sum_{k_1=1}^{\infty} \frac{c_{k_1,k_2}}{k_1 k_2} \prod_{i=1}^{2} \varphi_{i,k_i}(a_i+\delta_i) \right|$$

$$= A(\delta_1, \delta_2) + B(\delta_1, \delta_2) + C(\delta_1, \delta_2),$$

say.

Due to the assumptions (8.9.3) and (8.9.4), it suffices to prove that

$$A_{\delta_1,\delta_2} \leq \frac{1}{\lfloor \frac{1}{\delta_1} \rfloor} \sum_{k_1=1}^{\lfloor \frac{1}{\delta_1} \rfloor} \sum_{k_2=1}^{\infty} \sum_{r=k_1}^{\infty} \left| \frac{c_{r,k_2} - c_{r+1,k_2}}{k_2} \right| + \frac{1}{\lfloor \frac{1}{\delta_1} \rfloor \lfloor \frac{1}{\delta_2} \rfloor} \sum_{k_1=1}^{\lfloor \frac{1}{\delta_1} \rfloor} \sum_{k_2=1}^{\lfloor \frac{1}{\delta_2} \rfloor} |c_{k_1,k_2}|$$

$$+ \frac{1}{\lfloor \frac{1}{\delta_2} \rfloor} \sum_{k_2=1}^{\lfloor \frac{1}{\delta_2} \rfloor} \sum_{k_1=1}^{\infty} \sum_{s=k_2}^{\infty} \left| \frac{c_{k_1,s} - c_{k_1,s+1}}{k_1} \right|, \tag{8.9.12}$$

$$B(\delta_1, \delta_2) \leq 8 \sum_{k_2=1}^{\infty} \sum_{k_1=\lfloor \frac{1}{\delta_1} \rfloor+1}^{\infty} \frac{1}{k_2} \left| \Delta_{\{1\}}(c_{k_1,k_2}) \right| \tag{8.9.13}$$

and

$$C(\delta_1, \delta_2) \leq 8 \sum_{k_1=1}^{\infty} \sum_{k_2=\lfloor \frac{1}{\delta_2} \rfloor+1}^{\infty} \frac{1}{k_1} \left| \Delta_{\{2\}}(c_{k_1,k_2}) \right|. \tag{8.9.14}$$

Using triangle inequality, (8.9.5), (8.9.3) and (8.9.4), we get (8.9.12):

$$
\begin{aligned}
A(\delta_1,\delta_2) \le& \left|\sum_{k_1=1}^{\lfloor\frac{1}{\delta_1}\rfloor}\sum_{k_2=1}^{\lfloor\frac{1}{\delta_2}\rfloor}\frac{c_{k_1,k_2}}{k_1k_2}\Big(\big(\varphi_{1,k_1}(a_1)-\varphi_{1,k_1}(a_1+\delta_1)\big)\varphi_{2,k_2}(a_2)\Big)\right| \\
&+\left|\sum_{k_1=1}^{\lfloor\frac{1}{\delta_1}\rfloor}\sum_{k_2=1}^{\lfloor\frac{1}{\delta_2}\rfloor}\frac{c_{k_1,k_2}}{k_1k_2}\prod_{i=1}^{2}\big(\varphi_{i,k_i}(a_i)-\varphi_{i,k_i}(a_i+\delta_i)\big)\right| \\
&+\left|\sum_{k_1=1}^{\lfloor\frac{1}{\delta_1}\rfloor}\sum_{k_2=1}^{\lfloor\frac{1}{\delta_2}\rfloor}\frac{c_{k_1,k_2}}{k_1k_2}\big(\varphi_{1,k_1}(a_1)\big)\big(\varphi_{2,k_2}(a_2+\delta_2)-\varphi_{2,k_2}(a_2)\big)\right| \\
&\le \frac{1}{\lfloor\frac{1}{\delta_1}\rfloor}\sum_{k_1=1}^{\lfloor\frac{1}{\delta_1}\rfloor}\left|\sum_{k_2=1}^{\lfloor\frac{1}{\delta_2}\rfloor}\frac{c_{k_1,k_2}\varphi_{2,k_2}(a_2)}{k_2}\right|+\frac{1}{\lfloor\frac{1}{\delta_1}\rfloor\lfloor\frac{1}{\delta_2}\rfloor}\sum_{k_1=1}^{\lfloor\frac{1}{\delta_1}\rfloor}\sum_{k_2=1}^{\lfloor\frac{1}{\delta_2}\rfloor}|c_{k_1,k_2}| \\
&+\frac{1}{\lfloor\frac{1}{\delta_2}\rfloor}\sum_{k_2=1}^{\lfloor\frac{1}{\delta_2}\rfloor}\left|\sum_{k_1=1}^{\lfloor\frac{1}{\delta_1}\rfloor}\frac{c_{k_1,k_2}\varphi_{1,k_1}(a_1)}{k_1}\right| \\
&\le \frac{1}{\lfloor\frac{1}{\delta_1}\rfloor}\sum_{k_1=1}^{\lfloor\frac{1}{\delta_1}\rfloor}\sum_{k_2=1}^{\infty}\sum_{r=k_1}^{\infty}\left|\frac{\Delta_{\{1\}}(c_{r,k_2})}{k_2}\right|+\frac{1}{\lfloor\frac{1}{\delta_1}\rfloor\lfloor\frac{1}{\delta_2}\rfloor}\sum_{k_1=1}^{\lfloor\frac{1}{\delta_1}\rfloor}\sum_{k_2=1}^{\lfloor\frac{1}{\delta_2}\rfloor}|c_{k_1,k_2}| \\
&+\frac{1}{\lfloor\frac{1}{\delta_2}\rfloor}\sum_{k_2=1}^{\lfloor\frac{1}{\delta_2}\rfloor}\sum_{k_1=1}^{\infty}\sum_{s=k_2}^{\infty}\left|\frac{\Delta_{\{2\}}(c_{k_1,s})}{k_1}\right|.
\end{aligned}
$$

(8.9.13) follows from triangle inequality, Theorem 7.1.2(ii), (8.9.7) and (8.9.5):

$$
\begin{aligned}
&B(\delta_1,\delta_2) \\
&\le \sum_{k_2=1}^{\infty}\frac{1}{k_2}\left|\sum_{k_1=\lfloor\frac{1}{\delta_1}\rfloor+1}^{\infty}\frac{c_{k_1,k_2}}{k_1}\varphi_{1,k_1}(a_1+\delta_1)\right| \\
&\le 2\sum_{k_2=1}^{\infty}\frac{1}{k_2}\sum_{k_1=\lfloor\frac{1}{\delta_1}\rfloor+1}^{\infty}\left|\Delta_{\{1\}}(c_{k_1,k_2})\right|\cdot\sup_{n\ge\lfloor\frac{1}{\delta_1}\rfloor+1}\left|\sum_{k_1=\lfloor\frac{1}{\delta_1}\rfloor+1}^{n}\frac{\varphi_{1,k_1}(a_1+\delta_1)}{k_1}\right| \\
&\le 4\sum_{k_2=1}^{\infty}\frac{1}{k_2}\sum_{k_1=\lfloor\frac{1}{\delta_1}\rfloor+1}^{\infty}\left|\Delta_{\{1\}}(c_{k_1,k_2})\right|\cdot\frac{1}{\delta_1\cdot(\lfloor\frac{1}{\delta_1}\rfloor+1)} \\
&\le 8\sum_{k_2=1}^{\infty}\sum_{k_1=\lfloor\frac{1}{\delta_1}\rfloor+1}^{\infty}\frac{1}{k_2}\left|\Delta_{\{1\}}(c_{k_1,k_2})\right|,
\end{aligned}
$$

and a similar argument gives (8.9.14). This completes the proof of (8.9.9).

It remains to prove (8.9.10) and (8.9.11). It is not difficult to see that (8.9.10) is a consequence of Theorem 7.5.1, since (8.9.3), (8.9.4) and (8.9.5) imply that

$$\left|\sum_{k_2=1}^{\infty}\frac{c_{k_1,k_2}}{k_2}\varphi_{2,k_2}(a_2)\right| \leq \sum_{r=k_1}^{\infty}\sum_{k_2=1}^{\infty}\frac{1}{k_2}\left|\Delta_{\{1\}}(c_{r,k_2})\right| \to 0 \text{ as } k_1 \to \infty,$$

and

$$\sum_{k_1=1}^{\infty}\left|\Delta_{\{1\}}\left(\sum_{k_2=1}^{\infty}\frac{c_{k_1,k_2}}{k_2}\varphi_{2,k_2}(a_2)\right)\right| \leq \sum_{(k_1,k_2)\in\mathbb{N}^2}\frac{1}{k_2}\left|\Delta_{\{1\}}(c_{k_1,k_2})\right| < \infty.$$

Since a similar reasoning shows that (8.9.11) holds, the theorem is proved. □

The following two-dimensional analogue of Theorem 7.5.2 is a corollary of Theorem 8.9.3.

Theorem 8.9.4. *Let $(b_{k_1,k_2})_{(k_1,k_2)\in\mathbb{N}^2}$ be a double sequence of real numbers such that*

$$\lim_{\max\{n_1,n_2\}\to\infty} b_{n_1,n_2} = 0 \tag{8.9.15}$$

and

$$\sum_{(k_1,k_2)\in\mathbb{N}^2}\left\{\left|\frac{\Delta_{\{2\}}(b_{k_1,k_2})}{k_1}\right| + \left|\frac{\Delta_{\{1\}}(b_{k_1,k_2})}{k_2}\right| + \left|\Delta_{\{1,2\}}(b_{k_1,k_2})\right|\right\} < \infty. \tag{8.9.16}$$

Then the following assertions hold.

(i) *If $(x_1,x_2)\in[0,\pi]^2\backslash\{(0,0)\}$, then*

$$\sum_{(k_1,k_2)\in\mathbb{N}^2}\frac{b_{k_1,k_2}}{k_1k_2}\cos k_1x_1\cos k_2x_2$$

converges regularly.

(ii) *The function*

$$(x_1,x_2)\mapsto\sum_{(k_1,k_2)\in\mathbb{N}^2}\frac{b_{k_1,k_2}}{k_1k_2}\cos k_1x_1\cos k_2x_2$$

is continuous on $[0,\pi]^2\backslash\{(0,0)\}$.

(iii) *If*

$$\lim_{\substack{(x_1,x_2)\to(0,0)\\(x_1,x_2)\in(0,\pi)^2}} \sum_{(k_1,k_2)\in\mathbb{N}^2} \frac{b_{k_1,k_2}}{k_1k_2}\cos k_1x_1 \cos k_2x_2$$

exists, then

$$\sum_{(k_1,k_2)\in\mathbb{N}^2} \frac{b_{k_1,k_2}}{k_1k_2}$$

is regularly convergent.

(iv) *If*

$$\sum_{(k_1,k_2)\in\mathbb{N}^2} \frac{b_{k_1,k_2}}{k_1k_2}$$

is regularly convergent, then the function

$$(x_1,x_2)\mapsto \sum_{(k_1,k_2)\in\mathbb{N}^2} \frac{b_{k_1,k_2}}{k_1k_2}\cos k_1x_1 \cos k_2x_2$$

is continuous on $[0,\pi]^2$.

Example 8.9.5. Let $c_n = \sum_{k=n}^{\infty}(-1)^{k-1}\frac{1}{k(\ln(k+1))^2}$ for $n = 1,2,\ldots.$ Clearly,

$$\lim_{n\to\infty} c_n = 0, \quad \sum_{k=1}^{\infty}\left(|c_k - c_{k+1}| + \frac{|c_k|}{k}\right) < \infty,$$

and

$$\lim_{\substack{(\varepsilon,\delta)\to(0,0)\\(\varepsilon,\delta)\in(0,\pi)^2}} \sum_{(j,k)\in\mathbb{N}^2} \frac{c_jc_k}{jk}\cos j\varepsilon \cos k\delta \text{ exists.} \tag{8.9.17}$$

However, $\sum_{k=1}^{\infty}|c_k - c_{k+1}|\ln(k+1) = \infty$.

8.10 A convergence theorem for double sine series

The following convergence theorem for double sine series is a consequence of Theorems 8.8.1 and 8.9.4.

Theorem 8.10.1. *Let $(b_{k_1,k_2})_{(k_1,k_2)\in\mathbb{N}^2}$ be a double sequence of real numbers such that*

$$\lim_{\max\{n_1,n_2\}\to\infty} b_{n_1,n_2} = 0 \tag{8.10.1}$$

and

$$\sum_{(k_1,k_2)\in\mathbb{N}^2} \left\{ \left| \frac{\Delta_{\{2\}}(b_{k_1,k_2})}{k_1} \right| + \left| \frac{\Delta_{\{1\}}(b_{k_1,k_2})}{k_2} \right| + \left|\Delta_{\{1,2\}}(b_{k_1,k_2})\right| \right\} < \infty. \tag{8.10.2}$$

Then the following assertions hold.

(i) *If $(t_1,t_2) \in [0,\pi]^2$, then the double series*

$$\sum_{(k_1,k_2)\in\mathbb{N}^2} b_{k_1,k_2} \prod_{i=1}^{2} \sin k_i t_i$$

converges regularly to $g(t_1,t_2)$ (say).

(ii) *If $(x_1,x_2) \in [0,\pi)^2\backslash\{(0,0)\}$, then $g \in HK([x_1,\pi]\times[x_2,\pi])$ and*

$$(HK)\int_{[x_1,\pi]\times[x_2,\pi]} g = \sum_{(k_1,k_2)\in\mathbb{N}^2} b_{k_1,k_2} \prod_{i=1}^{2} \int_{x_i}^{\pi} \sin k_i t_i \, dt_i; \tag{8.10.3}$$

the double series on the right being regularly convergent.

(iii) *$g \in HK([0,\pi]^2)$ if and only if the double series $\sum_{(k_1,k_2)\in\mathbb{N}^2} \frac{b_{k_1,k_2}}{\prod_{i=1}^2 k_i}$ is regularly convergent. In this case,*

$$(HK)\int_{[0,\pi]^2} \left\{ \sum_{k_1=1}^{n_1}\sum_{k_2=1}^{n_2} b_{k_1,k_2} \prod_{i=1}^{2} \sin k_i t_i - g(t_1,t_2) \right\} d(t_1,t_2) \to 0$$

as $\min\{n_1,n_2\}\to\infty$. (8.10.4)

Proof. (i) If $(t_1,t_2)\in[0,\pi]^2$ and $\prod_{i=1}^2 \sin t_i \neq 0$. then the result follows from Theorem 8.1.13(iii). On the other hand, the result is clearly true if $(t_1,t_2)\in[0,\pi]^2$ and $\prod_{i=1}^2 \sin t_i = 0$.

(ii) Using Theorem 8.8.1 with $\prod_{i=1}^2 [a_i,b_i] = \prod_{i=1}^2 [x_i,\pi]$ $((x_1,x_2)\in[0,\pi)^2\backslash\{(0,0)\})$ and $\phi_{i,k}(t) = \sin kt$ $(k\in\mathbb{N}; i=1,2)$, we get the result.

(iii) We infer from (8.10.3) and Theorem 2.4.12 that $g \in HK([0,\pi]^2)$ if and only if

$$\lim_{\substack{(x_1,x_2)\to(0,0)\\(x_1,x_2)\in[0,\pi]^2}} \sum_{(k_1,k_2)\in\mathbb{N}^2} b_{k_1,k_2} \prod_{i=1}^{2} \int_{x_i}^{\pi} \sin k_i t_i \, dt_i \text{ exists,} \tag{8.10.5}$$

which is easily seen to be equivalent to the existence of the following limit

$$\lim_{\substack{(x_1,x_2)\to(0,0)\\(x_1,x_2)\in[0,\pi]^2}} \sum_{\Gamma\subseteq\{1,2\}} \sum_{(k_1,k_2)\in\mathbb{N}^2} \frac{b_{k_1,k_2}}{k_1k_2}\left(\prod_{i\in\Gamma}\cos k_i\pi\right)\left(\prod_{i\in\{1,2\}\setminus\Gamma}\cos k_ix_i\right).$$

Hence assertion (8.10.5) holds if and only if

$$\lim_{\substack{(x_1,x_2)\to(0,0)\\(x_1,x_2)\in[0,\pi]^2}} \sum_{(k_1,k_2)\in\mathbb{N}^2} \frac{b_{k_1,k_2}}{k_1k_2}\cos k_1x_1 \cos k_2x_2 \text{ exists,} \tag{8.10.6}$$

since Theorem 8.9.4(ii), (8.10.1) and (8.10.2) imply the existence of the limit

$$\lim_{\substack{(x_1,x_2)\to(0,0)\\(x_1,x_2)\in[0,\pi]^2}} \sum_{\emptyset\neq\Gamma\subseteq\{1,2\}} \sum_{(k_1,k_2)\in\mathbb{N}^2} \frac{b_{k_1,k_2}}{k_1k_2}\left(\prod_{i\in\Gamma}\cos k_i\pi\right)\left(\prod_{i\in\{1,2\}\setminus\Gamma}\cos k_ix_i\right).$$

It is now clear the first assertion of statement (iii) follows from parts (iii) and (iv) of Theorem 8.9.4.

Finally, Theorem 2.4.12, (8.10.3) and Theorem 8.9.4 yield

$$\begin{aligned}
&(HK)\int_{[0,\pi]^2} g \\
&= \lim_{\substack{(x_1,x_2)\to(0,0)\\(x_1,x_2)\in[0,\pi]^2}} (HK)\int_{[x_1,\pi]\times[x_2,\pi]} g \\
&= \lim_{\substack{(x_1,x_2)\to(0,0)\\(x_1,x_2)\in[0,\pi]^2}} \sum_{\Gamma\subseteq\{1,2\}} \sum_{(k_1,k_2)\in\mathbb{N}^2} \frac{b_{k_1,k_2}}{k_1k_2}\left(\prod_{i\in\Gamma}\cos k_i\pi\right)\left(\prod_{i\in\{1,2\}\setminus\Gamma}\cos k_ix_i\right) \\
&= \sum_{(k_1,k_2)\in\mathbb{N}^2} b_{k_1,k_2} \prod_{i=1}^{2}\int_0^{\pi} \sin k_i t_i \, dt_i.
\end{aligned}$$

□

Corollary 8.10.2. *Let $(b_{k_1,k_2})_{(k_1,k_2)\in\mathbb{N}^2}$ be a double sequence of real numbers such that $\lim_{\max\{n_1,n_2\}\to\infty} b_{n_1,n_2} = 0$ and*

$$\sum_{(k_1,k_2)\in\mathbb{N}^2} \left|\Delta_{\{1,2\}}(b_{k_1,k_2})\right| \Big(\ln(\max\{k_1+1,k_2+1\})\Big) < \infty.$$

Then the double series $\sum_{(k_1,k_2)\in\mathbb{N}^2} \frac{b_{k_1,k_2}}{\prod_{i=1}^2 k_i}$ *converges regularly if and only if*

$$\lim_{\substack{(\delta_1,\delta_2)\to(0,0)\\(\delta_1,\delta_2)\in[0,\pi]^2}} \int_{[\delta_1,\pi]\times[\delta_2,\pi]} \sum_{(k_1,k_2)\in\mathbb{N}^2} b_{k_1,k_2}\sin k_1t_1 \sin k_2t_2 \, d\mu_2(t_1,t_2) \text{ exists.}$$

Proof. According to Theorem 8.1.13, the double sequence $(b_{k_1,k_2})_{(k_1,k_2)\in\mathbb{N}^2}$ satisfies condition (8.10.2). Hence, for each $(t_1,t_2)\in[0,\pi]^2$, we apply Theorem 8.10.1 to conclude that the double series $\sum_{(k_1,k_2)\in\mathbb{N}^2} b_{k_1,k_2}\sin k_1t_1\sin k_2t_2$ converges regularly to $g(t_1,t_2)$ (say), and $g\in HK([0,\pi]^2)$ if and only if the double series $\sum_{(k_1,k_2)\in\mathbb{N}^2}\frac{b_{k_1,k_2}}{\prod_{i=1}^2 k_i}$ converges regularly.

It remains to prove that $g \in L^1([x_1,\pi]\times[x_2,\pi])$ whenever $(x_1,x_2)\in[0,\pi]^2\backslash\{(0,0)\}$. But this is an immediate consequence of the Riemann-Lebesgue Lemma, Lemma 7.3.3 and Theorem 8.3.3. □

The following example shows that Theorem 8.10.1 is a proper generalization of Corollary 8.10.2.

Example 8.10.3. Let $b_1 = 0$ and let

$$b_k = \sum_{j=k}^{\infty}(-1)^{\frac{\ln j}{\ln 2}}\frac{\chi_{\{2^r:r\in\mathbb{N}\}}(j)}{(\ln j)^{\frac{3}{2}}} \qquad (k\in\mathbb{N}\backslash\{1\}).$$

Using Theorem 8.10.1 with $m = 2$, we see that the function $(t_1,t_2)\mapsto\sum_{(k_1,k_2)\in\mathbb{N}^2} b_{k_1}b_{k_2}\sin k_1t_1\sin k_2t_2$ is Henstock-Kurzweil integrable on $[0,\pi]^2$. On the other hand, since $\sum_{k=1}^{\infty} b_k\sin kt$ is not the Fourier series of a function in $L^1(\mathbb{T})$ (cf. Example 7.4.15), the function $(t_1,t_2)\mapsto\sum_{(k_1,k_2)\in\mathbb{N}^2} b_{k_1}b_{k_2}\sin k_1t_1\sin k_2t_2$ cannot be Lebesgue integrable on $[0,\pi]^2$.

The following example sharpens Example 8.2.6.

Example 8.10.4. The double sine series

$$\sum_{(k_1,k_2)\in\mathbb{N}^2}\frac{\sin k_1t_1\sin k_2t_2}{(\ln(k_1+k_2+2))^2}$$

converges regularly for all $(t_1,t_2)\in[0,\pi]^2$. However, the function

$$(t_1,t_2)\mapsto\sum_{(k_1,k_2)\in\mathbb{N}^2}\frac{\sin k_1t_1\sin k_2t_2}{(\ln(k_1+k_2+2))^2}$$

is not Henstock-Kurzweil integrable on $[0,\pi]^2$.

Proof. For each $(k_1, k_2) \in \mathbb{N}^2$, let $b_{k_1,k_2} = \frac{1}{(\ln(k_1+k_2+2))^2}$. Then $\lim_{\max\{n_1,n_2\}\to\infty} b_{n_1,n_2} = 0$ and

$$\sum_{(k_1,k_2)\in\mathbb{N}^2} \left\{ \left|\frac{\Delta_{\{2\}}(b_{k_1,k_2})}{k_1}\right| + \left|\frac{\Delta_{\{1\}}(b_{k_1,k_2})}{k_2}\right| + \left|\Delta_{\{1,2\}}(b_{k_1,k_2})\right| \right\} < \infty.$$

By Theorem 8.10.1(i), the double sine series $\sum_{(k_1,k_2)\in\mathbb{N}^2} b_{k_1,k_2} \sin k_1t_1 \sin k_2t_2$ converges regularly for all $(t_1, t_2) \in [-\pi, \pi]^2$. On the other hand, since $\sum_{j=1}^{\infty}\sum_{k=1}^{\infty} \frac{b_{j,k}}{jk} = \infty$, we infer from part (iii) of Theorem 8.10.1 that the function

$$(t_1, t_2) \mapsto \sum_{(k_1,k_2)\in\mathbb{N}^2} b_{k_1,k_2} \sin k_1t_1 \sin k_2t_2$$

cannot be Henstock-Kurzweil integrable on $[0, \pi]^2$. □

Remark 8.10.5. Example 8.10.4 does not contradict Theorem 8.8.1 because

$$\sup_{n\in\mathbb{N}} \sup_{x\in[0,\pi]} \left|\sum_{k=1}^{n} \int_0^x \sin kt\, dt\right| = \infty.$$

8.11 Some open problems

In this section we will give some open problems relating to the following modification of Lemma 7.6.1.

Theorem 8.11.1. *If $f_1, f_2 \in L^1(\mathbb{T})$, then the double series*

$$\sum_{(k_1,k_2)\in\mathbb{N}^2} \frac{1}{k_1k_2} \int_{[-\pi,\pi]^2} \prod_{i=1}^{2} f_i(t_i) \cos k_it_i\, d\mu_2(t_1, t_2) \qquad (8.11.1)$$

converges regularly if and only if there exists $\phi \in HK([0, \pi]^2)$ such that

$$\phi(x_1, x_2) = \frac{1}{x_1x_2}\left\{\int_{[-x_1,x_1]\times[-x_2,x_2]} f_1 \otimes f_2\, d\mu_2\right\} \qquad ((x_1, x_2) \in (0, \pi]^2). \qquad (8.11.2)$$

Proof. According to Theorem 8.1.9, the assertion (8.11.1) holds if and only if the series

$$\sum_{k_i=1}^{\infty} \frac{1}{k_i} \int_{-\pi}^{\pi} f_i(t_i) \cos k_it_i\, d\mu_1(t_i) \text{ converges for all } i = 1, 2. \qquad (8.11.3)$$

By Lemma 7.6.1, assertion (8.11.3) holds if and only if for each $i = 1, 2$ there exists $\phi_i \in HK[0, \pi]$ such that

$$\phi_i(x) = \frac{1}{x}\int_{-x}^{x} f_i \, d\mu_1 \qquad (x \in (0, \pi)).$$

Now we infer from Theorem 5.6.5 and Fubini's Theorem that the last assertion is true if and only if (8.11.2) holds for some $\phi \in HK([0, \pi]^2)$. □

We observe that the proof of Theorem 8.11.1 depends on Theorem 5.6.5, for which no satisfactory analogue in higher dimensions is known (cf. Conjecture 5.6.7). Thus, it is natural to ask whether the following conjecture is true.

Conjecture 8.11.2. *Let $f \in L^1(\mathbb{T}^2)$ and assume that*

$$f(t_1, t_2) \sim \sum_{(k_1,k_2)\in\mathbb{N}_0^2} \lambda_{k_1}\lambda_{k_2}\Big\{ a_{k_1,k_2} \cos k_1t_1 \cos k_2t_2 + b_{k_1,k_2} \cos k_1t_1 \sin k_2t_2$$
$$+ c_{k_1,k_2} \sin k_1t_1 \cos k_2t_2 + d_{k_1,k_2} \sin k_1t_1 \sin k_2t_2 \Big\}.$$

Then the double series

$$\sum_{(k_1,k_2)\in\mathbb{N}^2} \frac{1}{k_1k_2}\int_{[-\pi,\pi]^2} f(t_1, t_2)\prod_{i=1}^{2} \cos k_it_i \, d\mu_2(t_1, t_2) \quad \textit{converges regularly}$$

if and only if there exists $\phi \in HK([0, \pi]^2)$ such that

$$\phi(x_1, x_2) = \frac{1}{x_1x_2}\int_{[-x_1,x_1]\times[-x_2,x_2]} f \, d\mu_2 \qquad ((x_1, x_2) \in (0, \pi]^2).$$

Although Theorem 8.7.6 provides a partial answer to Conjecture 8.11.2, its proof relies on Theorems 8.6.1 and 8.6.2 involving absolutely convergent double series. The following theorem gives another partial answer to Conjecture 8.11.2.

Theorem 8.11.3. *Let $(a_{k_1,k_2})_{(k_1,k_2)\in\mathbb{N}^2}$ be a double sequence of real numbers such that $\lim_{\max\{n_1,n_2\}\to\infty} a_{n_1,n_2} = 0$ and*

$$\sum_{(k_1,k_2)\in\mathbb{N}^2} \left\{ \left|\frac{\Delta_{\{2\}}(a_{k_1,k_2})}{k_1}\right| + \left|\frac{\Delta_{\{1\}}(a_{k_1,k_2})}{k_2}\right| + \left|\Delta_{\{1,2\}}(a_{k_1,k_2})\right| \right\} < \infty.$$

Then the following assertions hold.

(i) *If $(x_1, x_2) \in (0, \pi]^2$, then the double series*

$$\sum_{(k_1,k_2)\in\mathbb{N}^2} a_{k_1,k_2} \cos k_1x_1 \cos k_2x_2$$

converges regularly.

(ii) *There exists* $f \in HK([-\pi,\pi]^2)$ *such that*

$$f(x_1,x_2) = \sum_{(k_1,k_2)\in\mathbb{N}^2} a_{k_1,k_2}\cos k_1x_1\cos k_2x_2$$

for all $(x_1,x_2)\in(0,\pi]^2$.

(iii) *The double series* $\sum_{(k_1,k_2)\in\mathbb{N}^2}\frac{a_{k_1,k_2}}{k_1k_2}$ *converges regularly if and only if there exists* $\phi\in HK([0,\pi]^2)$ *such that*

$$\phi(x_1,x_2) = \cot\frac{x_1}{2}\cot\frac{x_2}{2}\left\{(HK)\int_{[0,x_1]\times[0,x_2]} f\right\}$$

for all $(x_1,x_2)\in(0,\pi]^2$. *In this case,*

$$(HK)\int_{[0,\pi]^2}\phi = \pi^2\sum_{(k_1,k_2)\in\mathbb{N}^2}\frac{a_{k_1,k_2}}{k_1k_2}.$$

Proof. Parts (i) and (ii) follow from Corollary 8.8.2.

(iii) Define the function $\phi : [0,\pi]^2 \longrightarrow \mathbb{R}$ by setting $\phi(x_1,x_2)=0$ $((x_1,x_2)\in[0,\pi]^2\backslash(0,\pi]^2))$, and

$$\phi(x_1,x_2) := \cot\frac{x_1}{2}\cot\frac{x_2}{2}\left\{(HK)\int_{[0,x_1]\times[0,x_2]} f\right\}\quad((x_1,x_2)\in(0,\pi]^2).$$

Since the inequality $\inf_{x\in[0,\pi]}\inf_{n\in\mathbb{N}}\sum_{k=1}^{n}\frac{\sin kt}{k}\geq 0$ (cf. [165, Chapter II, (9.4) Theorem]) and Fatou's Lemma yield

$$\sup_{n\in\mathbb{N}}\left|\int_0^\pi\sum_{k=1}^{n}\frac{\sin kt}{kt}\,dt\right| = \infty, \tag{8.11.4}$$

Theorem 8.8.1 cannot be applied directly. On the other hand, it is not difficult to check that the following claims hold.

Claim 1: If $x_1\in(0,\frac{\pi}{2})$ and $n\in\mathbb{N}$, then

$$\sum_{(k_1,k_2)\in\mathbb{N}^2}\left\{\left|\Delta_{\{1\}}(a_{k_1,k_2})\right|\max_{x_2\in[0,\pi]}\left|\int_0^{x_2}\frac{\sin k_2t_2}{k_2t_2}\,dt_2\right|+\left|\Delta_{\{1,2\}}(a_{k_1,k_2})\right|\right\}<\infty.$$

Claim 2: If $x_2\in(0,\frac{\pi}{2})$ and $n\in\mathbb{N}$, then

$$\sum_{(k_1,k_2)\in\mathbb{N}^2}\left\{\left|\Delta_{\{2\}}(a_{k_1,k_2})\right|\max_{x_1\in[0,\pi]}\left|\int_0^{x_1}\frac{\sin k_1t_1}{k_1t_1}\,dt_1\right|+\left|\Delta_{\{1,2\}}(a_{k_1,k_2})\right|\right\}<\infty.$$

Since claims 1 and 2 hold, it follows from Theorem 8.8.1 and Kurzweil's multiple integration by parts formula (Theorem 6.5.9) that if

$[x_1, \pi] \times [x_2, \pi] \in \mathcal{I}_2([0, \pi]^2)$ with $\max\{x_1, x_2\} > 0$, then $\phi \in HK([x_1, \pi] \times [x_2, \pi])$ and

$$(HK)\int_{[x_1,\pi]\times[x_2,\pi]} \phi = \sum_{(k_1,k_2)\in\mathbb{N}^2} \frac{a_{k_1,k_2}}{k_1 k_2} \prod_{i=1}^{2} \int_{x_i}^{\pi} \sin k_i t_i \cot \frac{t_i}{2}\, dt_i. \quad (8.11.5)$$

It remains to prove that $\phi \in HK([0, \pi]^2)$ if and only if the double series $\sum_{(k_1,k_2)\in\mathbb{N}^2} \frac{a_{k_1,k_2}}{k_1 k_2}$ is regularly convergent. Using Theorem 8.9.3 with $a_1 = a_2 = 0$, $b_1 = b_2 = \pi$, and $\varphi_{i,k}(x) = \int_x^{\pi} \sin kt \cot \frac{t}{2}\, dt$ ($i = 1, 2$; $k \in \mathbb{N}$; $x \in [0, \pi]$), we see that

$$\sum_{(k_1,k_2)\in\mathbb{N}^2} \frac{a_{k_1,k_2}}{k_1 k_2} \prod_{i=1}^{2} \left\{ \int_0^{\pi} \sin k_i t_i \cot \frac{t_i}{2}\, dt_i \right\} \text{ is regularly convergent}$$

if and only if

$$\lim_{\substack{(x_1,x_2)\to(0,0)\\(x_1,x_2)\in[0,\pi]^2}} \sum_{(k_1,k_2)\in\mathbb{N}^2} \frac{a_{k_1,k_2}}{k_1 k_2} \prod_{i=1}^{2} \int_{x_i}^{\pi} \sin k_i t_i \cot \frac{t_i}{2}\, dt_i \text{ exists.} \quad (8.11.6)$$

Using (8.11.5) and Kurzweil's multiple integration by parts formula (Theorem 6.5.9) again, we see that (8.11.6) is equivalent to the condition

$$\lim_{\substack{(x_1,x_2)\to(0,0)\\(x_1,x_2)\in[0,\pi]^2}} (HK)\int_{[x_1,\pi]\times[x_2,\pi]} \phi \text{ exists.} \quad (8.11.7)$$

Finally, it follows from Cauchy extension (Theorem 2.4.12) that (8.11.7) holds if and only if $\phi \in HK([0, \pi]^2)$. The proof is complete. □

The following example shows that Theorem 8.7.6 cannot be used to prove Theorem 8.11.3.

Example 8.11.4. We define a double sequence $(a_{k_1,k_2})_{(k_1,k_2)\in\mathbb{N}^2}$ of real numbers by setting $a_{k_1,k_2} = \sum_{r=k_1}^{\infty} \sum_{s=k_2}^{\infty} \frac{(-1)^{r+s}}{(r+s)^3}$ $((k_1, k_2) \in \mathbb{N}^2)$. Then the following statements are true.

(i) $(-1)^{k_1+k_2} a_{k_1,k_2} < 0$ for all $(k_1, k_2) \in \mathbb{N}^2$.
(ii) $\sum_{(k_1,k_2)\in\mathbb{N}^2} |a_{k_1,k_2}|$ converges.
(iii) $\sum_{(k_1,k_2)\in\mathbb{N}^2} \left\{ \left| \frac{\Delta_{\{2\}}(a_{k_1,k_2})}{k_1} \right| + \left| \frac{\Delta_{\{1\}}(a_{k_1,k_2})}{k_2} \right| + \left| \Delta_{\{1,2\}}(a_{k_1,k_2}) \right| \right\}$ converges.

According to (ii), there exists $\phi \in L^{\infty}([0, \pi]^2)$ such that

$$\phi(x_1, x_2) = \frac{1}{x_1 x_2} \int_{-x_1}^{x_1} \int_{-x_2}^{x_2} \sum_{k_1=1}^{\infty} \sum_{k_2=1}^{\infty} a_{k_1,k_2} \cos k_1 t_1 \cos k_2 t_2\, d\mu_2(t_1, t_2)$$

for all $(x_1, x_2) \in (0, \pi]^2$. However, Theorem 8.7.6 cannot be applied here because (i) holds.

8.12 Notes and Remarks

Section 8.1 is based on [121, 122]. Regularly convergent double series was first studied by Hardy in [50]; see also [18]. For a connection between regularly convergent multiple series and Henstock-Kurzweil integrals, see [38]. A more general version of Theorem 8.1.9 is given in [38], and a general approach of Theorem 8.1.12 can be found in [103, Theorem 2.2]. Theorem 8.1.13 is based on [103, Theorem 2.3].

Lemma 8.2.2 is the two-dimensional analogue of [106, Lemma 5.4]. Theorem 8.2.4, which is essentially [106, Theorem 5.5], refines a result of W.H. Young [157, p.155] concerning double Fourier series. A different proof of Example 8.2.6 can be found in [2]. For other results concerning multiple Fourier series, consult, for instance, [47, 128–131].

Theorem 8.3.3 is a special case of [103, Theorem 2.2]. A different approach to Theorem 8.3.5 can be found in [124, 126]. Theorem 8.3.6 is [126, Corollary 3].

Sections 8.4 and 8.5 are based on the paper [107]. More general versions of Theorem 8.4.3 are given in [123] and [19].

Sections 8.6, 8.7, 8.8, 8.9, 8.10 and 8.11 are based on the paper [106].

Bibliography

[1] T.M. Apostol, *Mathematical Analysis.* Second Edition, Addison-Wesley, Reading, MA, 1974.

[2] B. Aubertin and J.J. Fournier, Integrability of multiple series. Fourier analysis (Orono, ME, 1992), 47–75, Lecture Notes in Pure and Appl. Math., 157, Dekker, New York, 1994.

[3] V. Aversa and M. Laczkovich, Extension theorems on derivatives of additive interval functions, *Acta Math. Acad. Sci. Hungar.* **39** (1982), no. 1-3, 267–277.

[4] R.G. Bartle, Return to the Riemann integral. *Amer. Math. Monthly* **103** (1996), no. 8, 625–632.

[5] R.G. Bartle, *A Modern Theory of Integration.* Graduate Studies in Mathematics, 32. American Mathematical Society, Providence, RI, 2001.

[6] R.G. Bartle and D.R. Sherbert, *Introduction to Real Analysis.* Third Edition, John Wiley & Sons, Inc., New York, 2001.

[7] R.P. Boas, Integrability of trigonometric series. I. *Duke Math. J.* **18**, (1951). 787–793.

[8] R.P. Boas, *Integrability Theorems for Trigonometric Transforms.* Ergebnisse der Mathematik und ihrer Grenzgebiete, Band 38, Springer-Verlag New York Inc., New York 1967.

[9] B. Bongiorno, On the minimal solution of the problem of primitives. *J. Math. Anal. Appl.* **251** (2000), no. 2, 479–487.

[10] B. Bongiorno, *The Henstock-Kurzweil integral.* Handbook of measure theory, Vol. I, II, 587–615, North-Holland, Amsterdam, 2002.

[11] B. Bongiorno, Udayan B. Darji and W.F. Pfeffer, On indefinite BV-integrals. *Comment. Math. Univ. Carolin.* **41** (2000), no. 4, 843–853.

[12] B. Bongiorno, L. Di Piazza and K. Musial, A characterization of the weak Radon-Nikodym property by finitely additive interval functions, *Bull. Australian Math. Soc* **80** (2009), 476-485.

[13] B. Bongiorno, L. Di Piazza and K. Musial, A variational Henstock integral characterization of the Radon-Nikodm property. *Illinois J. Math.* **53** (2009), no. 1, 87–99.

[14] B. Bongiorno, L. Di Piazza and D. Preiss, A constructive minimal integral

which includes Lebesgue integrable functions and derivatives. *J. London Math. Soc.* (2) **62** (2000), no. 1, 117–126.

[15] B. Bongiorno, L. Di Piazza and V. Skvortsov, A new full descriptive characterization of Denjoy-Perron integral. *Real Anal. Exchange* **21** (1995/96), no. 2, 656–663.

[16] B. Bongiorno, L. Di Piazza and V. Skvortsov, On continuous major and minor functions for the n-dimensional Perron integral. *Real Anal. Exchange* **22** (1996/97), no. 1, 318–327.

[17] D. Bongiorno, On the problem of nearly derivatives. *Sci. Math. Jpn.* **61** (2005), no. 2, 299–311.

[18] D. Borwein and J.M. Borwein, A note on alternating series in several dimensions. *Amer. Math. Monthly* **93** (1986), no. 7, 531–539.

[19] G. Brown and Wang K. Y, On a conjecture of F. Móricz. *Proc. Amer. Math. Soc.* **126** (1998), no. 12, 3527–3537.

[20] A.M. Bruckner, R.J. Fleissner and J. Foran, The minimal integral which includes Lebesgue integrable functions and derivatives. *Colloq. Math.* **50** (1986), no. 2, 289–293.

[21] A.M. Bruckner, J. Mařík and C.E. Weil, Some aspects of products of derivatives. *Amer. Math. Monthly* **99** (1992), no. 2, 134–145.

[22] Z. Buczolich, Henstock integrable functions are Lebesgue integrable on a portion. *Proc. Amer. Math. Soc.* **111** (1991), no. 1, 127–129.

[23] Z. Buczolich, A v-integrable function which is not Lebesgue integrable on any portion of the unit square. *Acta Math. Hungar.* **59** (1992), no. 3-4, 383–393.

[24] Z. Buczolich, Characterization of upper semicontinuously integrable functions. *J. Austral. Math. Soc. Ser. A* **59** (1995), no. 2, 244–254.

[25] Z. Buczolich, Tensor products of AC_* charges and AC Radon measures are not always AC_* charges. *J. Math. Anal. Appl.* **259** (2001), no. 2, 377–385.

[26] Z. Buczolich, When tensor products of AC_* charges and Radon measures are AC_* charges. *Atti Sem. Mat. Fis. Univ. Modena* **49** (2001), no. 2, 411–454.

[27] P.S. Bullen, An unconvincing counterexample. *Int. J. Math Educ. Sci Technol* **19** (1988) no. 3, 455–459.

[28] Chew Tuan-Seng, B. Van-Brunt and G.C. Wake, First-order partial differential equations and Henstock-Kurzweil integrals. *Differential Integral Equations* **10** (1997), no. 5, 947–960.

[29] T. De Pauw and W.F. Pfeffer, The Gauss-Green theorem and removable sets for PDEs in divergence form. *Adv. Math.* **183** (2004), no. 1, 155–182.

[30] Luisa Di Piazza, Variational measures in the theory of the integration in $\mathbb{R}^m$. *Czechoslovak Math. J.* **51** (126) (2001), no. 1, 95–110.

[31] Luisa Di Piazza and V. Marraffa, The McShane, PU and Henstock integrals of Banach-valued functions. *Czechoslovak Math. J.* **52** (127) (2002), no. 2, 609–633.

[32] S. M. Edmonds, The Parseval formulae for monotonic functions. I. *Proc. Cambridge Philos. Soc.* **43**, (1947). 289–306.

[33] S. M. Edmonds, The Parseval formulae for monotonic functions. II. *Proc.*

Cambridge Philos. Soc. **46**, (1950). 231–248.

[34] S.M. Edmonds, The Parseval formulae for monotonic functions. III. *Proc. Cambridge Philos. Soc.* **46**, (1950). 249–267.

[35] S.M. Edmonds, The Parseval formulae for monotonic functions. IV. *Proc. Cambridge Philos. Soc.* **49**, (1953). 218–229.

[36] Claude-Alain Faure, A descriptive definition of some multidimensional gauge integrals. *Czechoslovak Math. J.* **45(120)** (1995), no. 3, 549–562.

[37] Claude-Alain Faure and J. Mawhin, The Hake's property for some integrals over multidimensional intervals. *Real Anal. Exchange* **20** (1994/95), no. 2, 622–630.

[38] Claude-Alain Faure and J. Mawhin, Integration over unbounded multidimensional intervals. *J. Math. Anal. Appl.* **205** (1997), no. 1, 65–77.

[39] R. Fleissner, On the product of derivatives. *Fund. Math.* **88** (1975), no. 2, 173–178.

[40] R.J. Fleissner, Multiplication and the fundamental theorem of calculus—a survey. *Real Anal. Exchange* **2** (1976/77), no. 1, 7–34.

[41] R.J. Fleissner, Distant bounded variation and products of derivatives. *Fund. Math.* **94** (1977), no. 1, 1–11.

[42] G.B. Folland, *Real Analysis.* Modern techniques and their applications. Second edition. Pure and Applied Mathematics (New York). A Wiley-Interscience Publication. John Wiley & Sons, Inc., New York, 1999.

[43] S Fridli, Hardy spaces generated by an integrability condition. *J. Approx. Theory* **113** (2001), no. 1, 91–109.

[44] R.A. Gordon, *The integrals of Lebesgue, Denjoy, Perron, and Henstock.* Graduate Studies in Mathematics, 4. American Mathematical Society, Providence, RI, 1994.

[45] T.L. Gill and W.W. Zachary, Banach spaces for the Feynman integral. *Real Anal. Exchange* **34** (2009), no. 2, 267–310.

[46] R.A. Gordon, The use of tagged partitions in elementary real analysis. *Amer. Math. Monthly* **105** (1998), no. 2, 107–117.

[47] L. Grafakos, *Classical Fourier Analysis.* Second edition. Graduate Texts in Mathematics, 249. Springer, New York, 2008.

[48] M De Guzman, A change-of-variables formula without continuity. *Amer. Math. Monthly* **87** (1980), no. 9, 736–739.

[49] G.H. Hardy, On the convergence of certain multiple series, *Proc. London Math. Soc. (2)* **1** (1903) 124–128.

[50] G.H. Hardy, On the convergence of certain multiple series. *Proc. Cambridge Philos. Soc.* **19** (1916-1919), 86-95.

[51] G.H. Hardy and J.E. Littlewood, Solution of the Cesàro summability problem for power-series and Fourier series. *Math. Z.* **19** (1924), no. 1, 67–96.

[52] G.H. Hardy and W.W. Rogosinski, Notes on Fourier series (IV): Summability (R_2). *Proc. Cambridge Philos. Soc.* **43** (1947), no.1, 19-25.

[53] G.H. Hardy and W.W. Rogosinski, Notes on Fourier series (V): Summability (R_1). *Proc. Cambridge Philos. Soc.* **45** (1949), no.1, 173-185.

[54] T. Hawkins, *Lebesgue's Theory of Integration.* Its origins and development. Reprint of the 1979 corrected second edition. AMS Chelsea Publishing,

Providence, RI, 2001.

[55] R. Henstock, Definitions of Riemann type of the variational integrals. *Proc. London Math. Soc. (3)* **11** (1961), 79–87.

[56] R. Henstock, Majorants in variational integration. *Canad. J. Math.* **18** (1966), 49–74.

[57] R. Henstock, A problem in two-dimensional integration. *J. Austral. Math. Soc. Ser. A* **35** (1983), no. 3, 386–404.

[58] R. Henstock, P. Muldowney and V.A. Skvortsov, Partitioning infinite-dimensional spaces for generalized Riemann integration. *Bull. London Math. Soc.* **38** (2006), no. 5, 795–803.

[59] E. Hewitt and Karl Stromberg, *Real and Abstract Analysis.* A modern treatment of the theory of functions of a real variable. Third printing. Graduate Texts in Mathematics, No. 25. Springer-Verlag, New York-Heidelberg, 1975.

[60] P. Heywood, Integrability theorems for trigonometric series. *Quart. J. Math. Oxford Ser. (2)* **13** (1962), 172–180.

[61] T.H. Hildebrandt and I.J. Schoenberg, On linear functional operations and the moment problem for a finite interval in one or several dimensions. *The Annals of Mathematics (2)* **34** (1933), 317-328.

[62] W.B. Jurkat and R.W. Knizia, A characterization of multi-dimensional Perron integrals and the fundamental theorem. *Canad. J. Math.* **43** (1991), no. 3, 526–539.

[63] J. Jarník and J. Kurzweil, A nonabsolutely convergent integral which admits transformation and can be used for integration on manifolds. *Czechoslovak Math. J.* **35(110)** (1985), no. 1, 116–139.

[64] J. Jarník and J. Kurzweil, A new and more powerful concept of the PU-integral. *Czechoslovak Math. J.* **38(113)** (1988), no. 1, 8–48.

[65] J. Jarník and J. Kurzweil, Perron-type integration on n-dimensional intervals and its properties. *Czechoslovak Math. J.* **45(120)** (1995), no. 1, 79–106.

[66] J. Jarník, J. Kurzweil and S. Schwabik, On Mawhin's approach to multiple nonabsolutely convergent integral. *Časopis Pěst. Mat.* **108** (1983), no. 4, 356–380.

[67] R.L. Jeffery, Functions defined by sequences of integrals and the inversion of approximate derived numbers. *Trans. Amer. Math. Soc.* **41** (1937), 171-192.

[68] K. Karták, Zur Theorie des mehrdimensionalen Integrals. (Czech) *Časopis Pěst. Mat.* **80** (1955), 400–414.

[69] K. Karták and J. Mařík, On representations of some Perron integrable functions. *Czechoslovak Math. J.* **19 (94)** (1969) 745–749.

[70] D.S. Kurtz and C.W. Swartz, *Theories of Integration.* The integrals of Riemann, Lebesgue, Henstock-Kurzweil, and Mcshane. Series in Real Analysis, 9. World Scientific Publishing Co., Inc., River Edge, NJ, 2004.

[71] J. Kurzweil, Generalized ordinary differential equations and continuous dependence on a parameter. *Czechoslovak Math. J.* **7(82)** (1957), 418–466.

[72] J. Kurzweil, On Fubini theorem for general Perron integral. *Czechoslovak Math. J.* **23(98)** (1973), 286–297.

[73] J. Kurzweil, On multiplication of Perron-integrable functions. *Czechoslovak Math. J.* **23(98)** (1973), 542–566.

[74] J. Kurzweil, *Nichtabsolut Konvergente Integrale.* (German) [Nonabsolutely convergent integrals] With English, French and Russian summaries. Teubner-Texte zur Mathematik [Teubner Texts in Mathematics], 26. BSB B. G. Teubner Verlagsgesellschaft, Leipzig, 1980.

[75] J. Kurzweil, *Henstock-Kurzweil Integration: its relation to topological vector spaces.* Series in Real Analysis, 7. World Scientific Publishing Co., Inc., River Edge, NJ, 2000.

[76] J. Kurzweil, *Integration between the Lebesgue integral and the Henstock-Kurzweil integral: its relation to local convex vector spaces.* Series in Real Analysis, 8. World Scientific Publishing Co., Inc., River Edge, NJ, 2002.

[77] J. Kurzweil and J. Jarník, Equi-integrability and controlled convergence of Perron-type integrable functions. *Real Anal. Exchange* **17** (1991/92), no. 1, 110–139.

[78] J. Kurzweil and J. Jarník, Differentiability and integrability in n dimensions with respect to α-regular intervals. *Results Math.* **21** (1992), no. 1-2, 138–151.

[79] J. Kurzweil and J. Jarník, Equivalent definitions of regular generalized Perron integral. *Czechoslovak Math. J.* **42(117)** (1992), no. 2, 365–378.

[80] J. Kurzweil and J. Jarník, Generalized multidimensional Perron integral involving a new regularity condition. *Results Math.* **23** (1993), no. 3-4, 363–373.

[81] J. Kurzweil and J. Jarník, Perron type integration on n-dimensional intervals as an extension of integration of stepfunctions by strong equiconvergence. *Czechoslovak Math. J.* **46(121)** (1996), no. 1, 1–20.

[82] J. Kurzweil, J. Mawhin and W.F. Pfeffer, An integral defined by approximating BV partitions of unity. *Czechoslovak Math. J.*.**41(116)** (1991), no. 4, 695–712.

[83] M. Laczkovich, On additive and strongly derivable interval functions. *Acta Math. Acad. Sci. Hungar.* **39** (1982), no. 1-3, 255–265.

[84] M. Laczkovich, Continuity and derivability of additive interval functions. *Acta Math. Acad. Sci. Hungar.* **39** (1982), no. 4, 393–400.

[85] Lee Peng Yee, *Lanzhou Lectures on Henstock Integration.* Series in Real Analysis, 2. World Scientific Publishing Co., Inc., 1989.

[86] Lee Peng Yee, The integral la Henstock. *Sci. Math. Jpn.* **67** (2008), no. 1, 13–21.

[87] Lee Peng Yee and Ng Wee Leng, The Radon-Nikodým theorem for the Henstock integral in Euclidean space. *Real Anal. Exchange* **22** (1996/97), no. 2, 677–687.

[88] Lee Peng Yee and Rudolf Výborný, *The Integral: An Easy Approach after Kurzweil and Henstock.* Australian Mathematical Society Lecture Series, 14. Cambridge University Press, Cambridge, 2000.

[89] Lee Tuo-Yeong, Multipliers for some non-absolute integrals in Euclidean spaces. *Real Anal. Exchange* **24** (1998/99), no. 1, 149–160.

[90] Lee Tuo-Yeong, The sharp Riesz-type definition for the Henstock-Kurzweil

integral. *Real Anal. Exchange* **28** (2002/03), no. 1, 55–70.

[91] Lee Tuo-Yeong, A full descriptive definition of the Henstock-Kurzweil integral in the Euclidean space. *Proc. London Math. Soc. (3)* **87** (2003), no. 3, 677–700.

[92] Lee Tuo-Yeong, Product variational measures and Fubini-Tonelli type theorems for the Henstock-Kurzweil integral. *J. Math. Anal. Appl.* **298** (2004), no. 2, 677–692.

[93] Lee Tuo-Yeong, A full characterization of multipliers for the strong ϱ-integral in the Euclidean space. *Czechoslovak Math. J.* **54(129)** (2004), no. 3, 657–674.

[94] Lee Tuo-Yeong, Some full characterizations of the strong McShane integral. *Math. Bohem.* **129** (2004), no. 3, 305–312.

[95] Lee Tuo-Yeong, A characterisation of multipliers for the Henstock-Kurzweil integral. *Math. Proc. Cambridge Philos. Soc.* **138** (2005), no. 3, 487–492.

[96] Lee Tuo-Yeong, The Henstock variational measure, Baire functions and a problem of Henstock. *Rocky Mountain J. Math.* **35** (2005), no. 6, 1981–1997.

[97] Lee Tuo-Yeong, Banach-valued Henstock-Kurzweil integrable functions are McShane integrable on a portion. *Math. Bohem.* **130** (2005), no. 4, 349–354.

[98] Lee Tuo-Yeong, A new characterization of Buczolich's upper semicontinuously integrable functions. *Real Anal. Exchange* **30** (2004/05), no. 2, 779–782.

[99] Lee Tuo-Yeong, Product variational measures and Fubini-Tonelli type theorems for the Henstock-Kurzweil integral. II. *J. Math. Anal. Appl.* **323** (2006), no. 1, 741–745.

[100] Lee Tuo-Yeong, Multipliers for generalized Riemann integrals in the real line. *Math. Bohem.* **131** (2006), no. 2, 161–166.

[101] Lee Tuo-Yeong, Proof of two conjectures of Móricz on double trigonometric series. *J. Math. Anal. Appl.* **340** (2008), no. 1, 53–63.

[102] Lee Tuo-Yeong, A measure-theoretic characterization of the Henstock-Kurzweil integral revisited. *Czechoslovak Math. J.* **58** (2008), no. 4, 1221–1231.

[103] Lee Tuo-Yeong, Some convergence theorems for Lebesgue integrals. *Analysis (Munich)* **28** (2008), 263–268.

[104] Lee Tuo-Yeong, A multidimensional integration by parts formula for the Henstock-Kurzweil integral. *Math. Bohem.* **133** (2008), 63–74.

[105] Lee Tuo-Yeong, Bounded linear functionals on the space of Henstock-Kurzweil integrable functions. *Czechoslovak Math. J.* **59** (2009), no. 3, 1005–1017.

[106] Lee Tuo-Yeong, Two convergence theorems for Henstock-Kurzweil integrals and their applications to multiple trigonometric series, to appear in Czechoslovak Math. J.

[107] Lee Tuo-Yeong, Some integrability theorems for multiple trigonometric series, to appear in Math. Bohem.

[108] Lee Tuo-Yeong, On a result of Hardy and Littlewood concerning Fourier series, *Anal. Math.* **36** (2010), no. 3, 219–223.

[109] Lee Tuo-Yeong and Lee Peng Yee, On necessary and sufficient conditions for non-absolute integrability. *Real Anal. Exchange* **20** (1994/95), no. 2, 847–857.

[110] Lee Tuo-Yeong, Chew Tuan Seng and Lee Peng Yee, Characterisation of multipliers for the double Henstock integrals. *Bull. Austral. Math. Soc.* **54** (1996), no. 3, 441–449.

[111] Lee Tuo-Yeong, Chew Tuan Seng and Lee Peng Yee, On Henstock integrability in Euclidean spaces. *Real Anal. Exchange* **22** (1996/97), no. 1, 382–389.

[112] Lszl Leindler, Comments regarding the Sidon-Telyakovskii class. *Anal. Math.* **34** (2008), no. 2, 137–144.

[113] Liu Genqian, The dual of the Henstock-Kurzweil space. *Real Anal. Exchange* **22** (1996/97), no. 1, 105–121.

[114] S. Lojasiewicz, *An Introduction to the Theory of Real Functions.* With contributions by M. Kosiek, W. Mlak and Z. Opial. Third edition. Translated from the Polish by G. H. Lawden. Translation edited by A. V. Ferreira. A Wiley-Interscience Publication. John Wiley & Sons, Ltd., Chichester, 1988.

[115] Lu Jitan and Lee Peng Yee, The primitives of Henstock integrable functions in Euclidean space. *Bull. London Math. Soc.* **31** (1999), no. 2, 173–180.

[116] Lu Jitan, Lee Peng Yee and Lee Tuo-Yeong, A theorem of Nakanishi for the general Denjoy integral. *Real Anal. Exchange* **27** (2001/02), no. 2, 669–672.

[117] M.S. Macphail, Functions of bounded variation in two variables. *Duke Math. J.* **8** (1941), 215–222.

[118] J. Mařík, Foundations of the theory of the integral in Euclidean spaces. (Czech) *Časopis Pěst. Mat.* **77**, (1952). 1–51, 125–145, 267–301.

[119] J. Mařík and C.E. Weil, Multipliers of spaces of derivatives. *Math. Bohem.* **129** (2004), no. 2, 181–217.

[120] J. Mawhin, Generalized multiple Perron integrals and the Green Goursat theorem for differentiable vector fields. *Czechoslovak Math. J.* **31** (1981), no. 4,614–632.

[121] F. Móricz, On the convergence in a restricted sense of multiple series. *Anal. Math.* **5** (1979), no. 2, 135–147.

[122] F. Móricz, Some remarks on the notion of regular convergence of multiple series. *Acta Math. Hungar.* **41** (1983), no. 1-2, 161–168.

[123] F. Móricz, Integrability of double lacunary sine series. *Proc. Amer. Math. Soc.* **110** (1990), no. 2, 355–364.

[124] F. Móricz, On the integrability of double cosine and sine series. I. *J. Math. Anal. Appl.* **154** (1991), no. 2, 452–465.

[125] F. Móricz, On the integrability of double cosine and sine series. II. *J. Math. Anal. Appl.* **154** (1991), no. 2, 466–483.

[126] F. Móricz, On the integrability and L^1-convergence of double trigonometric series. *Studia Math.* **98** (1991), no. 3, 203–225.

[127] F. Móricz, Integrability of double cosine-sine series in the sense of improper Riemann integral. *J. Math. Anal. Appl.* **165** (1992), no. 2, 419–437.

[128] F. Móricz, Lebesgue integrability of double cosine and sine series. *Analysis* **13** (1993), no. 4, 321–350.

[129] F. Móricz, Pointwise behavior of double Fourier series of functions of bounded variation.*Monatsh. Math.* **148** (2006), no. 1, 51–59.

[130] F. Móricz and A. Veres, On the absolute convergence of multiple Fourier series. *Acta Math. Hungar.* **117** (2007), no. 3, 275–292.

[131] F. Móricz and V. Totik, Pointwise convergence of multiple Fourier series using different convergence notions. *Anal. Math.* **12** (1986), 135–147.

[132] Piotr Mikusiński and K. Ostaszewski, *The space of Henstock integrable functions. II.* New integrals (Coleraine, 1988), 136–149, Lecture Notes in Math., 1419, Springer, Berlin, 1990.

[133] P. Muldowney, Ralph Henstock, 1923–2007. *Bull. Lond. Math. Soc.* **42** (2010), no. 4, 753-758.

[134] K.M. Ostaszewski, *Henstock Integration in the plane.* Mem. Amer. Math. Soc. 63 (1986), no. 353, viii+106 pp.

[135] K.M. Ostaszewski, The space of Henstock integrable functions of two variables. *Internat. J. Math. Math. Sci.* **11** (1988), no. 1, 15–21.

[136] W.F. Pfeffer, Lectures on geometric integration and the divergence theorem. School on Measure Theory and Real Analysis (Grado, 1991). *Rend. Istit. Mat. Univ. Trieste* **23** (1991), no. 1, 263–314 (1993).

[137] W.F. Pfeffer, *The Riemann Approach to Integration.* Cambridge Tracts in Mathematics, 109. Cambridge University Press, Cambridge, 1993.

[138] W.F. Pfeffer, On variations of functions of one real variable.*Comment. Math. Univ. Carolin.* **38** (1997), no. 1, 61–71.

[139] W.F. Pfeffer, The Lebesgue and Denjoy-Perron integrals from a descriptive point of view. *Ricerche Mat.* **48** (1999), no. 2, 211–223 (2000).

[140] W.F. Pfeffer, *Derivation and Integration.* Cambridge Tracts in Mathematics, 140. Cambridge University Press, Cambridge, 2001.

[141] W.F. Pfeffer, The Gauss-Green theorem in the context of Lebesgue integration. *Bull. London Math. Soc.* **37** (2005), no. 1, 81–94.

[142] O. Perron, Einige elementare Funktionen, welche sich in cine trigonometrische, aber nicht Fouriersche Reihe entwickeln lassen. *Math Ann* **87** (1922), 84–89.

[143] H.L. Royden, *Real Analysis.* Third Edition. Macmillan Publishing Company, New York, 1988.

[144] W. Rudin,*Real and Complex Analysis.* Third Edition. McGraw-Hill Book Co., New York, 1987.

[145] S. Saks, *Theory of the Integral.* Second revised edition. English translation by L. C. Young. With two additional notes by Stefan Banach Dover Publications, Inc., New York 1964.

[146] W.L.C. Sargent, On the integrability of a product. *J. London Math. Soc.* **23** (1948), 28–34.

[147] S. Schwabik, On non-absolutely convergent integrals, *Math. Bohem.* **121** (1996), 369–383.

[148] A. M. Stokolos, Some applications of Gallagher's theorem in harmonic analysis. *Bull. London Math. Soc.* **33** (2001), no. 2, 210–212.

[149] K.R. Stromberg, *Introduction to Classical Real Analysis.* Wadsworth International Mathematics Series. Wadsworth International, Belmont, Calif.,

1981.

[150] C.W. Swartz, *Introduction to Gauge integrals*, World Scientific, Singapore 2001.

[151] C.W. Swartz, Gauge integrals and series, *Math. Bohem.* **129** (2004), 324–332.

[152] B.S. Thomson, *Derivates of Interval Functions.* Mem. Amer. Math. Soc. 93 (1991), no. 452.

[153] B.S. Thomson, *Symmetric Properties of Real Functions.* Monographs and Textbooks in Pure and Applied Mathematics, 183. Marcel Dekker, Inc., New York, 1994.

[154] B.S. Thomson, σ-finite Borel measures on the real line. *Real Anal. Exchange* **23** (1997/98), no. 1, 185–192.

[155] B.S. Thomson, Some properties of variational measures. *Real Anal. Exchange* **24** (1998/99), no. 2, 845–853.

[156] A. J. Ward, On the derivation of additive interval functions of intervals in m-dimensional space. *Fund. Math.* **28** (1937), 265-279.

[157] W.H. Young, On the condition that a trigonometrical series should have the Fourier form. *Proc. London Math. Soc (2)* **9** (1911), 421–433.

[158] W.H. Young, On the integration of Fourier series. *Proc. London Math. Soc (2)* **9** (1911), 449–462.

[159] W.H. Young, On multiple Fourier series. *Proc. London Math. Soc (2)* **11** (1913), 133–184.

[160] W.H. Young, On a certain series of Fourier. *Proc. London Math. Soc (2)* **11** (1913), 357–366.

[161] W.H. Young, On the Fourier series of bounded functions. *Proc. London Math. Soc (2)* **12** (1913), 41–70.

[162] W.H. Young, On multiple integration by parts and the second theorem of the mean. *Proc. London Math. Soc. (2)* **16** (1918), 273–293.

[163] W.H. Young, On the fundamental theorem of integration for multiple integrals. *Proc. London Math. Soc. (2)* **22** (1924), 81–91.

[164] W.H. Young and G.C. Young, On the discontinuities of monotone functions of several variables. *Proc. London Math. Soc. (2)* **22** (1924), 124–142.

[165] A. Zygmund, *Trigonometric Series.* Vol. I, II. Third edition. With a foreword by Robert A. Fefferman. Cambridge Mathematical Library. Cambridge University Press, Cambridge, 2002.

Points, intervals and partitions

Functions, integrals and function spaces

Measures and outer measures

Miscellaneous

General index